Environmental Hydrology
and
Hydraulics

Eco-technological Practices for Sustainable Development

TL
409
G52
2006
WEB

Environmental Hydrology
and
Hydraulics

Eco-technological Practices for Sustainable Development

S.N. Ghosh
V.R. Desai
Department of Civil Engineering
Indian Institute of Technology
Kharagpur
India

Science Publishers

Enfield (NH)   Jersey   Plymouth

SCIENCE PUBLISHERS
An Imprint of Edenbridge Ltd., British Isles.
Post Office Box 699
Enfield, New Hampshire 03748
United States of America

Website: *http://www.scipub.net*

sales@scipub.net (marketing department)
editor@scipub.net (editorial department)
info@scipub.net (for all other enquiries)

Library of Congress Cataloging-in-Publication Data

Ghosh, S.N. (Some Nath)
  Environmental hydrology and hydraulics : eco-technological practices for sustainable
development / S.N. Ghosh, V.R. Desai.
    p. cm.
  Includes bibliographical references and index.
  Contents: Eco-hydrological background--Water uses--Hydraulic principles and
eco-friendly design approach--Water hazards and their management--Eco-technological
practices for sustainable development.
  ISBN 1-57808-403-2
  1. Water resources development. 2. Hydraulic engineering--Environmental aspects. 3.
Water conservation. 4. Ecohydrology. 5. Green technology. I. Desai V.R. (Venkappayya
Rangappayya), 1960-II.Title.

TC409.G52 2005

627--dc22   2005056311

ISBN 1-57808-403-2

© 2006, Copyright Reserved

All rights reserved. No part of this publication may be reproduced, stored
in a retrieval system, or transmitted in any form or by any means, electronic,
mechanical, photocopying, recording or otherwise, without the prior written
permission.

This book is sold subject to the condition that it shall not, by way of trade or
otherwise, be lent, re-sold, hierd out, otherwise circulated without the publisher's
prior consent in any form of binding or cover other than that in which it is
published and without a similar condition including this condition being imposed
on the subsequent purchaser.

Published by Science Publishers, Enfield, NH, USA
An Imprint of Edenbridge Ltd.
Printed in India.

Preface

The motivation for writing this book emanated from our firm belief that for sustainable development we need to conserve all our natural resources such as air, water etc. In addition, we need to emulate all the natural processes to the maximum possible extent in all our design endeavors aimed at achieving development through an improved infrastructure. Individuals and organizations at all levels should be made aware of the fact that water is a precious natural resource which is crucial to our survival. Water needs to be judiciously used in the context of an increasing population—not only to sustain essential requirements such as those for drinking and domestic usage but also for increased food production, industrial usage, power generation, navigational requirements, pisciculture, recreation, landscaping etc.

In view of the uncertainties associated with the global hydrological cycle over which human beings do not have any control, there is the problem of scarcity as well as excess of precipitation both spatially as well as temporally across the globe. Since precipitation is the primary source of water it is essential to harness/store this precious water resource for various usages at all the times as well as at all possible locations.

In the developed world, a large number of hydraulic structures have already been built to augment the water availability so that their overall water demand consistent with their industrial growth and their standard of living can be met. On the other hand, the developing/under-developed world consisting of low and medium income group of nations and accounting for about 85% of the global population have not been able to develop sufficient built-up capacity to augment/store their naturally available water resources. As a result of this most of their precipitation water flows into seas/oceans during peak flow season.

In view of the huge costs involved in major water resources projects as well as on account of their other associated operational, social and environmental problems, such as long gestation period, rehabilitation of project affected population, submergence of valuable forests, loss of rare plant/animal species and minerals, there is a strong international opinion against large scale water resource development. As an alternative to this, a greater emphasis is being given at all levels to small scale development and efficient use of water resources through appropriate technologies inclusive of revival of traditional technologies, water harvesting, low-cost and biological treatment/reuse of wastewater.

Already there are many books dealing with hydrology, hydraulics and hydraulic structures, which generally deal with larger problems of development, analysis, design and implementation of water resources. But there are not many books which deal with small-scale development of water resources consistent with the environmental concerns as well as application of relevant eco-friendly technologies.

Keeping all these factors into consideration, this book has been grouped into five chapters. Each of these chapters is briefly described here.

Chapter 1 provides the reader with a basic background of hydrology as well as all the hydrological processes. It also tries to establish the linkage between imbalance in these hydrological processes due to human intervention and its impact on ecology.

Chapter 2 deals with various uses of water in the order of importance to human civilization. Many relevant case studies from the developing as well as developed world are elaborated to describe water use for municipal requirement in urban/rural neighborhoods, agriculture, industries, hydropower generation, navigation, fisheries as well as recreation.

Chapter 3 explains the three fundamental conservation principles of hydraulics and their applications. Flow measuring instruments in conduits and open channels are also described prior to the listing of basic considerations for eco-friendly design of various water systems.

Chapter 4 discusses various types of qualitative/quantitative hazards arising out of water pollution, floods, landslides, collapse of dams and droughts. It also tabulates the information system requirement for mitigation of various disasters.

Chapter 5, the last among the Chapters, gives the details of traditional/ state-of-the-art water conservation/recharge practices in the developing/ developed world. Wastewater treatment/reuse by low-cost and localized biological processes are also described. Sustainable development through integrated water management practices (e.g., the construction of facilities which are almost self-sustaining in terms of their freshwater requirement and wastewater generation) are also explained.

The authors gratefully acknowledge the financial assistance by the Continuing Education Cell of IIT Kharagpur, in the preparation of this manuscript. The help provided by Sriyuts Mukesh Gupta, Abhinav Goel, Aditya, Tushar Ganvir, Ashwin and Lalit Jha in web searching as well as scanning is sincerely thanked. Likewise, the assistance by Sri Chinmoy Mukherji in the preparation of the figures is worth mentioning here. We thank them from the bottom of our heart. Lastly the authors are highly indebted to their family members Smt. Savita Desai and Dyuti Ghosh, Sriyuts Nachiket Desai and Archisman Ghosh, for their exemplary patience and cooperation shown during these years.

Nov. 2005   **S.N. Ghosh**
                                                       V.R. Desai
                                                       IIT, Kharagpur

Contents

Chapter **1**

Eco-hydrological Background

1.1 ENVIRONMENTAL HYDROLOGY IN GENERAL

Hydrology may be broadly defined as 'water science'. Hydrology is defined as the study of the occurrence and movement of water above, on and below the surface of the earth as well as the properties of water, and water's relationship to the biological and non-biological components of the environment.

The goal of physical hydrology is to explain the phenomena of water-flow in the natural environment by application of physical principles. Solutions to many hydrological problems require an understanding of the dynamics of water motion. The motion of water is through many processes, which occurs in a cyclic manner, through what is known as the *hydrologic cycle* or the *water cycle*.

1.2 HYDROLOGIC CYCLE AND ITS PROCESSES

The fundamental concept of hydrology is the hydrologic cycle: the global-scale, endless re-circulatory process linking water in the atmosphere, on the continents, and in the oceans. This cyclical process is usually thought of in terms of reservoirs (i.e., oceans, atmosphere, etc.) and the volumetric flows of water between them. Within the hydrologic cycle, the dynamic processes of water vapor formation and transport of vapor and liquid in the atmosphere are driven by solar energy, while precipitation and many of the various flows of water at or beneath the earth's surface are driven primarily by gravitational and capillary forces. The area of land in which water flowing across the land surface drains into a particular stream or river and ultimately flows through a single point or outlet on that stream or river is called the *catchment* or the *basin* or the *watershed* in American English. Catchments are delineated on the basis of land-surface topography. The boundary of a catchment is called a *divide* or a *ridge*.

The hydrologic cycle is a continuous process by which water is purified by evaporation and transported from the earth's surface as well as from the ocean surface to the atmosphere and back to the land and oceans. Figure 1.1 is a representation of the hydrologic cycle. All of the physical, chemical and biological processes involving water as it travels through various paths in the atmosphere, over and beneath the earth's surface and through growing plants, are of interest to those who study the hydrologic cycle. There are many pathways that water may take in its continuous cycle of falling as rain or snow and returning to the atmosphere. It may be captured for thousands of years in polar ice caps. It may flow into the rivers and finally to the seas and oceans. It may soak into the soil to be evaporated directly from the soil surface as it dries or be transpired by growing plants. It may percolate through the soil to groundwater reservoirs or aquifers, to be stored or it may flow to wells or springs or back to streams by seepage. The residence time for water may be either a few minutes, or it may take thousands of years. Refer to Table 1.1 for typical values of residence time for different phases.

Table 1.1   Estimate of the World's Waters and their Residence Time

Parameter	Surface area $(km^2 \times 10^6)$	Volume $(km^3 \times 10^6)$	Volume (%)	Equivalent Depth (m)[a]	Residence time
Oceans and seas	361	1,370	94	2500	~4,000 years
Lakes and reservoirs	1.55	0.13	<0.01	0.25	~10 years
Swamps	<0.1	<0.01	<0.01	0.007	1-10 years
River channels	<0.1	<0.01	<0.01	0.003	~2 weeks
Soil moisture	130	0.07	<0.01	0.13	2 weeks–1 year
Groundwater	130	60	4	120	2 weeks–10,000 years
Ice caps and glaciers	17.8	30	2	60	10-1,000 years
Atmospheric water	504	0.01	<0.01	0.025	~10 days
Biospheric water	<0.1	<0:01	<0.01	0.001	~1 week

[a]Computed assuming uniformly distributed storage over the entire surface of the earth.
[Nace, 1971]

People tap the water cycle for their own uses. Water is diverted temporarily from one part of the cycle by pumping it from the ground or by drawing it from a river or lake. It is used for a variety of activities such as households, businesses and industries, for transporting wastes through sewers, for irrigation of farms and parks and for production of hydroelectric power.

After use, water is returned to another part of cycle. It is either discharged downstream or allowed to seep into the ground. Used water is normally poorer in quality, even after treatment, which often poses a problem for downstream users.

Figure 1.1 Hydrologic Cycle Indicating Global Annual Quantities

[Source: http://www.unesco.org/science/waterday2000/Cycle.htm]

The various processes involved in the hydrologic cycle are as under.

Water vapor enters the atmosphere by *evaporation* and *transpiration*. Evaporation is the process of water in oceans, lakes, rivers, etc. changing into water vapor, while transpiration is the discharging of water vapor into the atmosphere from living vegetation such as leaves, grass, etc. Many a time, evaporation and transpiration are together known as *evapotranspiration*.

Once water vapor enters the atmosphere, it rises and cools. As the water vapor cools, *condensation* or the change from water vapor into liquid water begins to form small drops of water. These small droplets of water are what you look at when you see a cloud. As these droplets bounce around and hit one another, they stick together and make larger drops.

When the drops of water become too heavy to be held up, they fall back to the earth as *precipitation*. Depending on the temperature, it can fall as drizzle, rain, snow, sleet, and many other forms of precipitation.

Once the precipitation hits the ground, it begins to seep into the ground. This process is called *infiltration*. But the soil can hold only a limited quantity of water. And when the ground becomes saturated, the excess water drains into lakes, rivers, oceans, etc. This excess water is called *runoff*. Then the hydrologic cycle starts all over again.

Groundwater is that part of water which is below the ground level. It can be in the form of either storage in the subsurface water bearing strata known as *aquifers* or it can also flow as *groundwater flow* or *subsurface flow* or *subsurface runoff*.

1.2.1 Distribution of Water on the Earth

- Total ~ 1,358 million km^3
- Percentages
  - 97.20% oceans
  - 2.15% ice caps and glaciers
  - 0.65% lakes, streams, groundwater, atmosphere

At its simplest level the water cycle is the balance of evaporation and precipitation. The energy from the sun causes water on the earth to evaporate, this vapor moves through the atmosphere and falls as rain, sleet or snow. The fate of precipitation varies—it may:

1. Never reach the groundwater that is intercepted by buildings or plants, may evaporate rapidly.
2. Land on an impervious surface e.g., pavement or road. Such surface runoff will be channelled away in drains and sewers to rivers and ultimately to the sea or ocean.

3. Reach the soil and permeate it by infiltration. Of this, some will percolate until it reaches an impervious rock layer and form a pool of groundwater. Some will be taken up by the roots of plants and transported to the leaves for use in photosynthesis.

1.2.2 Water Availability and Use

Annual renewable water supply per capita varies greatly around the world: Iceland 670,000 m^3, Canada 112,000 m^3, United States 9000 m^3, Egypt 30 m^3 , Kuwait ~0 m^3, Bahrain ~0 m^3.

Water is a crucial environmental resource in many areas of the world. It does not respect political boundaries and consequently disputes exist around the world between neighboring countries about the use and contamination of shared water supplies e.g., in the Middle East, control of the Jordan and the large aquifer under the West Bank and in Turkey, Iraq and Syria over management of the Tigris and Euphrates.

Runoff does not automatically equate to availability because it can be spasmodic e.g., 90% of rain in India falls between June-September. Stable runoff is that which is available year round.

Regional variations in the water cycle are perturbed and sometimes intensified by drought cycles, which put extra pressure on semiarid zones. Undisturbed ecosystems can survive drought but human introduction of non-native animals and plants reduces resistance.

Types of water use are:

- Withdrawal: taken from lake, river or aquifer for non-destructive purpose and returned for re-use.
- Consumption: fraction withdrawn but lost in transmission, evaporation, absorption, and chemical transformation and therefore unavailable for re-use.
- Degradation: change in quality due to contamination, pollution.

Global use of water over the last century has increased by an order of magnitude. Agricultural use is inefficient and highly consumptive with up to 90% of the water never reaching the crops. Industrial use is largely for cooling power plants. If the withdrawn water is not contaminated then it can be re-used. Domestic water use varies according to the availability and the cost.

Many examples exist around the globe of how the intervention of man can disrupt and potentially destroy elements of the water cycle:

- The Ogallala Aquifer lies under 8 states in USA. Once contained 2000 km^3 in porous rock layers (more than all the surface freshwater on the earth). In 1930 average depth was 20 m but it reduced to 3 m by 1987 due to withdrawal for irrigation. Can't be recharged quickly enough. Groundwater depletion can lead to subsidence, sinkholes and salt-water intrusion (coastal areas).

- The Aral Sea was the world's fourth largest lake—shallow and saline. Located in the deserts of Uzbekistan and Kazakhstan (former USSR). It has two river inflows and no outlet but in 1918 it was decided to withdraw water from the inflowing rivers for cotton production. By 1990 the Aral Sea lost 40% of surface area and 66% volume. Fishing villages are now 40 km from the water and salty dust is devastating crops and causing health problems.

Sea level rise over last 100 years of 1-2 mm/yr or more—may reflect the apparent temperature rise due to thermal expansion plus the extraction of groundwater being delivered to the sea and the melting of glaciers. Arctic Sea ice is shrinking and ice shelves on the Antarctic Peninsula are retreating. Most climate models predict a more humid world with evaporation, precipitation and runoff enhanced i.e., a speeding-up of the hydrological cycle. Most temperature change is expected at high latitudes. Observations of increased water vapor in the stratosphere in some areas have been observed with increased rainfall over mid-latitudes and reduced rainfall over N. Africa and the Middle East but historical data is sparse and all must be considered in relation to the long-term cyclical variability in climate.

A hydrologist studies the fundamental transport processes to be able to describe the quantity and quality of water as it moves through the cycle by evaporation, precipitation, stream flow, infiltration, groundwater flow, and other components. The engineering hydrologist, or water resources engineer, is involved in the planning, analysis, design, construction and operation of projects for the control, utilization and management of water resources. Water resources problems are also the concern of meteorologists, oceanographers, geologists, chemists, physicists, biologists, economists, and political scientists, specialists in applied mathematics and computer science, and engineers in several fields.

1.3 PRECIPITATION-RUNOFF-INFILTRATION-EVAPORATION ANALYSIS

1.3.1 Precipitation

Precipitation is most commonly measured daily at a fixed time in the morning, and the measured quantity, expressed as depth in millimeters, is attributed to the previous day. The 'standard' rain gauge varies in orifice diameter and height above the ground in different countries, but most have adopted the World Meteorological Organization (WMO) guidelines of a 150 to 200 cm^2 collector area positioned 30 cm to 1 m above the ground. If rainfall occurs during the morning, and a rain gauge is read during a storm, the total quantity in that storm will be attributed to two separate days. This is an important factor to take into account when assessing the frequency of large falls of rain.

Another type of rain gauge is the recording or autographic gauge which traces amount of rainfall against time on a strip chart. Some newer gauges employ tipping buckets which record, via a magnetic switch, counts or tips in a digital form, on either solid-state or magnetic-tape loggers. These rain gauges are used to analyze the intensity of rainfall and are particularly useful in assessing storm profiles, infiltration rates, surface runoff or rainfall erosivity.

Storage gauges are rain gauges which are left in remote areas for periods of time between visits ranging from 1 week to 1 month. Where evaporation is high, oil can be used to inhibit losses. Calibrated dipsticks are used to measure rainfall between visits. Once again, small solid-state recorders make it possible for storage gauges to be measured at infrequent intervals to give daily rainfalls. The usefulness of all rain gauges is limited by the effectiveness of protection against vandalism.

Radar has been used with some success to measure rainfall in cases where high costs can be justified in order to obtain real time data, as in the case of flood forecasting. Radar has also been used to detect hail formation in the tea-growing area of Kericho in Kenya. Clouds can be seeded to curtail the growth of large hailstones, which can damage a high-value cash crop.

With the ever-increasing quality of satellite imagery it is possible to correlate cloud type, density and thickness with rainfall, using 'ground truth' stations. Once again the cost of such a computer-based exercise has to be weighed against the value of the rainfall data.

Telemetering of rainfall information has been developed in certain remote areas to minimize the transport costs of gathering rainfall data. This, and other remote or automated techniques (including the use of automatic weather stations), is not likely to replace the widespread system of volunteer observers using simple standard rain gauges until realistic monetary values can be placed on sets of reliable rainfall data. However, as technological advances reduce capital installation costs, and as fuel prices continue to rise, the gap between manual and automated systems in terms of total costs may not be so great.

1.3.1.1 *Estimation of Rainfall Over an Area*

Rainfall over an area is usually estimated from a network of rain gauges. In theory, these gauges should either be placed in a random sampling array, or be set out in a regular or systematic pattern. A combination of sampling techniques, as in a stratified random sampling network, is often used to increase the efficiency of sampling i.e., to decrease the number of gauges required to estimate the mean rainfall with a given precision (McCulloch, 1965).

In practice, gauges are usually placed near roads and permanent settlements for convenience of access, rather than according to a strict sampling array. There is a danger that bias can be introduced into the estimated mean rainfall over an area. This was first pointed out by Thiessen in 1911, who advocated the construction of *Thiessen polygons* (polygons generated around a rain gauge whose sides are the perpendicular bisectors of lines connecting neighboring rain gauges) to give a weighting inversely proportional to the density of rain gauges.

In mountainous areas, for example, fewer gauges tend to be placed in the inaccessible areas which often have the highest rainfall. Careful inspection is needed, therefore, to determine whether the mean is affected by serious bias, and whether this can be corrected using the Thiessen method.

The most rational way of estimating mean rainfall is by constructing *isohyets* (lines joining places of equal rainfall) on a map of the area in question. This takes into account such factors as the increase in rainfall with altitude on the windward side of hills and mountains, the rain shadow on the leeward side of hills and mountains and the aspect of topographic barriers in relation to prevailing winds. The assumption is made that rainfall is a continuous spatial variable i.e., if two rain gauges record 100 mm and 50 mm respectively, there must be some place between the two gauges which have received 75 mm.

Once the isohyets are drawn, mean areal rainfall is calculated by computing the incremental volumes between each pair of isohyets by adding the incremental amounts and dividing it by the total area.

Because of the labor involved in measuring the area between isohyets, several automated techniques have been devised. These include fitting various 'surfaces' to the rainfall data, calculating Thiessen polygon areas automatically or applying finite element techniques (Edwards, 1972; Shaw and Lynn, 1972; Lee et al, 1974; Chidley and Keys, 1970; Hutchinson and Walley, 1972). These techniques are basically designed to accommodate irregularly spaced rain gauges in areas where spatial variation is important, or to provide techniques that can readily be adapted to the computer processing of rainfall data.

In semi-arid areas, the temporal and spatial variability of rainfall is such that rainfall networks are rarely dense enough to reflect adequately either the mean or the variance of areal rainfall. Even where relatively dense networks (e.g., 1 gauge per 700 km^2) are installed as part of research programmes (Edwards et al, 1979), there are difficulties in constructing isohyetal maps, and considerable changes in the seasonal pattern of rainfall can be discerned.

Generally speaking, random sampling networks give a better estimate of the variance of mean rainfall, and systematic sampling networks give a better (i.e., unbiased) estimate of the mean. In some cases, as in network design, the former is more important since it leads to estimates of the precision of the mean or, conversely, to determining the number of gauges necessary for estimating the mean with the required precision. For most general purposes, however, a systematic spatial coverage can be relied upon to give unbiased estimates of the mean.

In this context it is useful to distinguish between an estimate's precision i.e., its repeatability in a sampling sense, and its accuracy, which is a function of both the sampling technique and the estimate of 'true' rainfall at a point.

Rain gauges are of many different heights above the ground. Variations in the cross-sectional area of the gauge from its nominal value, which may be due to poor construction or damage, overexposure of gauges to high wind speeds across the orifice, shelter of the gauge by vegetation, and common observers' mistakes, all contribute to inaccuracies in point rainfall measurements. Normally these are small compared to the seasonal and annual variability of rainfall but, in certain cases, these potential sources of error have to be taken into account. Such cases include the measurement of rainfall above forest canopies and in areas of high wind speed.

1.3.1.2 Frequency Distribution of Annual, Monthly and Daily Rainfall

Agriculturalists and pastoralists require statements concerning the reliability of rainfall. A convenient means of making such statements is provided by *confidence limits*. These are defined as the estimates of the risk of obtaining values for a given statistic that lie outside prescribed limits (Manning, 1956). The limits commonly chosen are 9:1 and 4:1. With 9:1 limits a figure outside the limits is to be expected once in 10 occasions, and half of these occasions (i.e., one in 20) are to be expected below the lower limit and half above the upper limit. Thus the 9:1 lower confidence limit of annual rainfall would represent the level of rainfall which is expected not to be reached once in 20 years. Similarly, the 4:1 lower confidence limit is expected not to be reached once in 10 years.

In order to establish values for such limits, it is necessary to assume a theoretical frequency distribution to which the sample record of data can be said to apply. Manning (1956) assumed that the distribution of annual rainfall in Uganda was statistically normal. Jackson (1977) has stressed that annual rainfall distributions are markedly 'skewed' in semi-arid areas and the assumption of a normal frequency distribution for such areas is inappropriate. Brooks and Carruthers (1953) make general statements that three-yearly rainfall totals are normally distributed, that annual rainfall is slightly skewed, that monthly rainfall is positively skewed and leptokurtic, and that daily rainfall is 'J'-shaped, bounded at zero. They go on to suggest that empirical distributions be used, such as log-normal for monthly rainfall, and an exponential curve (or a similar one) fitted to cumulative frequencies for daily rainfall.

For annual rainfall series which exhibit slight skewness and kurtosis, Brooks and Carruthers (1953) suggest that adjusting the normal distribution is easier and more appropriate than using the Pearson system of frequency curves. As the degree of skewness and kurtosis increases, log-normal transformations should be used.

These comments apply equally well to tropical rainfall where annual totals exceed certain amounts. For example, Gregory (1969) suggests that normality is a reasonable assumption where the annual rainfall is more than 750 mm. Kenworthy and Glover (1958) suggest that in Kenya normality can be assumed only for wet-season rainfall. Gommes and Houssiau (1982) state that rainfall distribution is markedly skewed in most Tanzanian stations.

Inspection of the actual frequency distribution, at given stations, and simple tests for normality can quickly establish whether or not the normal distribution can be used. Such tests include the comparison of the number of events deviating from the mean by one, two or three standard deviations with the theoretical probability integral. If not, a suitable transformation must be chosen before confidence limits or the probabilities of receiving certain amounts of rainfall can be calculated.

Maps have been prepared for some regions, particularly East Africa (East African Meteorological Department, 1961; Gregory, 1969), showing the annual rainfall likely to be equalled or exceeded in 80% of years. These are extremely useful for planning purposes although, as Jackson (1977) points out, these refer to average occurrences over a long period of years. Statements such as 'the rainfall likely to be equalled or exceeded in 4 out of 5 years' must be qualified by 'on average' to indicate that it does not rule out the occurrence of, say, 3 years of 'drought' rainfall in a row.

1.3.1.3 Frequency Distribution of Extreme Values

When dealing with the frequency distribution of maxima or minima, it is necessary to use other empirical frequency distributions which give a more satisfactory fit to the observed data. There are no rigid rules governing which type of distribution is most appropriate to a particular case, and a variety of empirical frequency or probability distributions are available in standard statistical textbooks (Table 1.2). As a general guide, extreme distributions are concerned with the exact form of the 'tail' of the frequency distribution. Because such occurrences are rare events, it is unusual to have a sufficiently long record for the shape of the asymptotic part of the curve to be defined with any certainty. Brooks and Carruthers (1953), for example, feel that the Gumbel distribution, which is commonly used in flood frequency prediction, tends to underestimate the magnitude of the rare rainfall events.

Table 1.2 Common Types of Frequency Distributions used for Hydrological Events

Type	Characteristics	Example
1. Binomial	Discrete events in two categories	Number of 'dry' months in each year
2. Poisson	'J'-shaped distribution of discrete events; possible number of occurrences very small	Frequency of heavy rainstorms
3. Normal	Symmetrical, bell-shaped continuous distribution (Gaussian)	Annual rainfall in wet regions
4. Adjusted normal	Like 'normal distribution' but slightly skew	Annual rainfall
5. Log-normal	Positively skew and leptokurtic	Annual rainfall in semi-arid areas, monthly rainfall
6. Pearson Type I	Bounded both ends, bell or 'J'-shaped, skew	Monthly rainfall with very dry months
7. Pearson Type III	Bounded one end, bell or 'J'-shaped, skew (log-normal is a special case)	Frequencies of wind speed
8. Extreme distribution Type I	Asymptotic, unbounded (Gumbel)	Flood frequency, rainfall intensity/frequency
9. Extreme distribution Type III	Asymptotic, bounded at a minimum value	Drought frequency

In most applications it is usual to linearize the distribution by calculating cumulative frequencies and then plotting on double logarithmic versus linear graph paper (extreme probability graph paper). If the actual distribution is close to the theoretical distribution postulated by Gumbel, the plotted points approximate to a straight line. An example of this is shown in

Figure 1.2 which demonstrates how a cumulative distribution can be linearized by the use of special probability paper (Ven te Chow, 1964). The same author points out that the abscissae and ordinates are normally reversed to show the probabilities as abscissae. Figure 1.3 is an example of the more usual presentation of data plotted on extreme probability paper. In this case, it is seen that the maximum daily rainfalls for June, in Lagos, Nigeria closely fit the Gumbel distribution.

In this context it is useful to mention that most analyses of frequencies of rainfall or other hydrological events are expressed in terms of the *recurrence interval* of an event of given magnitude. The average interval of time, within which the magnitude of the event will be equalled or exceeded, is known as the *recurrence interval* or *return period* which is the reciprocal of *frequency*. It is common, therefore, to refer to the 10-year or the 100-year flood.

The recurrence interval (N) should not be thought of as the actual time interval between events of similar magnitude. It means that in a long period, say 10,000 years, there will be 10,000/N events equal or greater than the N-year event. It is possible, but not very probable, that these events will occur in consecutive years (Linsley and Franzini, 1964).

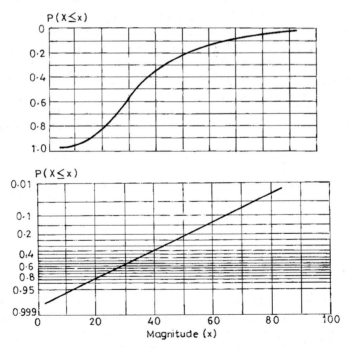

Figure 1.2 Linearization of a Statistical Distribution. Cumulative Probability Curve Plotted on Rectangular Coordinates.

[Source: Ven te Chow (1964)]

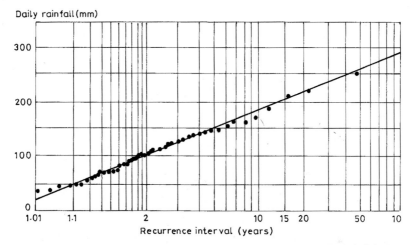

Figure 1.3 Gumbel Distribution for Maximum Daily Rainfall in Lagos, Nigeria for June During 1915-1926.
[Source: Ayoade (1977)]

Most of the record falls of rain are associated with intense tropical cyclones. Figure 1.4 shows the world's highest recorded rainfalls. The totals recorded at Cilaos (Reunion) were the result of an intense tropical cyclone in 1952, which was funnelled up a steep valley rising to 3000 m (Lockwood 1972). The previous world record for 24 hours was also due to the combination of tropical cyclone and land relief. On that occasion, the maximum rainfall occurred at Baguis in the Philippines following the passage of a typhoon in 1911. Less intense, but still remarkable falls of rain commented upon by Jackson (1977) are also shown on the graph.

Rainfall maxima such as those in Figure 1.4 only give an idea of the expected magnitude of the highest falls, with very long recurrence intervals. For general design purposes, statements about amount, duration and frequency of rainfall are required in order to compare the risk of failure of spillways, culverts and bridges against economic criteria. Such analyses can also be applied to soil erosion problems.

For a given duration, rainfall events can be ranked in either a partial duration series or an extreme value series. Thus, for daily rainfall, events can be selected so that their magnitudes are greater than a certain base value, or the maximum rainfall in any year can be chosen. Either series can be plotted on probability paper to yield recurrence intervals or return periods for a given magnitude. The difference between the two series is that the partial duration series may include several events which occur close together in 1 year. For practical purposes, the two series do not differ much, except in the

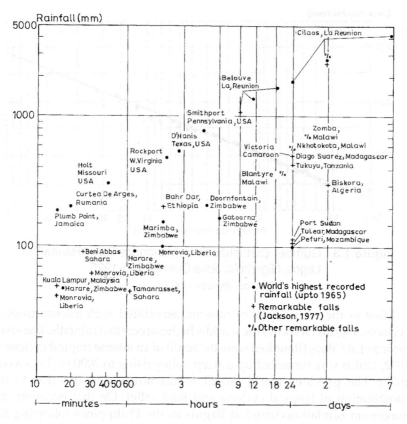

Figure 1.4 World's Highest Recorded Rainfall and Other Remarkable Falls.

[Source: http://www.fao.org/Wairdocs/ILRI/x5524E/x5524e0h.gif]

values of low magnitude (Ven te Chow, 1964). Alternatively, where information from recording rain gauges is available, rainfall intensity can be plotted against duration, and intensity-duration frequency curves can be constructed for different return periods.

1.3.1.4 Occult Precipitation

On mountain masses within the tropics, the summits are frequently covered in cloud in locations where the orographic uplift of moist air is sufficient to cause condensation. If the upper slopes are covered in forest, the surfaces of the leaves and branches of trees act as collectors for condensed water vapor. This phenomenon is referred to as *occult precipitation*, because it does not relate to any rain-making process in the cloud itself and rain gauges placed outside the forest area will not record any rainfall.

In southern and southwestern Africa, Marloth in 1904 and Nagel (1962) have shown that mist condensation or occult precipitation may contribute between 40 and 94% of the total precipitation on high ground. More recently, Edwards et al (1979) made some preliminary measurements on Mt. Kulal in northern Kenya and concluded that occult precipitation could be a significant addition to the groundwater store.

This mechanism may be important as a contribution to groundwater recharge and as a means of sustaining perennial basal springs on mountains in semi-arid areas. On the other hand, high rates of evaporation of intercepted water, and the deeper rooting zone of trees compared to shorter vegetation, may mean that any additional infiltrated water from occult precipitation will be rapidly used up by the trees.

Persistent and subjective reports of springs drying up following deforestation, and flowing again following afforestation, point to the value of investigating the detailed water balance of mountains in semi-arid areas. Unfortunately, the water balances are likely to be complex, with both infiltration capacities and rates of input of precipitation varying between forested and non-forested areas. Evaporation and transpiration rates are also likely to differ significantly and, because these components would need to be measured directly, projects aiming to achieve detailed water balances would require careful instrumentation.

1.3.2 Runoff

The volume of storm runoff depends to a large extent on the antecedent soil moisture conditions and the intensity of rainfall. Typical annual runoff volumes in semi-arid areas, expressed as a percentage of rainfall or *runoff coefficient*, would be less than 10%, but individual storms can produce much higher runoff volumes (~70%) when rain falls in an already saturated catchment. Runoff varies widely according to the seasonal distribution of rainfall, catchment characteristics (shape, size, steepness), vegetation type and density, in addition to the two basic criteria mentioned above. Generally speaking, the wide variation of annual totals makes it difficult to predict total catchment yield from rainfall alone. Most attempts to regress runoff against rainfall lead to unacceptable scatter.

While total runoff or stream flow gives the yield of water from a particular catchment in a particular year, it is closely dependent upon variations in annual rainfall. As in the case of rainfall, a set of runoff data can be subjected to a frequency analysis and statements can be made about the probability of occurrence of certain values.

More usually, however, interest is centered on the extremes of exceptionally high or low annual runoff events and their duration. In these cases, partial duration series or annual maximum (or minimum) series plotted on extreme probability paper will give the recurrence intervals of particular events.

To obtain the duration of annual flows above or below certain limits, an entirely different technique is required. This relies on rainfall being a stochastic process, where the annual rainfall in a particular year is independent of the rainfall in preceding years. Although irregular trends and spells of wetter-than-average or drier-than-average rainfall are commonplace, most attempts to discern reproducible cycles or harmonic patterns in rainfall have failed. In the absence of demonstrable cycles there is no reason to assume that the irregular patterns are other than those, which could be expected to occur in an entirely random series from time to time.

It is possible to extend a data set having characteristic statistics of mean and variance by generating a similar sequence of random values. This can be repeated many times and the actual occurrence of runs of values above or below certain limits can be analyzed. A good example of this technique in practical use is described by Kidd (1983), who used it to generate synthetic or theoretical inflows into Lake Malawi in order to predict outflow down the Shire River and the effects of controlling outflow with a barrage. Table 1.3 lists the runoff coefficient for the rivers around lake Malawi.

Often, as in the case of designing small water supplies, the seasonal variations in daily flow are important. The best way to represent these is by means of a *flow-duration curve*. This is a cumulative frequency curve of daily flows, expressed as a percentage over as long a period as possible. Thus, it is possible to estimate the percentage of time during which a given flow is equalled or exceeded.

In the case of low flows duration of flow is particularly important. It is often necessary to know what the minimum flow would be over a certain time with a given level of occurrence. Thus for a water supply scheme, 30 days might be the absolute maximum that either live stock or people could survive on storage. The 5-year, 30-day minimum flow can be compared with the demand to see what supplementary measures (boreholes, shallow wells) are necessary, on average, in a 5-year period.

This statistic and other comparable statistics (e.g., 10-year, 60-day minimum flow) can be calculated by a frequency analysis of overlapping 'n'-day flows in a given runoff data. This is best performed on a computer because of the amount of data involved. Where a long dry season occurs, as

Table 1.3 Annual Runoff Coefficient – Lake Malawi, Africa

River	Mean annual rainfall (mm)	Mean annual runoff (mm)	Runoff coefficient (%)
Ruo	1280	395	31
Kwakwazi	1240	332	27
Likangala	1430	499	35
Domasi	1730	882	51
Naisi	1280	312	24
Linthipe	880	133	15
Lilongwe	930	155	17
Lingadzi	810	82	10
Bua	900	83	9
Bua	900	73	8
Dwangwa	740	37	5
Luweya	1480	500	34
Luchelemu	1090	260	24
Lunyangwa	1210	232	19
N. Rumphi	1320	661	50
Kambwiya	1370	625	46
N. Rukuru	910	227	25

[Source: http://www.fao.org/Wairdocs/ILRI/x5524E/x5524e04.htm#3.5%20runoff]

in much of tropical Africa, a visual inspection of the recession curves of the streams is usually sufficient to gain some idea of the frequency of very low flows of specified duration.

It is necessary to know the peak flow or flood for the design of any structure intended to pass a given volume of water per unit time (e.g., spillways on dams, culverts, canals and bridges). Many empirical and analytical techniques are available to predict floods. It is outside the scope of this book to discuss more than the general principles. Readers interested in this topic will find a comprehensive treatment of the subject in NERC (1975), in spite of the direct relevance of this report to British conditions. The proceedings of the symposium on 'Flood hydrology' (TRRL, 1977) deal specifically with African conditions, and apply similar techniques of flood prediction to both urban and rural catchments.

Basically, a choice can be made between empirical techniques based on catchment characteristics, statistical techniques (where there is an abundance of reliable data), and unit-hydrograph techniques (which require some knowledge of rainfall characteristics, soil types, channel slope and shape of the unit hydrograph).

In many areas, there are insufficient streamflow records for statistical techniques to be applied with confidence. Usually statements are required about ungauged catchments where a large degree of uncertainty is bound to arise. Choice of method depends to a large extent, therefore, on the amount and type of data available. It is still common to find the so-called *'rational method'* being applied where sufficient data for other methods are lacking (Prabhakar, 1977). In this method, maximum runoff (Q_{max}) is related to average rainfall intensity (I) and catchment area (A) by means of an empirical runoff coefficient (K):

$$Q_{max} = K \cdot I \cdot A \tag{1.1}$$

Values of K are derived locally, often by trial and error, and values are given in Table 1.4.

Table 1.4   Empirical Values of the Factor K in the Rational Formula

Type of catchment	Empirical values of factor K
Rocky and impermeable	0.30 to 1.00
Slightly permeable, bare	0.60 to 0.80
Slightly permeable, partly cultivated or covered with vegetation	0.40 to 0.60
Cultivated absorbent soil	0.30 to 0.40
Sandy absorbent soil	0.20 to 0.30
Heavy forest	0.10 to 0.20

Source: Prabhakar (1977).

Another method, used when data are lacking, is the *'envelope curve'* method in which recorded peak flows are plotted against catchment area. Although a remarkable degree of conformity is often found, it is not possible with this method to relate magnitude to frequency of occurrence; but from experience, it is expected to give approximately the 25-year flood.

Implicit in many analyses of streamflow is a division between *surface runoff*, or *storm runoff*, and *base flow*. As the name implies, surface runoff is the direct runoff from the soil surface to the stream course, concentrated by the shape of the catchment and the drainage network into a flood wave, which reaches a maximum and then attenuates as it travels down the main river. Base flow, on the other hand, is conventionally taken as the contribution to streamflow from groundwater. After the passage of the flood wave, the catchment slowly drains until the flow in the river is related to the amount of water held in storage within the catchment. The *'recession curve'*, the slowly falling limb of the hydrograph, therefore, reflects the rate of release of water from the groundwater store, which is itself dependent upon a combination of hydraulic head and saturated permeability of the catchment. The exponential shape of a recession curve allows 'recession constants' to be calculated. These in turn can be used to predict the amount of water in the groundwater store, since the physical process of releasing water from storage can be simulated by a *linear reservoir* (in systems terminology) whose outflow is directly related to storage:

$$Q_t = Q_o K_r^t \qquad (1.2)$$

where Q_t = flow at time t after Q_o, and K_r = recession constant.

The storage (S_t) remaining in a basin at time t is given by:

$$S_t = -Q_t / \ln K_r \qquad (1.3)$$

The recession constant can be used, therefore, both to separate stormflow and baseflow (by extrapolating backwards in time from the recession limb of the hydrograph) and to estimate the groundwater storage at a given time. The latter technique is often used in water balance calculations, as will be seen later.

1.3.2.1 *Measurement of Streamflow*

Although hydrometric networks have improved over the past 30 years, the majority of gauging stations are still on large rivers or in areas of high agricultural potential, where information is required for water supply and / or irrigation.

In the rangelands not only are networks sparse, but the preponderance of shifting controls (e.g., sandy river beds), coupled with infrequent visits, lowers the reliability of the streamflow records.

In ungauged catchments, which are often of very small size, it is frequently necessary to estimate either storm runoff or streamflow. Once again there is a choice of methods, depending on the type of information required and the availability of data from nearby stations within the same hydrological region. These range from the statistical, analytical and empirical techniques referred to in the last section to mathematical models of various degrees of complexity, which are dealt with later in this chapter. However, the actual measurement of streamflow is better than the most sophisticated indirect methods, although usually a combination of the two is used to derive estimates of frequency or reliability.

Measuring the flow in a river is straightforward if the river is shallow enough for wading. The basic problem is that of obtaining a unique relationship between height of the water above a datum and discharge. Weirs and flumes are structures, which stabilize the channel section to give this unique relationship. For very small streams with a *discharge* (i. e., flow per unit time) up to 0.12 cumecs, portable "v"-notch weirs can often be installed (British Standards Institution, 1965), which will measure discharge for a given depth of flow over the weir. Where the benefits to be accrued do not warrant the construction of weirs or flumes, the natural channel has to be 'calibrated'. This is achieved by making a series of measurements of discharge, using a current meter, at different '*stages*' (i.e. at different heights of water above the zero-flow datum). These values can be plotted on a graph of discharge against stage, to give a smooth curve (the 'stage-discharge' or 'rating' curve). This curve can be extended using the Manning formula, which relates discharge to the slope of the channel and its conveyance (Ven te Chow, 1964). Once the rating curve is established, continuous or periodic measurement of a stage can be readily converted into discharge.

The choice of channel section is important in order to ensure, as far as possible, that the flow remains uniform, and that a '*control*' governs the stage-discharge relationship throughout its range by eliminating the effects of downstream channel features. Such controls can be rock bars or constrictions of the stream channel, which are characterized by having pools or smooth reaches upstream. Where such natural constrictions do not exist, long, straight reaches with stable beds should be sought. It is possible with the latter, however, that the control may be drowned out at high flows causing a change in the stage-discharge relationship.

Where a river is carrying a lot of sediment, changes in the channel cross-section will occur from time to time. These will affect the stage-discharge relationship, and will call for frequent measurements of discharge with a current meter to ascertain what error is being introduced into the discharge calculations by assuming a single rating curve. Often several rating curves will have to be used where a channel is eroding or silting up rapidly.

For a full discussion of the techniques for measuring streamflow the Norwegian Agency for International Development (NORAD) has produced a series of manuals on operational hydrology (Tilren, 1979) for the use of the Tanzanian Ministry of Water, Energy and Minerals. Other users would find these manuals full of sound practical information, which could be applied throughout tropical regions of the world.

1.3.2.2 Application of Hydrology in the National Weather Service

An example for the application of hydrology in weather prediction in USA can be cited here. There are 13 River Forecast Centers in the USA that are responsible for forecasting the water levels in the nation's major rivers and their tributaries. Each River Forecast Center is assigned a certain area of responsibility. Figure 1.5 shows the area of coverage of LMRFC—Lower Mississippi River Forecast Center. This Center covers an area of about 542,000 km^2.

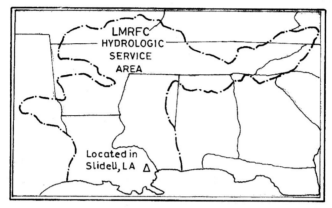

Figure 1.5 The Coverage Area of Lower Mississippi River Forecast Center (LMRFC) in USA.

[Source: http://www.srh.noaa.gov/lmrfc/about/#intro]

The following is a simplistic view of how to make a river stage forecast by using the basic steps below:

1. Data collection of stages, reservoir flows and mean area precipitation and future rainfall.
2. Data entry into the database.
3. Data pre-processing—Quality control of the data to make sure figures are correct.
4. Input into a hydrologic model. This contains all pertinent information at each forecast point, such as the present moisture in the top and bottom layers of the soil, to the climatology of how it reacts to rainfall.

The hydrologist is then able to obtain a forecast of the discharge and then relate this to a stage. River stage forecasts are made daily and are sent out to the Weather Service Forecast Offices. The River Forecast Centers (RFCs) make 3-day and also 5-day forecasts. Long-range forecast for 7 days and up to 30 days are made along the Mississippi River weekly by the LMRFC. When flooding is imminent, forecast is made for the time and height of the crest of the stage and also when the river is expected to go above flood stage. Another product sent out by the RFCs for use by the Weather Offices is the

Flash Flood Guidance. The Flash Flood Guidance is an indicator of the amount of rainfall required to produce flooding.

A *storm hydrograph* is a graph of river stage or discharge at a gauging site versus time. But before going into more detail of the hydrograph, let's define some terms. A storm hydrograph is made up of two components, viz., the base flow and surface runoff. Base flow is that rainfall which seeps (infiltrates) into the soil and moves laterally to the stream channel, reaching the stream channel after several days or more. Surface runoff consists of the rainfall and/or snowmelt, which travels overland to the stream channel plus rainfall directly deposited on the channel. Surface runoff reaches the stream channel quickly and is the main contributor of the total flow during flooding. Some main factors affecting runoff are:

1. Rainfall amount—High rainfall amounts produce more runoff than low rainfall amounts.
2. Rainfall intensity—For the same total amounts of rain, more runoff will occur with rain falling in short periods of time as opposed to rain falling in a longer period of time. For example, one inch of rain falling in 30 minutes will produce more runoff than one inch of rain falling in 24 hours. The lower runoff amounts are due to the water having a longer time to seep (infiltrate) into the ground.
3. Soil type—More runoff will occur with clay soils while sandy soils are able to absorb more rainfall.
4. Soil moisture—When the top layer of soil is moist, there will be more runoff than if the soil moisture content is low.
5. Vegetation—Vegetative cover may slow the runoff from rainfall. As vegetation takes in water, the runoff is retarded.
6. Topography—Runoff varies as the terrain varies. A mountainous terrain will have a faster runoff rate than one of a flat terrain.
7. State of ground—Rainfall over frozen ground produces more runoff than rainfall over non-frozen ground. Frozen ground is more impervious to rainfall.

1.3.2.3 Flooding

Flooding kills more people and causes more damage per year on average than any other natural disaster including hurricanes and tornadoes. There are many different forms of flooding, but only the most common and most destructive kinds will be covered here briefly. They are:

1. *Rainfall Flooding*: This is the most common form of flooding. When it rains enough to cause runoff, the runoff flows into the rivers and the

water levels begin to rise. The river will rise above its banks and flood the surrounding area when more water than the river can hold flows into it. This type of flooding can happen within less than an hour after the rain begins (usually in mountainous regions) or even days after the rain has stopped (in flat regions). It can also happen in a place where there was no rain at all! The rain can fall over an area far upstream and be carried downstream, therefore flooding areas where no rain had fallen.

2. *Snow Melt Flooding*: This type of flooding usually occurs in the cold regions, but water from melted snow can influence river heights all the way down to the delta. When temperatures begin to rise in the spring and there is an excessive amount of snow on the ground, the snow begins to melt and flow into the rivers. This type of flooding is especially dangerous when temperatures rise well above freezing (i.e., 10°C and above) combined with additional rainfall.

3. *Ice Jam Flooding*: This type of flooding only happens in cold climates. Thick ice forms on the surface of rivers during periods of continuous cold weather. When the water level rises quickly underneath the ice, the ice heaves up, cracks, and is carried down the river. The large chunks of ice get stuck under bridges or in sharp bends in the river. The ice blocks the flow of water and the water flows over its banks and floods the surrounding area. This type of flooding is usually confined to smaller areas.

4. *Dam Break Flooding*: This is usually the most destructive type of flooding. Dams can hold back literally millions of cubic meters of water which form huge man-made lakes that can be used for many things such as recreation (boating, swimming, fishing, etc.), flood management (holding excess water which would normally go down river and flood), and city water management (holding water for people to use for drinking, showering, etc.). But when the dams are not constructed or maintained correctly, it creates the potential for disaster. When a weakened dam breaks, it sends a huge wall of water down the river which can destroy entire towns in a matter of minutes. Fortunately, this very destructive type of flooding is also the least common.

1.3.3 Interception and Infiltration

Foresters have always observed the phenomena of interception and throughfall, but few have been prepared to acknowledge that these components were of great significance in the context of the hydrologic cycle. A recent study has established that interception can play a major role in the water balance of catchments where the aerodynamic component of the energy balance is large relative to net radiation. In the tropics this is not likely to be the case, but there are indications that, even in forests lying not too far from the equator, evaporation of intercepted water can be a significant feature.

Intercepted moisture, stored in the canopy, is the first component of the hydrological cycle to be lost directly back to the atmosphere. In areas of high wind speed, with aerodynamically 'rough' canopies, this loss can be very rapid and, in areas where the canopy is frequently wetted, the total quantity of intercepted water lost by evaporation can be a significant proportion of the total rainfall. Interception of raindrops by canopies is also a major factor in reducing soil erosion. This has an indirect effect on the hydrological cycle, in that, by conserving surface soil, infiltration is maintained.

In areas of shorter vegetation interception storage is likely to be small, and the rate of loss may not exceed the potential evaporation rate. Thus in rangelands, interception storage is unlikely to be a measurable quantity in the water balance. Many dry-season grazing areas, however, depend on perennial springs for water supply. These springs emanate from hillsides covered in vegetation, where interception protects the slopes through which the springs are being recharged. Throughfall, stem flow and drip (including the occult precipitation described earlier) form the precipitation input. In dense canopies throughfall is of minor importance, and hence the rate at which water is received by the soil surface is within its infiltration capacity. Surface runoff is practically nil on the heavily forested slopes, and deep percolation is often rapid, through fractured or weathered bedrock. These advantages in terms of recharge, however, have to be offset against the transpiration of deep-rooted, perennial vegetation. This tends to produce moisture deficits in the root zone. These deficits inhibit deep infiltration until sufficient rain falls to saturate the root zone.

Over most of the rangelands in tropical Africa infiltration is the key process in the hydrological cycle. *Infiltration capacity* as termed by Horton or *infiltrability* (Hillel, 1971) which is defined as the rate at which water can enter the soil. Horton recognized both a maximum and a minimum infiltration capacity of soils; the maximum being at the onset of rainfall with the capacity decreasing as the impact of raindrops changes the surface structure of the soil.

When the rate of rainfall exceeds the infiltration capacity of the soil, the result is surface runoff, unless the surface water is stored in depressions (surface or depression storage). Thus, management for soil conservation is directed both at maintaining high infiltration rates by preserving a good surface cover, and at increasing surface storage by pasture furrows, cut-off ditches, and range-pitting or water-spreading.

One of the additional problems in grazing areas is that of *'puddling'*. Puddling is the structural change associated with mechanical stress while soils are in a moist condition, and results in the destruction of large pores in the soil through which water percolates. This has the effect of decreasing infiltration capacity and increasing surface runoff. Further mechanical sorting results, together with the removal of organic material and, often, surface sealing. In fact, it was found that from different types of land use in a catchment in Machakos, Kenya the greatest loss of soil was from degraded grazing areas. Much of this can be attributed to loss of vegetation from overgrazing and the downward spiral of poor vegetation leading to high rates of soil loss and surface runoff.

With some soils, subsoil permeability may be the limiting factor in determining infiltration rates. While this points to surface management being less important, experimental evidence shows that the compaction of well structured surface layers in tropical forest soils, such as that following clearing, causes infiltration capacity to drop by half.

Although infiltration can be seen to be a key process in the cycle, very few experimental data are available to quantify infiltration capacity. It is known to be spatially highly variable, and for this reason isolated measurements are of little practical benefit. Indirect evidence of the improvement of infiltration, following the re-establishment of grass cover, can be found in Edwards and Blackie (1981), who report the results of the Atumatak catchment experiment in Karamoja, Uganda from 1959 to 1970. Two adjacent degraded catchments were chosen for this experiment. One was fenced, cleared of secondary bush and subjected to a controlled grazing-density scheme. The other continued under uncontrolled grazing conditions. Soil moisture tension blocks were installed at a number of sites and, from an analysis of the measurements, it was clear that, in the post-clearing phase, infiltration penetrated down to a depth of 60 cm at most sites. A corollary to the recovery in infiltration in the cleared catchment was the reduction in storm runoff to half that in the 'control' catchment - a striking example of rapid recovery following a modest management programme.

1.3.3.1 Groundwater

The infiltration characteristics of the land surface and the rainfall intensity and duration determine the rate at which the soil moisture store is replenished. After consumptive use by vegetation, a small proportion drains under gravity to the groundwater store. The occurrence of subsurface water can vary according to soil type, the nature of the underlying parent material and the depth of weathering. A classification of subsurface water is given in Figure 1.6. Water is held in each zone as a result of gravitational surface tension and chemical forces. There are no sharp boundaries, except at the capillary fringe in coarse-grained sediments. The water table is, in fact, a theoretical surface, and can be demonstrated approximately by the level of water in wells, which penetrate the saturated zone. The water table can be defined as the level at which the fluid pressure of the pores, in a porous medium, is exactly atmospheric. It can exist in the rock pores to very great depths of about 3000 m, but in dense rock the pores are not interconnected and the water will not migrate.

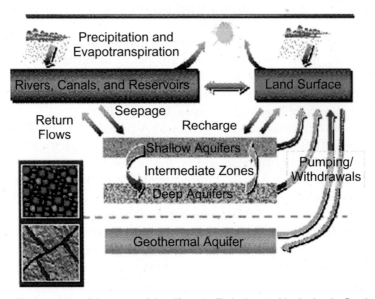

Figure 1.6   Diagram of Aquifers in Relation to Hydrologic Cycle.
[Source: Anon., 2003a]

Only a small proportion of the zone of saturation will yield water to wells. The water-bearing portions are called aquifers. Many types of formation can serve as aquifers, a key requirement being the ability to store water in pores. Table 1.5 gives a typical range of values of porosity, and it is clear that the unconsolidated deposits (chiefly sands and gravels) are the most important aquifers. Pores in the silts and clays are too small. Volcanic rocks are often good aquifers because of the many types of openings, which contribute to their permeability. Igneous and metamorphic rocks are generally regarded as poor aquifers but the Precambrian basement rocks are near the surface and are deeply weathered. As in the case these form shallow aquifers of low yield (0.5 to 1 l/sec), sufficient for small domestic supplies and stock watering. The areal extent of the basement complex rocks on the old erosion surfaces makes this type of aquifer extremely important in the context of tropical Africa. Joints and fractures in the crystalline basement hold some water, but storage is usually quite small and recharge is from the surface weathered zone. Therefore, yields tend to be less than or equal to those from the shallow surface aquifers.

Table 1.5 Typical Porosity Values for Different Formations

Aquifer type	Porosity (%)
Unconsolidated deposits	
Gravel	25-40
Sand	25-50
Silt	35-50
Clay	40-70
Rocks	
Fractured basalt	5-50
Karst limestone	5-50
Sandstone	5-30
Limestone, dolomite	0-20
Shale	0-10
Fractured crystalline rock	0-10
Dense crystalline rock	0-5

[Source: http://www.fao.org/Wairdocs/ILRI/x5524e04.htm#3.4%20groundwater]

Aquifers may be classified as confined and unconfined, according to whether or not the water is separated from the atmosphere by impermeable material. Confined aquifers often give rise to artesian wells, where the pressure within the aquifer is sufficient to produce flowing wells at the surface. Unconfined aquifers, of course, have water whose upper surface is at atmospheric pressure. The term semi-confined is used for the intermediate condition, where the confining layer is not completely impermeable. Often lenses of unconfined water are encountered above the water table, held by isolated layers of impermeable material. These are termed perched aquifers and the upper surface is a perched water table. Figure 1.6 gives a diagram of aquifers in relation to hydrologic cycle.

Groundwater may be discharged at the surface or into bodies of surface water. Springs are the most noticeable manifestation of this, but seepage into rivers and lakes is also an important part of the hydrological cycle. Well digging or borehole drilling induces artificial discharge. The quantity of discharge is a function of the porous medium, which is related to the size and interconnection of the pores. The storage term for unconfined aquifers is the 'specific yield' defined as the volume of water released from storage per unit surface area of aquifer, per unit decline in water table. The storage coefficient in confined aquifers is the 'storativity'—defined as the volume of water released per unit surface area of aquifer, per unit decline in the component of hydraulic head normal to that surface (Freeze and Cherry, 1979).

In the context of the hydrological cycle, groundwater flow passes from recharge areas to discharge areas. Fluctuations in the water table introduce transient effects in the flow system. However, it can often be simulated as a steady-state system if the fluctuations in water table are small in comparison with the total vertical thickness of the system.

Discharge areas in semi-arid climates can be mapped by the direct field observation of springs and lines of seepage, or by the occurrence of phreatophytes and other distinctive vegetative patterns. An analysis of the lithology, topography and existing borehole data will also give information on the recharge-discharge regime and allow the application of steady-state water balance equations to the surface and subsurface components within the recharge and discharge areas.

This approach is often applied to determine the recharge potential within a catchment area and, although livestock requirements are generally very small in comparison with the recharge components, it is a useful exercise to balance all the measured components in order to determine gross errors, or to assess the possible effects of a management programme.

1.3.4 Evaporation and Transpiration

The problem of measuring evaporation from open water surfaces, and transpiration from different types of vegetation, has been a central problem in hydrology for many years. In terms of the hydrological cycle and the water balance, evaporation and transpiration make up the second largest component. Errors in estimating evaporative loss, therefore, assume great significance, for example, in the calculation of groundwater recharge.

Difficulties in understanding the physical nature of the evaporation process, together with ambiguous results from the various types of instrument designed to measure evaporation directly (such as evaporation pans and evaporimeters), led to the development of empirical techniques for estimating evaporation, using generally available climatic data by Thornthwaite in

1948, by Blaney and Criddle in 1950 and by Turc in 1955. These techniques were recognized and acknowledged to give only approximate estimates, but in the absence of simple-to-apply, more theoretically sound methods they provided a useful means of calculating irrigation need and consumptive water use by crops.

Advances in micrometeorology have produced more sophisticated techniques for measuring evaporation. Generally speaking, these are still research techniques requiring far more instrumentation or experimental data than are normally available.

Perhaps the best compromise is the semi-empirical but physically based formulae of Penman developed in 1948, 1952 and subsequently (Penman, 1956; Penman, 1963), or its many derivatives (Monteith, 1965; Thom and Oliver 1977). This embodies the concepts of 'potential transpiration' (PET) from vegetation plentifully supplied with water, and of 'open-water evaporation' (E_O) from an extensive open-water surface. The original formula of Penman developed in 1948 is a combination of the energy balance and aerodynamic methods of measuring evaporation. The energy quantities available for evaporation and for heating the soil-plant-atmosphere system can be equated as follows:

$$R_n = \lambda\, E + K_a + G \qquad (1.4)$$

where R_n = net radiation, λ = product of mass density of water and latent heat of evaporation of water, E = daily lake evaporation, K_a = sensible heat transferred to the air, and G = sensible heat transferred to the soil.

In tropical regions G becomes small in relation to R_n over a day, and may be neglected. R_n can be measured or estimated from incoming solar radiation or from hours of bright sunshine (Grover and McCulloch, 1958), and the problem becomes that of partitioning R_n between sensible heating of the air and the latent heat flux.

The ratio $K_a/(\lambda\, E)$ is known as the *Bowen ratio* (β). Penman derived an estimate for β by introducing an empirical aerodynamic term E_a and eliminating the need to measure surface temperatures. Evaporation from an open-water surface (E_O) is then given by:

$$E_O = \frac{\Delta H + \gamma E_a}{\Delta + \gamma} \qquad (1.5)$$

where

$$E_a = f(u)\,(1 + u_2/160)\,(e_a - e_d), \qquad (1.6)$$

and

$$H = (R_n - G)/\lambda \qquad (1.7)$$

Here, Δ = the slope of the curve relating saturation vapor pressure to air temperature at mean air temperature,

γ = the psychrometric constant,

f(u) = an empirical constant,

u_2 = mean wind speed at 2 m height above the ground (km/d), and

e_a, e_d = saturation vapor pressure at air temperature and dew point respectively.

On the basis of the Lake Hefner experimental results of the US Navy in 1952, Penman modified the aerodynamic term to:

$$E_a = f(u)\,(0.5 + u_2/160)\,(e_a - e_d) \tag{1.8}$$

justifying the correction on the grounds that the new term gave better agreement with evaporation from a large body of water (Penman, 1956). To estimate PET, Penman first used a reduction factor, which varied seasonally. Averaged over the whole year, a value of 0.75 was derived for western Europe. At a later stage, making use of measurements of the albedo grass, and reinstating the original aerodynamic term to take into account the aerodynamic roughness of short vegetation, a one-step formula was introduced (Penman,1963):

$$PET = \frac{\Delta H + \gamma E_{at}}{\Delta + \gamma} \tag{1.9}$$

where,

$E_{at} = f(u)(1 + u_2/160)(e_a - e_d)$, and

$H = (R_n - G)/\lambda$, with R_n—measured over grass.

This formula has been used to provide an index of evaporation. In practice, it can be expected to give reasonable estimates of ET within the accuracy of the other components of water balance, where water supply to the root zone is not a limiting factor. Simplified calculation methods for the Penman formula can be found in McCulloch (1965), Berry (1964) and Doorembos and Pruitt (1977). Further modifications of the approach can be found in Monteith (1965) and Thom and Oliver (1977).

Where water supply to the root zone is a limiting factor, as in most of the semi-arid tropics, actual evaporation (AE) is considerably less than potential evaporation (PE). The direct methods of measuring AE require complex and expensive equipment, and the indirect methods, such as the water balance of watertight catchments, do not give short-period water use unless soil moisture measurements are available.

An alternative approach has been developed by Bouchet in 1963, and embodies the concept of complementary evaporation. Broadly speaking, this concept states that the difference between AE at a dry site and PE at a wet site, subject to the same radiation input (where water supply is not limiting), is the same as the difference between PE estimated at the dry site (i.e., with the same radiation input but lower humidities and higher air temperatures) and PE calculated for a wet site. This can best be illustrated by a simple diagram, in which annual evaporation is plotted against annual rainfall. The theoretical values of PE are seen to decrease with increasing rainfall at the drier site (location A) until they reach a value PE_0, which is a function of radiation input, temperature and humidity, where soil moisture is not limiting (location B). Variations in the theoretical values of AE are the precise opposite, increasing with increasing rainfall until they reach the same limiting value of PE_0.

Brutsaert and Stricker (1979) used the Priestley-Taylor equation (Priestley and Taylor, 1972) to calculate PE_0. The formula, which was developed to estimate evaporation in the absence of advection, reads:

$$PE_0 = \frac{1.26 \Delta R_r}{\Delta + \gamma} \qquad (1.10)$$

They used the Penman formula to estimate the local potential evaporation (PE_B). AE can now be found from the Bouchet relationship:

$$AE = 2PE_0 - PE_B \qquad (1.11)$$

This approach was tested by Stewart et al (1982), together with an improved Penman formula incorporating a larger aerodynamic term (Thom and Oliver, 1977). They found, using data from 120 tropical stations that the method is only valid when used with climatological data recorded at sites further than 50 km from a coast. At these stations the modified Brutsaert and Stricker formula gave encouraging results, which support the concept of complementary evaporation. Clearly the approach has great potential in areas of sparse data.

1.3.4.1 *Potential Evapotranspiration*

One aspect of the soil-water budget that involves significant uncertainty and ambiguity is estimating potential evapotranspiration. Just the concept of potential evapotranspiration is ambiguous by itself, as discussed in the next section. Due to limited meteorological data, two simple methods for estimating potential evapotranspiration were considered for the Niger basin

study, the Priestley-Taylor and Thornthwaite methods. For the short term
simulation (July 1983 to December 1990), a global net radiation data set
obtained from NASA facilitated making potential evapotranspiration
estimates using the Priestley-Taylor method. For reasons discussed later in
this book, the Priestley-Taylor method is considered superior to the
Thornthwaite approach; however, it is simpler to apply the Thornthwaite
approach to long term average conditions and to selected historical periods
because the global net radiation data used in this study are only available
from July 1983 to June 1991. It would be nice to have consistent methods for
estimating potential evapotranspiration over different time periods so that
fair comparisons can be made. Because the Thornthwaite method is more
easily applied over different historical time periods, determining whether
there are significant differences between predicted runoff using the
Priestley-Taylor and Thornthwaite methods is an important question. The
conclusion is that there are significant differences and the Priestley-Taylor
approach is better. For this study, the average net radiation over the eight-
year period when net radiation data were available was taken to be the long
term average net radiation. Both Priestley-Taylor and Thornthwaite
methods perform poorly in arid regions and the significance of this is briefly
discussed.

1.3.4.2 Potential Evaporation vs Potential Evapotranspiration

Thornthwaite in 1948, first used the concept of potential evapotranspiration
as a meaningful measure of moisture demand to replace two common surro-
gates for moisture demand temperature and pan evaporation. Potential
evapotranspiration refers to the maximum rate of evapotranspiration from a
large area completely and uniformly covered with growing vegetation with
an unlimited moisture supply. There is a distinction between the term potential evapo*trans*piration and potential evaporation from a free water surface
because factors such as stomatal impedence and plant growth stage influence evapotranspiration but do not influence potential evaporation from
free water surfaces.

Brutsaert (1982) notes the remarkable similarity in the literature among
observations of water losses from short vegetated surfaces and free water
surfaces. He poses a possible explanation that the stomatal impedance to
water vapor diffusion in plants may be counterbalanced by larger roughness values. Significant differences have been observed between potential
evapotranspiration from tall vegetation and potential evaporation from free
water surfaces. The commonly used value of 1.26 in the Priestley-Taylor

equation was derived using observations over both open water and saturated land surfaces. For the most part, the term potential evapotranspiration will be used predominantly in this chapter and, as used, includes water loss directly from the soil and/or through plant transpiration.

An additional ambiguity in using the potential evapotranspiration concept is that potential evapotranspiration is often computed based on meteorological data obtained under non-potential conditions. In this study, temperature and net radiation measurements used for calculating potential evapotranspiration in dry areas and for dry periods will be different than the values that would have been observed under potential conditions. The fact that the Thornthwaite and Priestley-Taylor methods have exhibited weak performance at arid sites is related to this ambiguity because the assumptions under which the expressions were derived break down. Poor performance in arid regions is highly relevant to the Niger Basin study in Africa, because of large arid areas in the northern part of the basin. This problem will be addressed a bit further during the detailed discussions of each method.

Although not used directly in this chapter, a brief review of the widely used Penman equation serves as a good starting point for discussing the estimation of potential evapotranspiration.

1.3.4.3 Estimating Actual Evapotranspiration

To estimate the actual evapotranspiration (AET) in the soil-water budget method many investigators have used a *soil-moisture extraction function* or *coefficient of evapotranspiration* (β') which relates the actual rate of evapotranspiration to the potential rate of evapotranspiration based on some function of the current soil moisture content and moisture retention Properties of the soil

$$AET = \beta' \cdot PET \tag{1.12}$$

1.4 THE WATER BALANCE

Water balance or *water budget* or *hydrologic budget* indicates the mass balance in water input and output for a system over a certain time period. The general expression describing the water balance of a watertight catchment over a given period is:

$$P = Q + AET + \Delta S + \Delta G \tag{1.13}$$

where, P is precipitation and Q is the streamflow respectively and can usually be measured directly; AET is actual evapotranspiration; and $\Delta S, \Delta G$ are changes in soil moisture and groundwater storage respectively.

Where no bias is present in any of the measured terms this expression can be used to determine the value of any one term, by difference, over a given period. There will be a random error present, of course, which is dependent on the precision of the instruments, the efficiency of the sampling networks and the extent to which ΔS or ΔG can be measured in the catchment. Figure 1.7 gives the annual surface water balance for different continents while Figure 1.8 depicts the average daily water balance for the state of Mississippi in the USA.

In regions, which experience a long dry season the differences from year to year between soil moisture storage at the end of the dry season, are usually very small. Similarly, groundwater storage tends to return to a minimum storage state at the end of the dry season. Exceptions occur when the groundwater store drains out excessively during a drought period. On return to wet conditions a proportion of the storage input is required to top up the stores to a more usual state. Conversely, during a very wet period the groundwater storage is not depleted so fully, and there is a surplus of recharge, which tends to drain rapidly. These features are recognizable in the baseflow recession of the stream. By using the composite recession curve an estimate of the groundwater storage can be made.

As mentioned previously, the actual evaporation (AE) term is the term most commonly estimated by the water balance method (Blackie et al, 1979). By choosing a 'water year', i.e., a period of approximately 1 year running from the end of one dry season to the end of the next, consecutive estimates of annual AE can be obtained. For shorter-period estimations of AE detailed soil moisture records are required, and it is more usual to distribute the annual total according to an empirical seasonal model in the absence of direct measurements of AE. Water balance has also been used to estimate groundwater recharge (Lloyd et al, 1967) in an arid area of northeast Jordan. The authors estimated that, on average, 8.2% of the annual rainfall reached the aquifers—which is in agreement with two previous estimates obtained using different techniques.

If the development of a particular aquifer involves the abstraction of a quantity of water even approaching its recharge potential, the water balance should be used as an additional check to avoid overexploitation of the resource. Generally speaking, unless a large well field is contemplated for an urban area, the abstraction rates are likely to be well below the recharge potential. Israel is one country where abstraction is dangerously close to the rate of recharge, and careful monitoring of pumping is carried out along the coastal aquifers to prevent the intrusion of saline groundwater from the

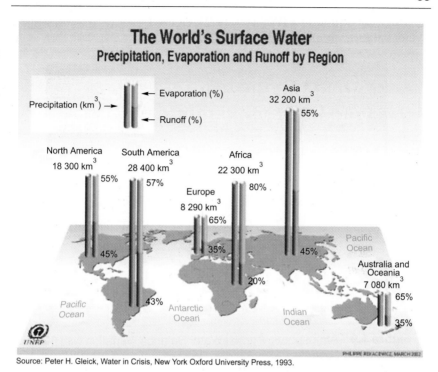

Source: Peter H. Gleick, Water in Crisis, New York Oxford University Press, 1993.

Figure 1.7 Annual Surface Water Balance for Different Continents.

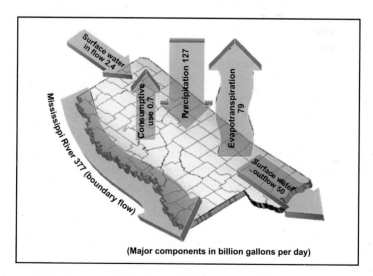

Figure 1.8 Average Daily Water Balance for Mississippi State, USA.

[Source: Anon., 2003b]

Mediterranean interface (Webster, 1971). Figure 1.9 shows the atmospheric water balance variation with the latitude.

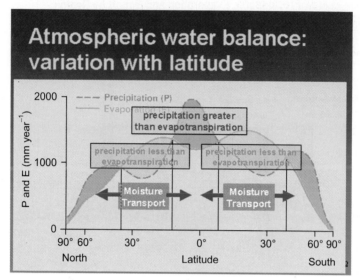

Figure 1.9   Atmospheric Annual Water Balance: Variation with the Latitude

[Source: http://mesoscale.agron.iastate.edu/agron406/406-4/sld022.htm]

The water balance equation also provides the basic framework for mathematical modeling of the hydrological cycle. Such models are usually stochastic-conceptual (see Clarke, 1973, for a full discussion of the use of mathematical models in hydrology): 'stochastic' because the chance of occurrence of the variables is taken into consideration and the concept of probability is introduced; and 'conceptual' because the form of the model is suggested by consideration of the physical processes acting on the input variable(s) to produce the output variable(s).

The unit-hydrograph approach to flood prediction, on the other hand, may be termed a deterministic-empirical model, if all variables are regarded as free from random variation ('deterministic') and the model is formulated without reference to the physical processes ('empirical') (O'Donnell, 1966; Dooge, 1965).

Where discharge records are short, and longer records of rainfall are available, a stochastic-conceptual model can be formulated in which the parameters are estimated for the period of common data. The longer rainfall records can then be used to extend the discharge period. Kidd (1983) gives a good example of these techniques applied to African conditions, whereas Blackie (1972) demonstrates a deterministic-conceptual model applied to the water balance of East African catchments.

Simple deterministic models can often be formulated to solve unknown variables in the water balance (Lloyd et al, 1967), but in order to introduce any element of probability (i.e., reliability), in e.g. estimating recharge or actual evaporation, it is necessary to introduce stochastic techniques unless exceptionally long periods of records are available.

Here again, the type of model must be suited to the end product desired, to the quantity and quality of data available, and to a cost-benefit analysis of exploiting a water resource with or without knowledge of the risks involved. The need for immediate practical benefits and the lack of data usually call for the use of mathematical models in most applications to livestock development. As investment costs increase, it is wise to examine the implications of proceeding without some examination of the medium- and long-term effects of an expensive system of exploitation. An understanding of the hydrological cycle, therefore, and the quantification of the various components in the form of a water balance, should be an essential prerequisite of all large-scale management plans.

1.4.1 The Soil-Water Budget

Where detailed data about soil layers, depth to groundwater, and vegetation are not available, hydrologists have often resorted to simple bucket models and budgeting schemes to model near-surface hydrology. The soil-water budget is a simple accounting scheme used to predict soil-water storage, evaporation, and water surplus. A typical budgeting time step is one day. Surplus is the fraction of precipitation that exceeds potential evapotranspiration and is not stored in the soil. The simple model used here does not distinguish between surface and subsurface runoff, so surplus includes both. For the Niger project, the main purpose of calculating the water budget is to estimate surplus, which serves as input to groundwater and surface water flow models. With this in mind, the basic equation for calculating surplus is:

$$S = P - E - \frac{\Delta w}{\Delta t} \qquad (1.14)$$

In Equation 1.14, S is surplus, P is precipitation, E is evaporation, $\frac{\Delta w}{\Delta t}$ is the rate of change of soil moisture w with time t. This model does not consider horizontal motion of water on the land surface or in the soil. Snowmelt was ignored in the water budget computations for this study because temperatures throughout the study region remain above freezing throughout the year.

A major source of uncertainty in evaluating Equation 1.14 is estimating the evaporation. Estimation of evaporation is based upon knowledge of the potential evapotranspiration, available water-holding capacity of the soil, and a moisture extraction function. The concepts of water-holding capacity and the method for evaluating Equation 1.14 are discussed here.

1.4.2 Water-holding Capacity

In order to calculate the soil-water budget, an estimate of the soil's ability to store water is required. Soil scientists to define the water storage capacities of soils under different conditions use several terms. The *field capacity* or *drained upper limit* is defined as the water content of a soil that has reached equilibrium with gravity after several days of drainage. The field capacity is a function of soil texture and organic content. The *permanent wilting point* or *lower limit of available water* is defined as the water content at which plants can no longer extract a health sustaining quantity of water from the soil and begin to wilt. Typical suction values associated with the field capacity and wilting point are –10 kPa (–0.1 bars) and –1500 kPa (–15 bars) respectively. Like water content, field capacity and permanent wilting point are defined on a volume of water per volume of soil basis. Given these two definitions, the *water available for evapotranspiration after drainage* (or the *available water-holding capacity*) is defined as the field capacity minus the permanent wilting point. Table 1.6 gives some typical values for available water-holding capacity.

Table 1.6   Water Retention Properties for Agricultural Soils

(values taken from ASCE, 1990)

Texture class	Field capacity	Wilting point	Available capacity
Sand	0.12	0.04	0.08
Loamy sand	0.14	0.06	0.08
Sandy loam	0.23	0.10	0.13
Loam	0.26	0.12	0.14
Silt loam	0.30	0.15	0.15
Silt	0.32	0.15	0.17
Silty clay loam	0.34	0.19	0.15
Silty clay	0.36	0.21	0.15
Clay	0.36	0.21	0.15

For budgeting calculations, it is useful to know the *total* available water-holding capacity in a soil profile. This value is typically expressed in mm and can be obtained by integrating the available water-holding capacity over the effective depth of the soil layer. A one-meter soil layer with a uniform available water-holding capacity of 0.15 has a total available water-holding capacity of 150 mm. Here, the term water-holding capacity means total available water-holding capacity in mm. Soil-water storage (mm) is denoted by w and water-holding capacity is denoted by w*. A large water-holding capacity implies a large annual evapotranspiration and small annual runoff relative to a small water-holding capacity under the same climatic conditions.

Dyck (1983) has provided a summary of some moisture extraction functions used by different investigators. Mintz and Walker (1993) have also illustrated several moisture extraction functions. Many researchers agree that soils show the general pattern of behavior that moisture is extracted from the soil at the potential rate until some critical moisture content is reached when evapotranspiration is no longer controlled by meteorological conditions. Below this critical point, there is a linear decline in soil moisture extraction until the wilting point is reached. Shuttleworth (1993) has noted that the critical moisture content divided by the field capacity is typically between 0.5 and 0.8. The type of moisture extraction function just described is commonly applied to daily potential evaporation values. A simpler function, $\beta' = w/w^*$, has been applied to monthly values.

There are several drawbacks to using simple soil moisture extraction functions. Mintz and Walker (1993) have cited field studies that show β' may vary with potential evapotranspiration for a given soil wetness and it may also vary with leaf-area index. In addition, the spatial variation of water-holding capacity is difficult to determine. A new and possibly better approach to determining the relationship between plant transpiration and potential evapotranspiration is to correlate β' with satellite-derived indices of vegetation activity so that β' will reflect plant growth stage and the spatial vegetation patterns. Gutman and Rukhovetz (1996) have investigated this possibility. Their approach still requires an estimate of potential evapotranspiration.

1.4.3 Budgeting Soil-Moisture Storage to Yield Surplus

Soil-moisture budget calculations are commonly made using monthly or daily rainfall totals because of the way data are recorded. Computing the water balance on a monthly basis involves the unrealistic assumption that rain falls at constant low intensity throughout the month, and consequently surplus estimates made using monthly values are typically lower than those made using daily values. Particularly in dry locations, the mean potential evaporation for a given month may be higher than the mean precipitation, yet there is observed runoff, and budgeting with monthly values may yield zero surplus. For this reason, the use of daily values is preferred over monthly values when feasible. In this study, daily rainfall records were provided by FAO for a number of stations in the Niger basin; however, the spatial coverage of these stations is sparse in some areas, and it is difficult to interpolate daily rainfall over space.

Because of the difficulty in estimating daily rainfall at regularly spaced computational points, and because a consistent basis for comparison is needed between the short term case (for which daily data are available) and the long term case (only monthly data are available), monthly data were used in this study. However, the use of monthly data does not yield enough surplus to match observed river flows in some areas of the Niger basin even with the assumption of zero overland flow and stream losses. To resolve this problem, a modification was made to the commonly used bucket model in which it is typically assumed that no surplus is generated until the soil is completely saturated with water; this assumption is not consistent with situations where the rainfall rate exceeds the infiltration rate of the soil. In the modified model, a fraction of the precipitation is extracted and declared runoff, before remaining precipitation is passed to the soil. This extraction of precipitation is represented by the term $P.\alpha_{i-1}$ in Equation 1.15 below. This scheme generates more runoff, enough to satisfy mass balance constraints in most areas of the Niger basin. Without this term, the model will predict zero runoff during many months of the year when some observed streamflow actually occurs. The runoff extraction term roughly accounts for "event" or "quick" flow that cannot be modeled using monthly averaged values. Equation 1.15 describes how soil moisture storage is computed.

$$w_i = w_{i-1} + P - P \cdot \alpha_{i-1} - PE \cdot \beta'_{i-1} \tag{1.15}$$

In Equation 1.15, w_i is the current soil moisture, w_{i-1} is the soil moisture in the previous time step, P is precipitation, PE is potential evaporation, α is the runoff extraction function, and β' is the soil-moisture extraction function. With monthly data, computations are made on a quasi-daily basis by assuming that precipitation and potential evapotranspiration for a given day are equal to their respective monthly values divided by the number of days in the—month. Several conditions apply when evaluating Equation 1.15. If w_i drops below zero, then set w_i equal to 0.01; if $w_i > w^*$

where w^* is the water-holding capacity, then the surplus for that day is $w_i - w^* + P \cdot \alpha$.

The soil-moisture extraction function $\beta' = w/w^*$ was used here. Since there is no precedent for the use of a runoff extraction function (α), the formulation of this function was more speculative and deserves further study. In the mean time, a simple relationship, $\alpha = w/w^*$, was used and yielded enough surplus to satisfy mass balance constraints in the surface flow routing model.

1.4.4 Balancing Soil Moisture

If the initial soil moisture is unknown, which is typically the case, a balancing routine is used to force the net change in soil moisture from the beginning to the end of a specified balancing period (N time steps) to zero. To do this, the initial soil moisture is set to the water-holding capacity and budget calculations are made up to the time period (N+1). The initial soil moisture at time 1, [w(1)] is then set equal to the soil moisture at time (N+1), [w(N+1)] and the budget is re-computed until the difference [w(1) − w(N+1)] is less than a specified tolerance.

1.4.5 The Spatial Soil-Water Budget

The soil-water budget is most easily applied at single points in space dictated by the location of climate stations where water-holding capacity can be measured or estimated, but the result of these point calculations must be interpolated over space in order to get a surplus volume. An alternative approach taken in the study was to use pre-computed climate, net radiation, and water-holding capacity grids, augmented with climate station measurements, to calculate the soil-water budget at each point on a 0.5 by 0.5 grid. Using a single value for precipitation, temperature, and net radiation in each 0.5 cell seems reasonable at the monthly time scale. The water-holding capacity may vary considerably within each 0.5 cell, and the value used in budgeting calculations is only an average property of the cell.

1.5 WATER BODIES

Water bodies are described by a plethora of different names in English— rivers, streams, ponds, bays, gulfs, and seas to name a few. Many of the definitions of these terms overlap and thus become confusing when one attempts to pigeon-hole a type of water body. Read on to find out the similarities (and differences) between terms used to describe water bodies.

We will begin with the different forms of flowing water. The smallest water channels are often called *brooks* but *creeks* are often larger than brooks but may either be permanent or intermittent. Creeks are also sometimes known as *streams* but the word stream is quite a generic term for any body of flowing water. Streams can be intermittent or permanent and can be on the surface of the earth, underground, or even within an ocean (such as the Gulf Stream).

A *river* is a large stream that flows over land. It is often a perennial water body and usually flows in a specific channel, with a considerable volume of water. The world's shortest river, the D River, in Oregon, USA is only 36.6 m long and connects Devil's Lake directly to the Pacific Ocean.

A *pond* is a small lake, most often in a natural depression. Like a stream, the word *lake* is quite a generic term—it refers to any accumulation of water surrounded by land—although it is often of a considerable size. A very large lake that contains salty water, is known as a *sea* (except the Sea of Galilee, which is actually a freshwater lake).

A sea can also be attached to, or even be part of, an ocean. For example, the Caspian Sea is a large saline lake surrounded by land, the Mediterranean Sea is attached to the Atlantic Ocean, and the Sargasso Sea is a portion of the Atlantic Ocean, surrounded by water.

Oceans are the ultimate bodies of water and refers to the four oceans— Atlantic, Pacific, Arctic and Indian (although sometimes the Antarctic Ocean or the Southern Ocean is referred to—it is the ocean south of $50°$ South latitude, surrounding Antarctica). The equator divides the Atlantic Ocean and Pacific Oceans into the North and South Atlantic Ocean and the North and South Pacific Ocean.

Coves are the smallest indentations of land by a lake, sea, or ocean. A *bay* is larger than a cove and can refer to any wide indentation of the land. Larger than a bay is a *gulf*, which is usually a deep cut of the land, such as the Persian Gulf or the Gulf of California. Bays and gulfs can also be known as *inlets*.

Any lake or pond directly connected to a larger body of water can be called a *lagoon* and a *channel* is a narrow sea between two landmasses, such as the English Channel.

1.6 REFERENCES

Anon., (2003a), "Diagram of aquifers in relation to hydrologic cycle".

Anon., (2003b), "Average daily water balance of Mississippi state, USA".

ASCE (American Society of Civil Engineers), 1977, *Sedimentation Engineering*. ASCE, New York.

ASCE, (1990), *Evapotranspiration and Irrigation Water Requirements*, Jensen, M.E., R.D. Burman, and R.G. Allen (editors), ASCE Manuals and Reports on Engineering Practice No. 70.

Ayoade J.O., (1977), A preliminary study of the magnitude, frequency and distribution of intense rainfall in Nigeria. In: *Proc. Symp. on Flood Hydrology*, Nairobi 1975. TRRL Suppl. Rep. 259. Transport and Road Research Laboratory, Crowthorne, UK.

Berry G., (1964), The evaluation of Penman's natural evaporation formula by electronic computer. *Austr. J. Appl. Sci.* 15(1): 61-64, Australia.

Blackie J.R., (1972), The application of a conceptual model to some East African catchments. M.Sc. thesis, Imperial College, London, UK.

Blackie J.R, Edwards K.A and Clarke R. T., (1979), Hydrological research in East Africa. *East Afr. Agric. For. J.* 43 (Special Issue)

British Standards Institution (BSI), (1965), *Methods of measurement of liquid flow in channels, British Standard 3680. Part 4A: Thin plate weirs and flumes. British Standards Institution*, London, UK.

Brooks C.E.P. and Carruthers N., (1953), *Handbook of statistical methods in meteorology*. HMSO, London.

Brutsaert W. and Stricker H., (1979), An advection-aridity approach to estimate actual regional evapotranspiration. *Water Resour. Res.* 15: 443-450.

Brutsaert, W., (1982), *Evaporation into the Atmosphere: Theory, History, and Applications*, D. Reidel Publishing Company, Dordrecht, Holland.

Chidley, T.R.E. and Keys K.M., (1970), A rapid method of computing areal rainfall. *J. Hydrol.* 12: 15-24.

Chow, V.T., D.R. Maidment, and L.W. Mays, (1988), *Applied Hydrology*, McGraw-Hill, Inc., New York, NY. USA.

Clarke R.T., (1973), *Mathematical models in hydrology*. Irrigation and drainage paper 19. FAO, Rome Italy.

Dooge J.C.I., (1965), Analysis of linear systems by means of Laguerre functions. *J. Siam Control Ser.* A2: 396-408.

Doorembos J. and Pruitt W.O., (1977), "Crop water requirements", *Irrigation and drainage paper 24*. FAO, Rome, Italy.

Dyck, S., (1983), "Overview on the Present Status of the Concepts of Water Balance Models," *IAHS Publ. 148*, Wallingford, 3-19.

East African Meteorological Department, (1961), *Probability maps of annual rainfall in East Africa*. East African Community, Nairobi, Kenya.

Edwards, K.A., (1972), "Estimating areal rainfall by fitting surfaces to irregularly spaced data". *Proceedings of the WMO/IAHS symposium on the distribution of precipitation in mountainous areas*. Vol II: 565-587.

Edwards, K.A. and Blackie J.R., (1981), Results of the East African catchment experiments 1958-1974. In: eds Lal R. and Russell E. W. *Tropical agricultural hydrology*. John Wiley, Bath, UK, 163-188.

Edwards, K.A., Field C.R. and Hogg I. G. G., (1979), *A preliminary analysis of climatological data from the Marsabit District in northern Kenya*. UNEP/MAB Integrated Project in Arid Lands, Tech. Report B-1, Nairobi, Kenya.

Freeze R.A. and Cherry J.A., (1979), *Groundwater*. Prentice-Hall, Englewood Cliffs, New Jersey, USA

Gleick, Peter, H., (1993), Water in crisis, Oxford University Press, New York, USA.

Glover, J. and McCulloch, J.S.G., (1968), The empirical relationship between solar radiation and hours of sunshine. *Q.J.R. Meteorol. Soc.* 82:172-175.

Gommes, R. and Houssiau M., (1982), Rainfall variability, types of growing season and cereal yields in Tanzania. In: *Proceedings of the technical conference on climate in Africa.* WMO/OMM No. 596. World Meteorological Organisation, Geneva. 312-324.

Gregory S., (1969), Rainfall reliability. In: *Environment and land use in Africa.* Thomas and Whittington, Methuen, London, UK. pp. 57-82.

Gutman, G., and L. Rukhovetz, (1996), "Towards Satellite-Derived Global Estimation of Monthly Evapotranspiration Over Land Surfaces," *Adv. Space Res.,* 18, No. 7, 67-71.

Hillel D., (1971), *Soil and water: Physical principles and processes.* Academic Press, New York.

http://www.fao.org/Wairdocs/ILRI/x5524E/x5524e04.htm (2003)

http://mesoscale.agron.iastate.edu/agron406/406-4/sld022.htm

http://www.srh.noaa.gov/lmrfc/ about/#intro

http://www.unesco.org/science/waterday2000/Cycle.htm

Hutchinson, P. and Walley W.J., (1972), "Calculation of areal rainfall using finite element techniques with altitudinal corrections". *Bull. IAHS.* 17: 259-272.

Jackson, I.J. (1977), *Climate, water and agriculture in the tropics.* Longman, London, UK.

Kenworthy, J.M. and Glover, J., (1958), The reliability of the main rains in Kenya. *East Afr. Agric. J.* 23: 276-72.

Kidd, C.H.R. (1983), *A water resources evaluation of Lake Malawi and the Shire River.* UNDP/WMO Report. Project MLW/77/012. UNDP, Lilongwe, Malawi.

Lee, P.S., Lynn P.P. and Shaw E.M. (1974), Comparison of multiquadric surfaces for the estimation of areal rainfall. *IAHS Hydrol. Sci. Bull.* 19(3): 303-317.

Linsley, R.K. and Franzini J.B., (1964), *Water resources engineering.* McGraw-Hill, New York, USA.

Linsley, R.K., Kohler, M.A. and Paulhus, J.L.H., (1998), *Hydrology for engineers,* McGraw Hill, New York, USA

Lloyd, J.W., Drennan D.S.H. and Bennell B M. (1967), A groundwater recharge study in northeastern Jordan. *J. Instn. Water Engrs.* Paper No. 6962: 615-631.

Lockwood, J.G., (1972), *World climatology.* Edward Arnold, London, UK.

Manning H.L., (1956), The assessment of rainfall probability and its application in Uganda agriculture. *Proc. R. Soc. London, Ser.* B 144: 460-480.

McCulloch, J.S.G., (1965), Tables for the rapid computation of the Penman estimate of evaporation. *East Afr. Agric. For. J.* 30: 286-295.

Mintz, Y., and G.K. Walker, (1993), "Global Fields of Soil Moisture and Land Surface Evapotranspiration Derived from Observed Precipitation and Surface Air Temperature," *J. Appl. Meteor.,* 32: 1305-1334.

Monteith J.L., (1965), Evaporation and environment. *Symp. Soc. Exp. Biol.* 19: 205-234.

Nace, R. L. (ed.)., (1971), *Scientific framework of the world water balance.* UNESCO Tech. Papers Hydrol. 7. UNESCO, Paris, France.

Nagel, J.F., (1962), Fog precipitation measurement on Africa's south-west coast. *NOTOS* (Pretoria Weather Bureau). pp. 51-60, South Africa.

NERC (Natural Environment Research Council)., (1975), *Flood studies report* (Five volumes). NERC, Whitefriars Press, London, UK.

O'Donnell, T., (1966), Methods of computation in hydrograph analysis and synthesis: Recent trends in hydrograph synthesis. In: *Proc. Tech. Meeting 21. Cent. Org. for Appl. Sci. Res.* Netherlands.

Penman, H.L., (1956), Evaporation—an introductory survey. *Neth. J. Agric. Sci.* 4: 9-29.

Penman, H.L. (1963), *Vegetation and hydrology.* Tech. Commun. 53. Commonwealth Bureau of Soils. Common. Agric. Bur., Farnham Royal, UK.

Prabhakar, D.R.L. (1977), Description of current flood prediction methods in East Africa. In: *Proceedings of the Symposium on Flood Hydrology,* Nairobi 1975. TRRL Suppl. Rep. 259. Transport and Road Research Laboratory, Crowthorne, U.K. pp. 205-226.

Priestley, G.H.B. and Taylor R J., (1972), On the assessment of surface heat flux and evaporation using large-scale parameters. *Mon. Weather Rev.* 100: 81-92.

Shaw E.M. and Lynn, P.P., (1972), Areal rainfall evaluation using two surface fitting techniques. *IAHS Hydrol. Sci. Bull.* 17(4): 419-433.

Shuttleworth, J.W., (1993), "Evaporation," Handbook of Hydrology, D.R. Maimdent Editor, McGraw-Hill, Inc., USA.

Singh, V.P., (1992), *Elementary Hydrology,* Phentice Hall, New Jersey, USA.

Stewart, J.B. Batchelor, C.H. and Gunston, H., (1982), Actual evaporation: Preliminary study of the usefulness of the method of estimation proposed by Brutsaert and Stricker. Report of the Institute of Hydrology. Wallingford, UK.

Subramanya, K., (1994), *Engineering Hydrology,* 2nd Ed., Tata McGraw Hill, New Delhi, India.

Thom, A.S. and Oliver H R., (1977), On Penman's equation for estimating regional evaporation. *Q. J. R. Meteorol. Soc.* 103: 345-357.

Tilren, O.A., (1979), *Manual on procedures in operational hydrology* (Five volumes). Norwegian Agency for International Development, Oslo, Norway.

Todd D.K., (1959), *Groundwater hydrology.* John Wiley, New York, USA.

TRRL (Transport and Road Research Laboratory), (1977), TRRL Supplementary Report 259. In: *Proceedings of the Symposium on Flood Hydrology,* Nairobi 1975. TRRL, Crowthorne, U.K.

Ven te Chow, (1964). *Handbook of applied hydrology.* McGraw-Hill, New York, USA.

Webster, K.C., (1971), *Water resources development in Israel.* Tahal Consulting Engineers Ltd. Water Planning for Israel Ltd., Tel Aviv, Israel.

Chapter 2

Water Uses

2.1 GENERAL

Worldwide, agriculture has been the single biggest drain on water supplies, accounting for about 69 percent of all use. About 23 percent of water withdrawals go to meet the demands of industry and energy, and just 8 percent to domestic or household use. Patterns of use vary greatly from country to country, depending on levels of economic development, climate and population size. Africans, for instance, devote 88 percent of the water they use to agriculture, mostly irrigation, while highly industrialized Europeans allot more than half their water to industry and hydropower production.

In 1990, 250 million hectares of land were under irrigation worldwide; supplying a third of the world's harvested crops and agriculture was the primary use for water in two out of three countries. Agricultural water use is particularly high in arid areas such as the Middle East, North Africa and the southwestern United States, where rainfall is minimal and evaporation so high that crops must be irrigated most of the year. Afghanistan and Sudan apply an estimated 99 percent of all the water they use to agriculture.

Domestic water use—including drinking, food preparation, washing, cleaning, gardens and service industries such as restaurants and Laundromats—accounts for only a small portion of total use in most countries. The amount of water people apply to household purposes tends to increase with rising standards of living, and variations in domestic water use are substantial. In the United States, each individual typically uses more than 700 liters each day, for domestic tasks. In Senegal, the average individual uses just 29 liters daily to meet domestic requirement, which is less than one-twentieth of that in the USA. Domestic needs account for a greater share of overall use in countries, rich or poor, that have little agriculture or industry. In both Kuwait and Zambia, nearly two out of every three liters of water are used in households.

Industry, a category that includes energy production, uses water for cooling, processing, cleaning and removing industrial wastes. Nuclear and fossil-fueled power plants are the single largest industrial users, applying staggering amounts of water to the job of cooling. While most of the water used for industrial purposes is returned to the water cycle, chemicals and heavy metals often contaminate it, or its temperature is increased to the detriment of water ecosystems. Industrial use varies from less than 5 percent of withdrawals in dozens of developing countries to as much as 85 percent in Belgium and Finland. Only in Europe, where reliance on irrigation is relatively low, does industrial water use equal the sum of water applied to agriculture and domestic uses.

In complex society water use is varied, from municipal and agricultural to mining and other industrial uses. Herein broad categories of water use and its cost is discussed with description within each category. For example, municipal water use includes diverse activities, from plant watering to toilet flushing.

2.2 WATER SUPPLY FOR RURAL AND URBAN NEIGHBORHOODS

Water supply is one of the most important facets to the life of man, if not the most important. Concern for water and its adequate supply is the most overbearing problem faced by peoples all over the world. The problem and its causes in its perspective are discussed in the following articles.

Over 1 billion people—about one-fifth of the world's population—lack access to safe drinking water. A brief review of the most recent figures on access to safe drinking water provided by the World Bank and the UN indicate that the majority of safe drinking water problems appear to be concentrated in just five nations as given in Table 2.1.

Table 2.1 Major Countries having Populations Without Access to Safe Drinking Water

Country	People without access to safe drinking water (estimated number in millions)
China	124–348
India	182–355
Indonesia	75–77
Pakistan	58
Nigeria	67–72
Total	513–900

(The high estimates come from the World Bank's World Development Report 1997 and World Development Indicators 1997, and the low estimates come from the United Nations The State of World Population 1997)

Any international effort to greatly enhance the gross number of people with access to safe drinking water must focus upon and fully involve these key countries which account for the majority of the people without safe drinking water. The scenario of population growth, leading to degradation of water resources, leading to disease and low child survival rates, leading to population growth is most illustratively borne out in urban centers, where population densities are the highest. The United Nations has noted that:

Water-related disease is particularly acute in these urban communities. In 1985, at least 25 per cent of urban communities (and 58 per cent of rural communities) were without clean water for sanitation needs. As the countryside around cities like Manila and Panama City loses its tree cover, so too the domestic water supplies decline in quantity and quality. In turn, human communities face a growing threat of contaminated-water pandemics. Similarly, public-health programmes in Bangkok, Nairobi, Lagos and Abidjan, among other conurbations of the humid tropics, are being set back because their water supplies are declining in the wake of deforestation in upland catchments (UNFPA, 1991).

According to the World Bank's World Development Indicators 1997 between 1985 and 1993, 24 countries have seen their urban population's access to safe drinking water decrease. These countries are: Colombia, Ecuador, Ethiopia, The Gambia, Guatemala, Guinea, Guinea-Bissau, Haiti, Jamaica, Jordan, Madagascar, Mali, Mauritania, Mozambique, Nepal, Nicaragua, Saudi Arabia, Tanzania, Togo, Trinidad and Tobago, United Arab Emirates, Uruguay, Venezuela, and Zimbabwe.

It should come as no surprise that most of the countries that saw a decline in urban access to safe drinking water also have a high rate of urbanization. It should also come as no surprise that many are in Africa, considering that Africa, with an urban growth rate of 4 per cent per year, is urbanizing faster than any other world region.

Meanwhile, thirteen countries have seen their rural population's access to water decrease between 1985 and 1993. They are Bolivia, Botswana, Cameroon, El Salvador, Ethiopia, Gabon, Madagascar, Nigeria, Saudi Arabia, Trinidad and Tobago, Tunisia, Tanzania, and United Arab Emirates. The following articles provide some case studies from India and Nepal in the developing world and USA in the developed world.

2.2.1 Strategies for Sustainable Water Supply for All in India

2.2.1.1 Water Supply in India—Scenario Around the Year 2000

India has achieved significant strides in the water supply sector. Yet in 1991, only about 82% of the urban households covering 85% of the urban population had access to safe drinking water. Against the national average target of supply of 140 liters per capita per day (lpcd) of water, the per-capita consumption is too low and ranges from 165 lpcd in few larger towns to about 50 lpcd in smaller towns. In fact, the availability in the urban slums is around 27 lpcd. The situation is critical in many large cities. For instance, in Chennai—the fourth largest city of the country, it is estimated that only 2 million of the 3.7 million residential consumers within the service area of the Water Supply and the Sewerage Board are connected to the system. On an average, they receive a supply of about 36 lpcd while the rest within the service area use the public taps which serve about 240 persons per tap.

2.2.1.2 Financial Requirements for Water Supply

Water supply and sanitation is a state subject and the Government of India essentially supports funding of projects in this sector through the Plan provisions. The Plan provisions for this sector have steadily increased from 0.65 per cent of the total outlay in the Second Five Year Plan (1956-61) to 1.81 per cent in the Seventh Five Year Plan (1985-90). However, due to financial inadequacy, overpopulation the coverage achieved in urban areas was only 85 per cent with safe drinking water and 45 per cent with sanitation.

The resources required to achieve 100 per cent coverage with safe drinking water and 75 per cent with sanitation are massive. The Planning Commission of India has estimated additional investment needs for water supply for the period 1996 to 2001, to be about US$ 3000 million.

2.2.1.3 Factors Inhibiting Development of Sustainable Water Supply Systems in India

There are a number of factors that inhibit achieving sustainable water supply in India and they are as follows:

Poor Financial Status of Local Bodies: Provision of water supply in cities and towns has remained the primary responsibility of the Urban Local Bodies (ULBs). The financial base of the ULBs over the last four decades has become increasingly fragile and the ULBs are finding it difficult to maintain even the existing lower level of water supply services or meet essential expenditure on staff out of the revenues that they are able to collect from tax sources. The task of development, maintenance of the distribution system and collection of water charges generally remained with the local bodies. The Plan funds of urban development are spent through a variety of State level agencies such as State or institutions like Water Supply and Sewerage Boards, Public Health Engineering Departments etc., the assets created are transferred to the local bodies for operation and maintenance without assuring them the consequent Plan assistance.

Unrelated Links between Cost, Price and Consumption: In India, the pricing mechanism is rarely used to guide the decision of the consumer, as to how much to consume or to balance supply with demand. The existing water (connection) charges and water tariffs are highly subsidised. Revision of water charges and water tariff have remained indifferent to the inflation rate. Quite often, the State governments, being the guarantors of loans received from the financial institutions for implementation of water supply systems, come to the rescue of the state level agencies for repayment of loans and the water tariff structures intended to be revised remains untouched due to various socio-economic and political reasons. The ratio between the water tariff and water charges collected and expenditures incurred on operation and maintenance in some of the States has been found to range between 30 per cent and 46 per cent.

The Continuance of Poor Staff Strength for Maintenance Activities: The responsibilities of the local bodies include identification of sources, generating of potable water from those sources, distribution of water, fixation of tariffs and charges, collection of revenue and operation and maintenance. Studies reveal a substantial shortage in the staff strength of appropriate calibre to deal with the above complex issues. Poor cost recovery has been primarily attributed to irrational water charges, wastages and mismanagement. A 'National Policy towards Full Cost Recovery' in respect of water supply and sanitation sector was adopted in March 1993. A sense of public participation in terms of 'customer' rather than 'consumer' needs to prevail. A quick response system to the customer needs and problems coupled with an efficient metering system and transparent billing and collection would substantially improve the 'willingness to pay' of the public.

Absence of Regular Maintenance and Consequent Higher Operation and Maintenance Cost: There exists a general apathy of the ULBs towards the important issue of maintenance of water supply systems. To assist the local bodies, the Central Public Health Environmental Engineering Organization (CPHEEO) of the Government of India has formulated a 'Manual on Water Supply and Treatment' which lays down guidelines as to how the systems should be maintained. The guidelines also lay emphasis on keeping a set of plans giving details of the layout and the production/distribution lines; establishing a systematic programme for daily operations including an operation schedule for machinery and equipment; keeping data and record on all equipment—their condition, when repaired and replaced; maintenance of records on the analysis on the waste collected at various points and listing safety measures that are necessary for proper maintenance of the system; etc. However, in most of the towns these are more often not followed. Water supply projects are capital intensive and require longer repayment period more so due to initial capital costs and recurring operation and maintenance costs. Neglect of maintenance of assets created has led to decline in quality of services resulting in resistance from the users against any increase in tariff rates/user charges.

Substantial Losses and Leakages: A considerable portion of the treated water is lost through leakages in the transmission and distribution system. Reduction in loss due to leakages through leak detection equipment has been found to be one of the most vital measure to minimize service cost and maximize service, but this aspect is still largely ignored, there are no reliable estimates of the actual quantities lost. This has a direct linkage to the revenue generation and sustainability of the entire water supply system. The awareness about use of conservation methods/equipments to ensure efficient use of water is generally lacking. Detailed investigations carried out by National Environmental Engineering Research Institute (NEERI) have revealed that about 17 to 44 percent of the total flow in the distribution system is lost as unaccounted through leakages in main, communication and service pipes and leaking valves. The major portion of leakage (about 82 percent) occurs in the house service connection, through service pipes and taps. The remaining 18 percent is due to leakages in pipelines. Water supply is unmetered in major parts of urban areas and significant proportion is supplied to low-income areas through stand posts resulting in unaccountable losses.

High Administrative and Supervision Charges Burden on ULBs: Operation and maintenance of water supply and sewerage schemes is the responsibility of the ULBs. The administrative and supervision charges of the Water Supply and Sewerage Boards is quite high and varies between 18 and 22 percent and is calculated as a percentage of the total project cost.

Lack of Recycling Initiatives: In India, water is essentially used as a one time commodity. Often treated and un-treated water is used indiscriminately. There is substantial scope for segregated use of the water for appropriate uses and recycling of the wastewater for further use for gardening, industries, street cleaning, fire fighting, agriculture etc. This also brings in another important consideration that the same quality of water used for drinking purposes need not be misused for large number of other activities like flushing, washing besides other uses, and from that point of view, the possibilities of alternate water supply systems could also be kept in view for potable and un-potable water.

2.2.1.4 Initiatives in India towards Achieving Sustainable Water Supply for All

Recognizing the need for an integrated approach to the provision of those environmental services and policies that are essential for human life, the 'Global Plan for Action: Strategies for Implementation', urges the Governments at the appropriate levels, in partnership with other interested parties, to ensure that clean water is available and accessible to all human settlements as soon as possible. It recognizes that water resources management in human settlements presents an outstanding challenge for sustainable development which combines the challenge of securing for all the basic human needs for a reliable supply of safe drinking water and meeting the competing demands of industry and agriculture, which are crucial to economic development and food security, without compromising the ability of future generations to meet their water needs.

Emphasizing the need for a strong political commitment the Habitat Agenda stresses the need to pursue policies for water resources management that are guided by the broader consideration of economic, social and environmental sustainability of human settlements at large, rather than by sectoral considerations alone and establishment of strategies and criteria (biological, physical and chemical water quality) to preserve and restore aquatic ecosystems in a holistic manner, giving consideration to entire drainage basins and the living resources contained therein. It advocates for management of supply and demand for water in an effective manner that provides for the basic requirements of human settlements development, while paying due regard to the carrying capacity of natural ecosystems. More importantly, it suggests promotion of partnerships between the public and private sectors and between institutions at the national and local levels so as to improve the allocative efficiency of investments in water and sanitation and to increase operational efficiency. In regard to the institutional mechanisms, it impresses to implement the institutional and legal reforms necessary to remove unnecessary overlaps and redundancies in the functions and jurisdictions of multiple sectoral institutions and to ensure effective coordination among those institutions in the delivery and management of services. It advocates for the introduction of economic instruments and regulatory measures to reduce wastage of water and encourage recycling and reuse of waste water and to develop strategies to reduce the demand for limited water resources by increasing efficiencies in the agricultural and industrial sectors. It also emphasizes the need to involve women in decision-making process in regard to management of infrastructure systems at large.

In recognition of such concerns in advance, a number of new initiatives have been taken in the last seven years for achieving sustainable water supply systems in India. The details are given below:

Registration Charges on Water Connections and Collection of Advance Payments: In order to elicit the cooperation of the public in advance, it would be appropriate for the Development Authorities/Local Bodies to announce a registration scheme as being done in housing and telecommunications sector inviting applications from the public to register their names for new connections as and when the city authorities plan to implement water supply scheme. A non-refundable one time deposit can be collected from the houseowner who are willing to get the house service connection from the proposed water supply scheme. This will enable the local authorities to gain part of the resources required against the capital cost or generate revolving funds even before the service is made available. The Tirupur Municipality in Tamil Nadu has tried this proposal successfully at Tirupur. An amount of Indian Rupees 25 million from 10,000 house-owners at the rate of Indian Rupees (Rs) 2000 per house and Rs 4,000 from commercial institutions has been collected. Part of this amount has been deposited with the Tamil Nadu Water Supply and Drainage Board— the state-level implementing agency, for executing the work of laying additional distribution system of 110 km in the town.

Metering of Water Connection: The procedure of application of flat rates for un-metered connections is against the principles of allocative efficiency. Metering of water supplied to industrial and commercial premises have largely been accepted, metering domestic water supply has been a subject of debate for a long time. Findings of a study carried out in the state of Uttar Pradesh revealed that the consumption of domestic water supply came down to less than 50 per cent when metering was resorted to. In view of the fact that substantial percentage of the area does not have the facility of metering, flat rates are levied which are based on the location as well as the area of houses. In certain cases, even the size of the ferrule, which connects to the main water supply line, is taken as the parameter to decide the charges payable on flat rate monthly.

Realistic Pricing Policies: The pricing policies should be on the principle of full cost recovery. Water supply and sanitation agencies including local bodies need to be given full autonomy for determination of tariff with the provision for automatic annual increase to cover the average incremental operation and maintenance costs, depreciation charges, debt services, etc. Associations and involvement of local bodies and the resident population at large from the conceptual stage to implementation stage of the project and also in fixation of tariff will ensure sustainable operation of the service system. The local bodies, in general, adopt the rates that are prescribed for by the State as a whole as a minimum and they rarely exercise the power conferred on them to enhance such rates depending on the local conditions and requirements.

In addition, the financial institutions in India are increasingly making it a pre-condition to gain access to institutional credit. For instance, in respect of the water supply scheme in Jaypore, Orissa, the Housing and Urban Development Corporation (HUDCO) had emphasized the need for immediate hike in tariff rates both in respect of domestic, commercial and industrial rates followed by an annual automatic increase of 10 percent. In addition a one time connection charge of Rs. 4000 per connection was also insisted. Similar is the case of Kolhapur in Maharashtra.

Conservation through Rain Water Harvesting: Ground water is very valuable and to avoid its depletion measures to recharge the aquifers should be extensively practiced. In places where there is an acute scarcity of drinking water, there is an increasing need to adopt rain water harvesting methods to adequately recharge the aquifers. In the water starved city of Chennai, people resorted to rain harvesting in a big way during the North-East monsoon in 1993 following acute scarcity preceding the monsoon. The rainwater was collected in ground level sumps, which was subsequently used after filtering and boiling. A number of voluntary organizations took the case for wider adoption and today incorporation of provisions for rain water harvesting and aquifer charging in the building plans in Chennai is mandatory for approval as the Chennai Municipal Development Authority has incorporated a regulation for conservation of rain water in its building guidelines in the area of Chennai in view of the acute scarcity of water.

Private Sector participation in Water Supply: It is believed that owing to the capital intensive nature with long gestation period, private sector would not be interested in urban infrastructure provision, and that it may not be desirable to allow full private sector monopolies in this crucial sector. Varied forms of public-private partnership in water supply provision are being practiced in many developing countries with ranging options from large scale trucking , formation of water corporation, water vending kiosks and door-to-door service, coin-operated meters. Recently many progressive cities like Pune, Belgaum, Dewas, Bangalore, Tirupur etc. are coming up with privatization of water supply on a build-own-transfer (BOT), build-own-operate-transfer (BOOT) basis.

Unbundling of Water Supply Systems: A substantial scope for involvement of private sector in water supply provision requires unbundling of the operations. The private sector could be involved effectively in the source development from where bulk transfer of treated water could be effectuated through a bulk water purchase agreement by the ULBs. While the distribution has to remain under the control of the public sector, the collection of tariffs/user charges could be effectively delegated to the private sector with the allocation of commensurate profit, which would encourage them to maintain appropriate metering and reach of water to the individual households.

Participation of Community Groups: It is realized that the user participation, either as provider or for performance assessment, can be critical to the effectiveness of the smaller community level infrastructure services. In many of these, the provisions and operations at the local level can be better handled by the user or community groups. Important initiatives taken in this context are given below:

- Specific arrangements for involving user and community groups may be achieved by unbundling of the services in an effective manner. For example, depending on the technical considerations of scale economies, local distribution networks for water may be provided by the community groups and they may be supplied with bulk water by a water utility company.

- The City and Industrial Development Corporation (CIDCO), a public sector institution at Navi Mumbai (i.e., New Bombay) has had a successful experience with privatization efforts. The privatization experience includes maintenance of sewerage pumps, and water pumps, meter reading and billing, maintenance of parks and gardens, collection of CIDCO's service charges and so on. CIDCO has given the responsibility of collection of its service charges to the Senior Citizens Club (an association of retired persons) to whom it pays 1 per cent as commission. If CIDCO was to collect the charges on its own it would have cost it three times more.

- An important initiative is to make available the cooperative societies, the bulk water supply by the local agencies and in turn the members of the Group Housing to take care of the distribution of available water and collection of water charges.

- A unique example of community participation in the field of water supply has been the experience of the Baroda Citizens Council (BCC), constituted in 1966 by the partnership of University of Baroda, Baroda Municipal Corporation, the American Friends Service Society and the Gujarat Federation of Mills and Industries. Initially the BCC was involved in construction of family toilets, handpumps, sanitation facilities and training of health and hygiene workers and training of women volunteers in handpump maintenance. In this scheme the beneficiaries contributed upto 70 percent of the capital costs. The remaining came from contributions from bilateral and other donors, routed through the BCC. Usually the contribution per family were in the range of Indian Rupees (Rs) 500 per family for the Mark III hand pump and Rs 250 per family for a community drainage scheme. These handpumps and community drains are fully maintained by the community and money is collected from the users for procuring spare parts or for engaging mechanics. The BCC has moved on to find sustainable systems for financing the maintenance and upgradation of this infrastructure in the slums with the help of the Municipal Corporation and the UNICEF.

- A significant initiative has been taken through a centrally sponsored scheme namely 'Urban Basic Services for the Poor (UBSP)'. The programme primarily aims at integration of community efforts through a convergence approach involving the basic infrastructure services including health services with additional focus on livelihood for the people. The scheme is primarily operated in the slum areas and so far has addressed about 7 million slum dwellers as against total target group of 50 million slum dwellers. The scheme encourages formulation of neighborhood groups (of 20 houses) represented by Women Resident Community Volunteer who combine to form a community development society under a community organizer.

- A National Level Initiative for Sustainable Rural Water Supply taken by the Government is the establishment of the Rajiv Gandhi National Drinking Water Mission which was launched in August 1986 to accelerate the progress of drinking water supply in rural areas and to provide cost effective science and technology inputs to improve the programme implementation in active collaboration and cooperation with the states, local people and institutions. The Mission's objective is to provide safe drinking water free from chemical and biological contamination as also ensure provision of safe drinking water of 40 liters per person per day (lpcd) in all areas for all human being and additional 30 lpcd in Desert Development Programme areas for drinking water requirement of cattle. Habitations, which are not getting full supply of 40 lpcd, are treated as partially covered requiring augmentation facilities to bring them to the level of 40 lpcd.

- The Mission's major activities include the improvements in the quality of drinking water through the sub-missions on eradication of guineaworm, control of fluorosis, removal of excess iron and brackishness, removal of arsenic, water conservation and recharge of aquifers. In addition, other programmes on water quality surveillance, training of villagers and officers/staff involved in the programme, research and development, and information, education and communication for health awareness are being implemented in cooperation with the state/Union territory Governments, Panchayats and non-Governmental Organizations (NGOs) with special provisions for scheduled castes and scheduled tribes.

- Another innovative initiative in Gujarat has been the organization of 'Paani Panchayats' (Water Courts). An NGO called 'Shelter for Health Awareness' was asked to help in developing local users' water committees to increase the per capita availability of water. The forum was also asked to distribute available water and help in resolving disputes between the villagers on issues pertaining to water supply.

Role of Financing Agencies in Institutionalizing Change: The financing institutions have been substantially successful in sensitizing the ULBs on the need to evolve and implement infrastructure systems which are sustainable. In order to make the water supply schemes sustainable, while agreeing to extend financial assistance HUDCO emphasizes on the provisioning of i) principle of full cost recovery to be adopted, ii) adequate subsidy to be provided in a transparent manner to meet the basic minimum requirement of the poor, iii) efforts to be made for cost reduction by effective savings on manpower, energy consumption, reduction in leakages, improvement in billing and collection, etc., iv) concerned agencies including the local bodies, to be given full autonomy for determination of tariffs with the provision for automatic annual increase to cover costs, v) tariff fixation to be based on average incremental cost including O&M cost depreciation charges, debt dues etc., vi) state level institutions should associate the local bodies and the community at large to instill better sense of participation, vii) submitting to the fact that chances of success of privatization are greater in operation and maintenance, HUDCO agrees that privatization could be introduced for new installations initially, viii) compulsory 100 per cent metering, ix) elimination of Stand Post as far as possible and x) operation of escrow account.

Substantial capacity building programmes are needed for all the stakeholders, for them to play an important role in the sector. Such initiatives can result in the ULBs adopting the financial viability and user pay approaches in other sectors too.

2.2.2 The Case Study of 'Akshyadhara' in India

One of the several methods suggested for urban water supply in India is the Akshayadhara concept. "Akshaydhara"—adopted from Sanskrit, means pristine, perennial flow. The key element of this approach is the manipulation of shallow depth (<20 m) soil-aquifer system to effect renovation of storm- and domestic waste-water and its subsequent transmission to surface water bodies through natural subsurface flow, maintaining their pristine quality perennially.

The system is based on the fact that non-sanitary (grey water) component of the domestic wastewater is much larger (>70%) in volume and much less dangerous than the sanitary component (dark water). The two basic building blocks of the proposed *Akshyadhara* approach are based on the concept of soil-aquifer-treatment (SAT) of wastewater. These are:

- The percolation well for renovation and recharge of storm water and the non-sanitary component of domestic wastewater, and
- The infiltration basins for renovation of primary settled municipal wastewater.

A percolation well is a variant of the well known soak pit—modified to (i) effect high percolation rates, (ii) provide large volume for temporarily holding for storm water, and (iii) effect adequate anaerobic and aerobic microbial treatment of impurities contained in the domestic greywater and storm water. The SAT renovation of secondary treated municipal wastewater involving infiltration basins is already being practiced in several countries. A successfully operated SAT system is in operation in Ahmedabad on a pilot scale for over six months, using primary settled sewage.

2.2.2.1 *Things Required to Implement Akshaydhara*

- Separation of sanitary and non-sanitary wastewater components of domestic wastewater.
- Anaerobic digestion of the sanitary component locally employing a bio-digester. Such bio-digesters linked to public latrines and using

human excreta as feed stock are already in use in several parts of India. Liquid effluents from the bio-digester can be discharged either (i) into the city sewerage system or (ii) can be used locally for horticulture. In areas not covered by a sewerage system, the fraction may be disposed of into the unsaturated soil zone employing a separate septic system/ percolation well.

- The large volumes of non-sanitary wastewater component generated from individual/community housing units are also to be disposed of into the unsaturated soil zone locally using percolation wells, after passing through a small settling basin and sieves. Since the percolation well is functionally similar to a conventional septic system, the wastewater undergoes anaerobic oxidation followed by aerobic oxidation and nitrification as it passes through the unsaturated soil zone eventually reaching the water table.

- There is, however, an inherent risk of nitrate pollution of shallow groundwater in the above mode of domestic wastewater disposal (Wilhelm et al. 1994a, 1994b). To avoid or minimize this risk, the local groundwater is proposed to be pumped from within 10-20 m distance from the percolation well employing a shallow bore well (~30 m deep). The pumped water can be supplied for non-potable household uses through a separate plumbing system. The advantage of such a system is that the user is not required to limit the use of this renovated water. As more of this water is used, the more is returned back to the percolation well for renovation and recycling locally. Nitrate in this water will be denitrified on returning to the percolation well under the prevailing anaerobic conditions in presence of organic carbon present in the grey water. The pumping well thus serves two purposes: (i) pumps the renovated water for recycling, and (ii) confines the ground water zone assigned to renovation/recycling process.

- The scheme also envisages collection of storm runoff water and its disposal in percolation wells to achieve the two additional objectives, namely, (i) to reduce incidence of flooding of streets and, (ii) to provide groundwater recharge and dilution of the concentration of pollutants in groundwater from non-biodegradable inorganic contaminants. These may otherwise tend to accumulate locally during renovation and recycling of non-sanitary wastewater by the percolation well.

- The drinking water which constitutes a small component of the domestic consumption, but is required to be of the highest purity, is envisaged to be supplied from the community/municipal supply through a separate plumbing system. This supply may either be from treated surface water or deep groundwater source. To conserve this resource, supply of this water may be limited to approximately 20-30 liters per person per day to meet drinking and kitchen needs only.

- Thus, through a combination of (i) renovation and recycling of wastewater through the soil-aquifer system resulting in conservation of high quality potable water supply; and (ii) recharge of storm runoff water into the aquifers, the trend of declining groundwater levels may be reversed and the hydrostatic gradient restored towards the surface water bodies thereby rejuvenating/enhancing their base flow component.

- With local disposal/reuse of both sanitary and non-sanitary components of domestic wastewater, the pollution loading of the surface water bodies will be avoided. This coupled with increase in base flow component will rejuvenate our rivers and lakes, providing sustainable and aesthetic civic environment.

- In cities where non-segregated domestic wastewater is collected through sewerage system, the *Akshaydhara* approach envisages use of SAT wastewater renovation systems using infiltration basins.

- In these technologies, capital costs are expected to be small and the proposed systems employ essentially the same processes as nature does to purify the storm water and to recharge the groundwater through natural percolation. Since the rates of infiltration and recharge are required to be high, periodic system maintenance is necessary. However, maintenance requirements for the proposed system are not so elaborate as for the conventional system, the latter involving collection, transport and disposal of wastewater.

2.2.2.2 Benefits of 'Akshaydhara' Strategy

- Reduction in the cost of centralized sewage collection, treatment and disposal.
- Reduction in the cost of high quality water supply and its conservation thereby ensuring resource sustainability.
- The infrastructure and maintenance costs are shared between the city government and the residents' bodies, thereby involving the residents in maintaining hygienic conditions in the city. The residents get compensated for their share of the cost by virtually unlimited availability of household (non-potable) water.
- Rejuvenation and restoration of groundwater and surface water systems for health, aesthetic and recreation purposes.

- Possibility of evolving a new trade involving periodic de-sludging of bio-reactors and use of sludge as natural manure substitute.

2.2.2.3 *Some Concerns about Akshayadhara*

In the proposed mode of wastewater disposal, pockets of shallow underground aquifers are likely to receive some amount of pollution load. While most of this load is biodegradable in nature, some areas may have to be not only earmarked but also monitored for spread of pollution plume. This also calls for appropriate research and experimentation in terms of:

- Improving the soil-aquifer treatment technology for wastewater renovation.
- Increased understanding of pollutant removal/retardation during movement through soil aquifer medium; and
- Monitoring containment and movement of subsurface pollution plumes.

Akshaydhara is a seminal concept in the field of water resources management. This needs to be experimented upon and through appropriate technological innovations adapted for applications to real life situations. It may take several years to realize large scale application of the *Akshaydhara* concept.

2.2.2.4 *Future of Water Supply and Sanitation in India*

Urban water and sanitation infrastructure services do not pay for themselves and the government doesn't have the financial capacity to continue subsidising them. Many users who currently receive free or highly subsidised services could in fact afford to pay. A National Committee constituted in 1997 suggested extensive private-public partnerships in the field of water supply particularly in areas relating to source development, treatment and bulk supply with private agency, retail distribution and pricing with public sector; differential treatment of water for different uses, micro-level treatment to recycle water at the household level and metering of water supply to reduce leakage. It also suggests proper packaging of projects to reduce project cost and improve viability.

India is poised for substantial involvement of private sector in the field of water supply. There has already been a welcome trend in this direction in the fields of power and telecommunications. The realization and recognition of the possibilities for unbundling of water supply operations has opened up a whole new world of opportunities for the private sector to involve itself profitably. The emerging concepts of bulk purchase agreements in water supply is gaining substantial attention in recent years. A consensus is slowly emerging on the need for establishment of Urban Utilities Regulatory Board on the lines of Telecom Regulatory Authorities, either at the city level or at the state level, which may look into the larger issues of equity aspects of pricing/supply/distribution, production cost and leakage reduction ensuring quality of service as well as involvement of private sector with special reference to urban utilities is felt. It is hoped that shortly many city water supply schemes would be managed collaboratively through public-private partnerships.

Another area is to put people at center stage and involve the community in regard to development of safe drinking water systems and create the proper environment for developing an attitude of 'willingness to pay' through the resident welfare associations, Ward Committees, Councilors, city authorities etc. so that a deep sense of participation is ensured. In the unbundling process in the water supply provision and management sector the community can be allocated certain physical roles to play in this important sector.

2.2.3 People's Water Supply and Sanitation in Nepal

The case study of JAKPAS (People's Water Supply and Sanitation) is dealt with here.

2.2.3.1 Objectives of the JAKPAS Action

The basic objectives of JAKPAS were to set up and test a trial Rural Water Supply and Sanitation Fund Development Board to work in partnership with service organizations (SOs) and communities. The following were the specific objectives:

1. Test and refine a variety of services delivery options (institutional, social and technical approaches) to be taken at local level, including approaches which link women's organizations, education and income with sustainable water and sanitation service delivery;
2. Refine the eligibility criteria for the identification and selection of support agencies and schemes;
3. Based on the field testing of options help define the funding mechanism (financial and technical intermediacy of fund) that will manage the Board's resources, including its policy and operations manuals and organizational structure; and
4. Prepare pipeline of schemes for the first year implementation by the Board.

2.2.3.2 Description of the JAKPAS Action

The pilot project JAKPAS (people's water supply and sanitation) was implemented between March 1993 and June 1996 for the period of 39 months, under two Japanese Grant Facilities to the amount of US$ 2.45 million.

JAKPAS Project was managed by a small core of Nepali professional and support staff assisted by an international expatriate. Service Agencies (SAs) who were local consulting firms, individuals or NGOs, provided crucial support to the JAKPAS operations under contractual agreements.

The services of SOs were originally expected to be procured on a turnkey basis to prepare the proposal and then to implement it both together with the community. The SOs were not contracted to directly undertake a pre-designed program. The JAKPAS was expected to provide fund, technical advice and training to SOs and they were supposed to operate within the general framework of the objectives given by JAKPAS. Later it was determined that the SOs needed more technical support and monitoring of their activities.

SOs were selected by evaluation against a set of eligibility criteria and were contracted at three stages. Stage-1 Development Phase Contracts used to be bipartite contracts between JAKPAS and SOs, under which they facilitated community action planning process, built up the social capital of the communities, strengthened the management capability of the communities and prepared comprehensive implementation phase proposals. Stage-2 Implementation Phase Contracts were tripartite contracts between JAKPAS, SOs and communities (water user groups—WUGs) under which software and hardware activities outlined in the implementation phase proposals were implemented. Stage-3 Post Implementation Phase Contracts were again bilateral contracts between JAKPAS and SOs to deliver software activities to strengthen aspects relating to the sustainability of the investments made.

The JAKPAS implemented 113 schemes benefiting total population of 44,000 in two batches of scheme cycles and prepared a pipeline of 105 schemes (37,000 population) for financing in the first year of operation of the Board.

2.2.3.3 Results of JAKPAS

Strong points of the experience

Demand and beneficiary willingness to contribute to the costs: Contrary to the traditional belief that water delivery is a social service and therefore should be provided free of cost, and that sanitation services is difficult to improve, the pilot demonstrated that there exists a strong demand for water supply and sanitation services shown by the fact that communities have contributed, on average, 40% (in cash and kind) of scheme costs and 100% (in cash) of the scheme's first year operation and maintenance costs.

Existence and quality of SOs: Contrary to the general doubt about the capability of SOs, the pilot demonstrated that there are enough SOs to implement rural water supply and sanitation schemes, provided they are appropriately supported in improving their technical and institutional capabilities. They, however, are required to subject to more intensive monitoring of their performance of contractual agreements.

Timing of scheme cycle: The original estimated duration of the scheme cycle (12-18 months) was found to be short. More time is needed for the community development work required by the participatory approach, the technical activities, and the processing of proposals. Allowing for the effect of seasons, harvests, and holidays on site accessibility, availability of local labor, and building conditions had doubled the duration of the scheme cycle to 36 months consisting of a pre-development phase of 13 months for the selection of SOs and schemes and development phase of 10 months for preparing community action planning and implementation phase proposals and implementation phase of 13 months for the construction of the scheme.

Eligibility criteria: The original criteria for the selection of SOs and schemes were found to be generally appropriate. Experience showed that criteria should be clearly stated at the start of each batch of schemes and kept constant during the processing of that batch. It was however learnt that support organization should be considered ineligible until proven eligible.

Management of the program: The pilot demonstrated that a few qualified staff could administer many agreements with SOs by using efficient information system, maximizing the use of local consultants to handle specific tasks, employing clear criteria for scheme and support organization selection and payment by results. It has also shown that a satisfactory incentive structure is needed to motivate staff and sufficient autonomy from outside interference is a must to ensure transparency in resource allocation to both SOs and communities.

Perspective and impacts: The successful implementation of the pilot project— JAKPAS has resulted in the establishment of an autonomous institution, the Rural Water Supply and Sanitation Fund Development Board (RWSSFDB), in June 1996, under an IDA credit assistance of US$ 18.28 million. The basis of the Board's operation are comprehensive operation manuals with clear and transparent eligibility criteria for the selection of SOs and schemes, a comprehensive scheme implementation cycle, a mechanism for procurement and contracting of SOs and service agencies, legal framework for the community organization, service level and technical options affordable and manageable by the communities, funding mechanism, monitoring and evaluation procedure, etc., all successfully tested by JAKPAS.

JAKPAS, and its successor the Board had a remarkable effect on the sector as a whole. The service delivery mechanisms are well publicized and shared with other agencies working in the sector. The Board and the intervention of the World Bank in the sector paved the way for several donor agencies to adopt the Board modalities and new donors entering the sector now see the private sector service delivery as possible option that had previously not existed. The Board has also set an example that establishment of an autonomous institution is possible within the government framework. New institutions coming up in Nepal now have tendency to seek the autonomy enjoyed by the Board.

2.2.4 A Case from the Developed World: Tucson Valley in Arizona, USA

So far, case studies from the developing world have been discussed. Here a case study from the developed world from Tucson, USA is presented.

There have been three major groups of water users in Tucson: homes and businesses, agricultural interests, and industry (including mining). The increase in municipal water use during the 1980s was offset by decreasing agricultural use, with total water use in the Tucson area holding steady during this period at about 338.25 million cubic meters (see Figure 2.1). Agricultural use has risen since 1993. That, coupled with rising municipal use, pushed total water use to 397.29 million cubic meters by 1997.

Agriculture has historically consumed the largest share of water of any sector in the Tucson area. After reaching a plateau between 1955 and 1975, however, agricultural water use declined during the 1980s and early 1990s. Municipal water use has increased since 1984 as the population has grown—both in total acre-feet used and in percentage of total water use, and now consumes a larger share than does agriculture.

2.2.4.1 Municipal Water Use

Population growth has caused municipal water use to be the fastest growing water use sector. Total municipal water use in the Tucson Active

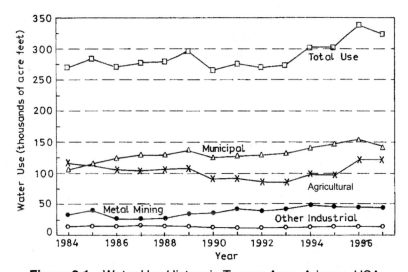

Figure 2.1 Water Use History in Tucson Area, Arizona, USA.

[Source: Water in the Tucson Area: Seeking Sustainability: a status report by the Water Resources Research Center, College of Agriculture, The University of Arizona]

1000 acre-ft = 1.2335 million m^3

Management Area (TAMA) increased from approximately 142.68 million cubic meters in 1985 to about 189.42 million cubic meters in 1997.

One hundred and fifty-one municipal water providers have been reported to be operating in the Tucson area. Of this number, 19 large providers serve over 96 percent of total municipal demand. The service areas of the major water providers are shown in Figure 2.2.

Tucson Water's total water usage rate in 1955 was 652 liters per capita per day (lpcd) including both residential and non-residential customers. It has remained fairly constant since 1985, ranging between 667 and 641 lpcd. People tend to use more water during hot, dry summers than during relatively cooler, wetter ones. More homes having swimming pools and other water-using features increases water usage while installing low-water use toilets can decrease consumption.

Residential customers (single family and multi-family) are considered municipal water users along with businesses and institutions. Water use characteristics generally differ for each category of demand, with residents consuming most of the water in the municipal category. Residential demand for Tucson Water has remained fairly consistent at about 417 lpcd from 1985 to 1995. The average residential consumption rate for other large providers is higher, averaging about 459 lpcd. A number of factors explain the higher consumption rates of these large providers, including the age of the housing

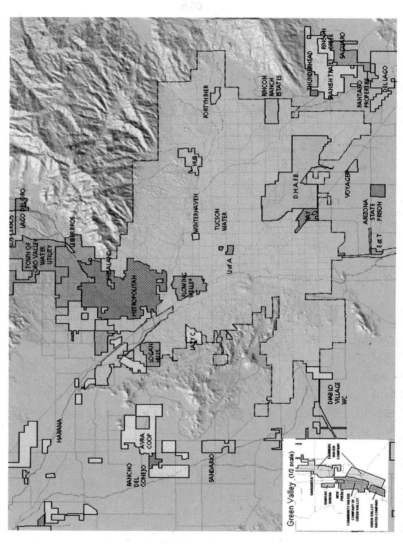

Figure 2.2 Physiographic Map of Tucson Area, Arizona, USA Showing the Service Areas of Major Water Providers

[Source: Water in the Tucson Area: Seeking Sustainability: A status report by the Water Resources Research Center, College of Agriculture, The University of Arizona]

within the service area, the availability and effectiveness of conservation programs, income levels and water rates. Table 2.2 provides the details of selected water providers in Tuscon Area.

Table 2.2 Selected Water Providers in the Tucson Area (1995 data)

Water provider	Population served	Groundwater deliveries (Million m^3)	Residential lpcd
Tucson Water	559,602	135.21	417
Metro Water	42,861	10.53	561
Oro Valley	23,229	7.02	440
Flowing Wells Irrigation District	14,951	3.50	481
Community Water Company of Green Valley	12,819	2.54	424
Avra Water Co-op	5,663	0.95	398
Ray Water	4,617	0.74	402

[Source: Water in the Tucson Area: Seeking Sustainability: A status report by Water Resources Research Center, College of Agriculture, Univ. of Arizona]

2.2.4.2 Residential Use

Single family residents account for approximately 75 percent of residential demand in the Tucson area. Multifamily residential demand makes up the balance. These include apartment complexes, duplexes, triplexes, townhouses and condominiums. Their use is typically about 60 percent of the single family residential use rate. Multifamily complexes use less water per person in part because landscaping is generally limited to common areas, and some water uses occur away from the residence, e.g., apartment dwellers are more likely to use car washes.

During the 1970s and 1980s new housing construction shifted towards multifamily dwellings. This increased construction of lower water-use housing promised to reduce gpcd rates. This trend, however, did not continue. Economic expansion and much lower mortgage rates caused single family home construction to rebound. In addition, the majority of multifamily units being constructed are more luxurious units which are more likely to have large turf areas, pools and other water-using amenities.

Older homes tend to consume more water both indoors and outdoors than newer homes. They generally use more water outdoors because of larger lots with more turf and landscaping. They use more indoors because they are less likely to have low-water use fixtures such as ultra low flush (ULF) toilets designed to use 6 liters per flush. Homes built after 1975 are less likely to have lawns, and homes built after 1989 are required to have ULF toilets.

2.2.4.3 *Non-residential Use*

Non-residential demand generally consists of turf facilities (golf courses, cemeteries, etc.), water features in public rights-of-way and commercial establishments. Non-residential demand for large providers averaged 155 lpcd in 1995. Tucson Waters' non-residential demand decreased by 23 lpcd from 1985 to 1995. For other large providers, however, non-residential demand increased by 42 lpcd. Tucson Water has been able to reduce its gpcd rate by switching some golf courses and other facilities to effluent. (Although considered water, effluent does not count in the official calculations.) Other large providers are serving an increasing number of golf courses but do not have access to reclaimed water. As a result, their water use appears greater. Some areas served by large municipal providers are experiencing a transition from bedroom communities to areas with more retail and commercial activity, a change that is reflected in their water use.

The 35 golf courses in the area account for about ten percent of municipal water use in TAMA. The total amount of water used on TAMA golf courses has increased from 14.39 million cubic meters in 1985 to 20.91 million cubic meters in 1997 (see Figure 2.3). Of the total amount of water used on golf courses in TAMA, the share of effluent increased from 24 percent in 1985, peaking at 38 percent in 1990, and has since fallen to 35 percent in 1997.

The number of holes of golf in TAMA has increased by 35 percent since 1985. Golf course design in the area has shifted towards more desert-like courses, incorporating fewer water hazards and significantly more low-water use plants along fairways instead of turf. In addition, the average number of hectares of turf per hole has decreased from 2.36 in 1985 to 1.92 in 1997. However, reductions in water use resulting from these changes have been offset by a large increase in the percentage of golf course turf which is overseeded with winter rye grass, from 21 percent in 1985 to 66 percent in 1997. The water use per hole of golf has increased over time since 1985, in part due to variations in weather.

Other large-scale turf facilities include parks, cemeteries and schools. For legal reasons Arizona Department of Water Resources (ADWR) divides turf facilities into industrial and municipal categories according to whether or not they are served by municipal water providers. All these turf facilities use about 24.6 million cubic meters of water, about 33 percent of which is effluent. The remainder is groundwater.

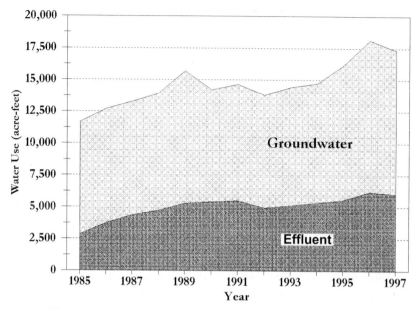

Figure 2.3 Water Use for Golf Courses in Tucson, USA.

[Source: Water in the Tucson Area: Seeking Sustainability: A status report by the Water Resources Research Center, College of Agriculture, The University of Arizona]

1000 acre-ft = 1.2335 million m³

Patterns of water use for retail and commercial establishments vary widely, but often include water for sanitary and landscaping needs. The number of employees often is a significant factor in business water use. Partly because water use patterns vary significantly among businesses, water audits tailoring conservation measures to particular needs are an effective strategy for reducing business water use.

2.2.4.4 Conservation Rules

Under the Groundwater Management Act, ADWR sets goals for per capita water use by municipal water providers, but cannot regulate individual customers. Large providers can choose among four water conservation programs. Most often selected is the Total Gross Per Capita Decrease (GPCD) programme, which sets targets for reductions in per capita water use for each water provider based on an analysis of conservation potential for that provider. Tucson Water and Metro Water have selected the non-per capita conservation program; they do not have to meet specific per capita water use targets, but must implement a range of conservation measures. Small providers, which account for four percent of total municipal water use, are regulated differently than large providers. Small providers are required to reduce waste and encourage conservation, but they generally lack the resources to implement conservation programs.

Other ways ADWR tries to balance supply and demand include requiring developers to demonstrate that renewable supplies are available to serve the development; that water use is consistent with TAMA's management plan and goal; and that the developer has the financial capability to construct needed water facilities. These Assured Water Supply (AWS) rules are designed to work with the conservation programs to reduce mining of groundwater.

2.2.4.5 Uses and Renewable Supplies

Historically, Tucson has been largely dependent on groundwater, a mostly non-renewable supply. Since mid-1970's, effluent use has increased and Central Arizona Project (CAP) water has arrived in the area. The following paragraphs describe the uses and renewable supplies available.

Supply from Central Arizona Project (CAP): CAP is potentially the largest renewable water source, although the only direct use of CAP water in the municipal sector was found to occur for treatment plant maintenance—about 0.25 million cubic meters in 1997. CAP's potential as a water source obviously was not fully realized.

Effluent use: Effluent use currently meets about five percent of municipal water demand. Of the 85.36 million cubic meters of effluent produced at the Ina Road and Roger Road wastewater treatment plants in 1998, about 15.99 million cubic meters was reused, with the rest discharged to the Santa Cruz River channel, where some 96 percent eventually recharges the aquifer within TAMA. Of the amount reused, approximately 1.48 million cubic meters was delivered directly to turf facilities and some 3.69 million cubic meters was delivered to the Cortaro Marana Irrigation District. This effluent only has secondary treatment and is mostly delivered downstream via gravity.

The City of Tucson processed approximately 10.71 million cubic meters of secondary treated effluent at its reclaimed water facilities located next to the Roger Road Wastewater Treatment Facility. This effluent receives further treatment (tertiary treatment) by filtration through sand filters or soil and additional disinfection. The reclaimed water then is delivered for use or is stored at the Sweetwater Underground Storage and Recovery Facility (recharge facility) to meet peak demands in the summer, primarily to irrigate golf courses.

Reclaimed water flows through a different set of pipes, separate from the potable water system. So far, $66 million has been spent building the system, including the reclamation facilities, the recharge facility and the distribution system. Tucson charges $3,862 per ha-m. for reclaimed water. Full cost for production and distribution of reclaimed water was reported as about $4,537 per ha-m, with $2,626 covering debt service and capital costs and $1,911 covering operation, maintenance and overhead costs. The price charged for reclaimed water is substantially lower than the wholesale price Tucson Water charges for potable water. This is done to further encourage the use of effluent.

Figure 2.4 shows Tucson's 137-km reclaimed water system, which has about 200 users. New users pay the cost of connecting to the system. This cost can be quite high, due to the expense of extending pipe to carry the effluent to the new user and modifying the user's delivery system to handle effluent.

Arizona Department of Water Resources (ADWR) incentives encourage the use of effluent. The most important incentive allows municipal providers to exclude effluent used on golf courses from their gpcd calculations (although the same total amount of water is actually used) which means conservation goals for individual customers can be higher.

2.2.4.6 Grey Water Reuse

Grey water is water recovered after various indoor household uses, excluding toilet use. Grey water includes water from clothes washers, bathroom sinks, showers, baths, dishwashers and sometimes the rinse side of the kitchen sink. A grey water reuse system can be set up to capture and recycle such water for uses not requiring drinking water quality water, e.g., landscape irrigation. Approximately 60 to 65 percent of the wastewater generated from residential indoor use is grey water. The average resident generates an estimated 114 liters of grey water per day. This is a significant source of water available to meet peak outdoor irrigation demands in the summer.

Although grey water use is common in rural areas and has been practiced by many people in urban areas for years, grey water reuse is technically illegal in many places in the United States. Plumbing codes generally require water coming from the drain to be discharged to the sewage system or a septic tank. In Arizona, a permit must be obtained from the Arizona Department of Environmental Quality (ADEQ) or the Pima County Department of Environmental Quality to operate a grey water reuse system. To issue a permit, ADEQ must approve the design and construction of the

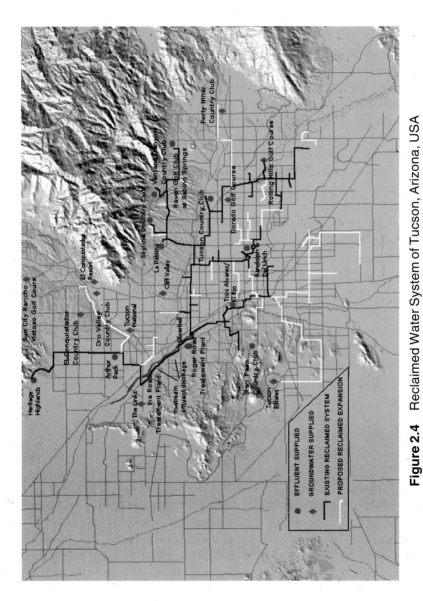

Figure 2.4 Reclaimed Water System of Tucson, Arizona, USA

[Source: Water in the Tucson Area: Seeking Sustainability: A status report by the Water Resources Research Center, College of Agriculture, The University of Arizona]

system. The system must include a settling tank to settle out grit in the water and also must have a filtration device. Water to be applied to the surface of the ground (defined as within two feet of the earth's surface) must be disinfected, meet water quality standards and be monitored. Daily testing of the samples may be required and can be expensive.

Concerns about public health are the biggest obstacles in legalizing grey water use. The quality varies depending on how it was used. Water from washing diapers, for example, will probably be more contaminated than water from a shower. Fecal coliform bacteria levels and nitrates have been of particular concern, although the threat is perhaps exaggerated. Salmonella and polio virus have been shown to last several days in grey water. Data accurately characterizing factors that determine grey water quality and assessing risks of use is limited. Studies are needed to develop guidelines for the safe reuse of residential grey water. The Water Conservation Alliance of Southern Arizona (Water CASA), in cooperation with ADWR, ADEQ and Pima County Department of Environmental Quality, is studying residential grey water reuse in Tucson. Study results will help determine if health risks increase with grey water reuse, and whether permitting standards can be loosened.

2.2.4.7 Indoor Water Use

Indoor uses remain fairly constant throughout the year. People may wash more clothes in the summer, but in the winter the bulk of clothes washed is greater. Similarly, other indoor water uses vary little during the course of the year. This constancy is reflected in relatively flat levels of sewage water flow.

Figure 2.5 shows that the largest indoor uses of water are toilets, showers and baths, and washing machines in all types of housing. Newer models of toilets are designed to use less water than older models. Until the early 1980s, most new toilets used five to seven gallons per flush. Water-conserving 3.5 gallons per flush toilets were the standard until the early 1990s when 1.6-gallon ultra low flush (ULF) toilets became available.

Replacing older toilets with ULF toilets is one of the best ways to save water indoors. The University of Arizona's Water Resources Research Center was conducting a study to determine which models of ULF toilet have not held up over time. This information can be used to rewrite plumbing codes, upgrade ULF toilet quality, and make the correct replacement parts easier to obtain.

Toilet dams and other water displacement devices can help save water in older model toilets. Toilet dams, which are placed in the tank to keep water

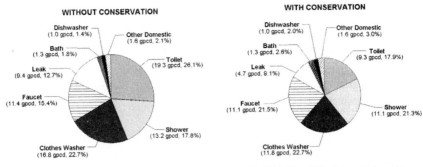

Figure 2.5 Composition of Indoor Water Use with and without Conservation

[Source: Water in the Tucson Area: Seeking Sustainability: A status report by the Water Resources Research Center, College of Agriculture, The University of Arizona]

from fully filling the tank, typically save about 3.8 liters per flush. Water-filled bags or plastic bottles also can be used in place of dams and typically save about the same amount.

Showers and baths are another large component of indoor water use. Showers and baths typically comprise about 20 to 25 percent of total indoor water use in older homes. Low-flow shower heads also can save water. Older shower heads typically use 13.3 liters per minute (lpm), while low-flow shower heads typically use 9.5 lpm or less.

Cloth washers typically account for between 20 and 25 percent of indoor water use. Clothes washers vary widely in their water use. Older models used about 208 liters per load, while more water-efficient models use around 159 liters per load. Newer, more efficient models, including horizontal axis machines, use about 114 liters per load.

Faucet use typically accounts for around 15 to 20 percent of indoor use. Faucet aerators reduce some water use by introducing more air into the stream, thereby increasing the water's wetting action. This can save approximately one gpm over older 13.3 lpm faucets. Other faucet uses, such as filling a glass or teapot, are unaffected by aerators.

Leaks average approximately 10 percent of indoor water use. Inspection of the home for leaks as part of a water audit is often an effective way to save water. Tucson Water offers free leak detection as part of its Zanjero Programme, which provides customers an analysis of the water use in their home, and information on how to lower their water use and water bills.

2.2.4.8 Outdoor Water Use

In the Tucson area, single-family residents use 30 to 50 percent of their water outdoors, for landscape watering, swimming pools, spas, evaporative cooling and other such uses. Outdoor water use varies over the year. In the summer before the monsoon rains starts, outdoor water use peaks as plants require more water to survive and evaporation from pools is at its greatest. In the winter, outdoor water use drops dramatically, especially during the winter rainy season from January through March. Bermuda grass is dormant during this season, and only about seven percent of landscapes have winter rye grass lawns.

Landscape irrigation is the largest category of outdoor water demand. Most landscaping in Tucson combines grass with desert plants. A 1992 random survey of Tucson Water customers found that about 43 percent of respondents had some grass in their landscaping. Eight percent of residents reported their landscaping was mostly grass while the other 35 percent had landscapes combining turf area with other plant materials.

Relying on a garden hose to water vegetation is the most prevalent form of irrigation in Tucson. Drip irrigation is the second most common method of irrigation and has gained significantly in popularity since the early 1980s. At that time, one percent of households had drip, compared to approximately 27 percent of households in the early 1990s. Approximately 22 percent of Tucson Water service area households reported having in-ground irrigation systems in the early 1990s. About eight percent of homes surveyed in the early 1990s did not irrigate their landscaping at all.

Swimming pools are less common than lawns in Tucson, but the percentage of homes with swimming pools has been increasing over time. A swimming pool typically uses three to five times as much water as the same area of turf. This is due in part to the fact that most private lawns are under-irrigated, and pool consumption includes not only evaporation but also filter back flushing and occasional draining for maintenance.

As is shown in Figure 2.6, the percentage of Tucson homes with pools is a function of when the home was built, increasing from about 15 percent in homes built prior to the mid-1950s, to about 22 percent from the mid-1950s though the 1960s, and then to nearly 30 percent in newer homes. In the mid to late 1990's, almost 20 percent of all homes in Pima County had pools. Outdoor water use can be optimized by certain techniques like xeriscaping, which is described below.

Xeriscaping Xeriscaping is using efficient landscape design and lower water use vegetation to create attractive landscapes—and equally

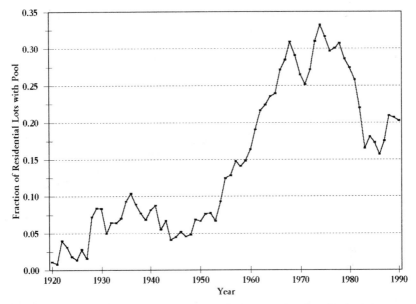

Figure 2.6 Yearwise Percentage of Houses having Swimming Pools in Tucson, USA

[Source: Water in the Tucson Area: Seeking Sustainability: A status report by the Water Resources Research Center, College of Agriculture, The University of Arizona]

important, to save water. The word "xeriscape" combines the Greek word "xeros", meaning dry, with "scape" from "landscape." Xeriscaping principles make use of "micro-climates" that exist in the landscape. Microclimates are defined according to the amount of sun and shade, the slope, and air movement that characterize a landscape.

The property is divided into low, medium and high water use areas, with the highest water use areas close to the house, in areas with the most shade. These are cooler areas, and xeriscaping would limit turf to these areas. Drought-tolerant plants and native vegetation are used in low water use zones to provide attractive landscape with a variety of colorful and interesting plants.

Water harvesting techniques might be applied to capture and store rainwater for use on plants or to channel runoff directly to vegetation. Drip irrigation can be installed to water individual plants, while sprinkler systems are used for turf. Refer to sections 2.3.5 and 2.3.6 later in this Chapter for more details. Soil can be improved and topped with mulch to hold water from rainfall as well as irrigation. Taken together these practices help residents save both water and money while creating beautiful and interesting landscapes.

One of the few ways to reduce pool water use is covering the pool when it is not in use to minimize evaporation. A survey of newer homes in Tucson revealed that approximately 60 percent of home pools have pool covers. But pool covers are used only about half the year. Usage is at a minimum during the summer swimming season to allow convenient and frequent access to the pool. Also covers are not used in the summer because they cause the water to become uncomfortably warm. Peak cover usage surprisingly is not in winter, but in the fall and spring, when pool users are trying to extend the swim season. Lower pool cover usage in the winter may reflect a desire to protect the cover from sun damage while evaporation rates are the lowest.

Water for evaporative cooling systems accounts for around five percent of outdoor water use in Tucson. (Evaporative cooling is classified as an outdoor water use because it results in water being consumptively used and not returned to the sewage system as is the case with other indoor uses.) In 1992, approximately 79 percent of homes in Tucson had evaporative coolers. At that time approximately 59 percent of households had only an evaporative cooler; 21 percent had both a cooler and an air conditioner; and 19 percent had only an air-conditioner. As air-conditioners have gained popularity in new construction, the percentage of Tucson homes with evaporative coolers, along with the amount of outdoor water devoted to evaporative cooling, has declined. Approximately 85 percent of new constructions surveyed in 1996 had only an air-conditioner, 11 percent had both an air-conditioner and an evaporative cooler, and four percent had only an evaporative cooler.

2.2.4.9 Higher Water Use Trends

While newer homes have more low-flow plumbing fixtures and appliances and are likely to have less turf and no evaporative cooler, some trends in new construction cancel out these conservation gains. For example, newer homes are more likely to have water-using amenities such as pools, spas and whirlpool tubs. Further, new apartment complexes and condominiums are more likely to have large amounts of turf and landscaping, as well as pools.

The number of homes with outdoor misting systems grew through the early and mid-1990s. These systems spray droplets of water into the air that evaporate to cool an area. Although manufacturers of misting devices claim water efficiency, few residential misting systems are as effective or water-efficient as advertised, and some are poorly designed. System emitters can scale up or corrode, producing drips instead of the intended mists.

The trend in cooling system design is towards greater water use. The latest models of evaporative coolers are designed to prolong the life of the cooler by draining water after a certain number of hours of operation. This prevents mineral content from building up and corroding metal cooler parts and scaling up the pads. Some new coolers, however, empty the pan automatically after only a few hours of operation. This is excessive considering Tucson-area water quality. In older coolers, water collects in the bottom of the cooler and can be drained using a bleed-off valve. Some new air-conditioners also use water. Manufacturers can achieve higher efficiency ratings by dissipating heat generated by the unit in an attached evaporative cooler.

2.2.4.10 Water Rates and Conservation

For decades water has been priced as if it were free. What people pay for is the cost of capturing the water, delivering it to them and making sure it is safe to drink. People who pump their own water pay to build and operate their wells, but they do not pay anyone for the water itself. ADWR, however, does charge well owners a small pumping fee, which goes primarily towards conservation and augmentation (water banking) programs. In some states people pay an annual fee for their pumping permit, which recognizes that the state owns the water and sets certain conditions for people to use it.

2.2.4.11 Components of Tucson's Water Distribution System

Sources of Water: Water pumped from an aquifer to pipes for distribution and delivery has been the source of much of Tucson's water. Tucson's newest available source of water is the CAP, which brings water from the Colorado River. Tucson also has a system for using treated wastewater on facilities such as golf courses and parks.

Pipelines: Large pipelines ("water mains") bring water from its source to central points, and smaller pipes distribute that water throughout the community.

Reservoirs: These are storage areas that hold water until used. Reservoirs are important for balancing supply and demand and for ensuring that an extra amount of water is in reserve for fire fighting. At one time, elevated storage tanks provided adequate supplies of water. With our large population, however, huge reservoirs are needed, with capacities ranging from one million to twenty million gallons of water.

Booster stations: Since Tucson is not flat nor completely situated downhill from its water supply, booster pumps are needed to boost the water to higher elevations.

Many people will conserve water if the price is very high. The point at which people will respond to higher bills varies greatly, depending mostly on personal income and the percentage of total household expense the water bill represents. Some people may find a $100 water bill acceptable while others may have problems paying a fourth as much.

Water rates can be modified to encourage conservation in two very different ways:

The rate structure can be designed to reward conservation and discourage excessive water use. For example, cost per gallon could increase as customers use more water; cost could increase at peak times of the day or year to reflect higher costs at that time; or cost could be higher for areas that are more expensive to serve. Rate levels can be raised to cover the cost of finding new future water supplies.

2.2.4.12 Changing Rate Structures

Tucson's first water rates in 1900 were flat rates. A flat rate means people pay the same for water no matter how much they use. For many years, Winterhaven had a flat rate for water, with the expectation that people would maintain lush lawns and landscapes, although this has changed and desert landscapes are now acceptable in that Tucson neighborhood. Today almost all communities nationally meter water usage and thus charge people more for increased water use.

Water bills from all water providers in the Tucson area have two basic parts—a basic rate determined by the size of the connection (this is a fixed charge applied whether or not water is used) and a commodity rate which is charged for every unit of water used over a minimum amount. For many water providers, the minimum amount is 2,000 gallons. Some water companies charge a higher commodity rate for water use above certain amounts (referred to as an increasing block rate or progressive rate structure). This type of rate structure is designed to discourage high-volume water use. Tucson Water, Metro Water and Avra Water Co-op have increasing block rates.

Another water rate structure variation designed to encourage water conservation is seasonal water rates. Seasonal rates usually involve charging a higher commodity rate during summer months than during winter months. Higher summer rates are designed to encourage more conservation when more water is needed to meet peak demand on the water system. Metro Water has had seasonal rates since 1995. Tucson Water had seasonal rates for all customer classes from 1977 to 1995 but removed seasonal rates for the single family residential and duplex-triplex classes in 1995.

Water rates are an important signal about the relative scarcity of water and the need to conserve. If rates do not keep up with inflation, the real price of water actually declines. People may take that as a signal that water is cheap and conservation is not important. Even if the real price of water is held constant, incomes in Pima County have increased at a faster rate than inflation. This is good news for the economy, but means that water bills shrink as a relative share of the overall budget. The incentive to conserve is reduced.

Tucson Water has pursued a water rate policy designed to lower peak demand and encourage water conservation. Tucson Water instituted increasing block rates and seasonal rates as a conscious effort to discourage excessive water use. Especially after the 1975-76 water controversy, many Tucson Water customers saw significant increases in their water bills, and water use decreased significantly as the message to conserve hit home. At this time the water conservation program, Beat the Peak, was initiated.

As is shown in Figure 2.7, between 1976 and 1993 rates were updated every year. For average water use customers, bills just barely kept ahead of inflation. At least in part because the real price of water stayed about the

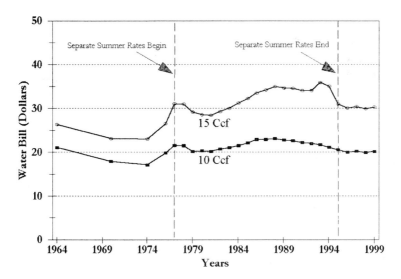

Figure 2.7   Water Rate Variation during 1964-1999

[Source: Water in the Tucson Area: Seeking Sustainability: A status report by the Water Resources Research Center, College of Agriculture, The University of Arizona]

1 Ccf = 100 ft^3

same, and incomes in County increased at a higher rate than the rate of inflation, water use increased. In 1993, there was a significant increase in the price of water, especially the summer rate. Public reaction, however, forced a redesign of the rates and a decline in the real price of water in the summer. The real (inflation adjusted) price of water has declined since then.

Water providers must get approval to change water rates. Municipal water utilities must get approval from the city or town council. Changing the rates is much more difficult for water companies that must get approval from the Arizona Corporation Commission. Going through this process is expensive because of the legal fees involved and costs of a rate hearing can easily reach hundreds of thousands of dollars, even for a small water company. This is required even if a company doesn't plan to raise more revenue from customers, but just change the rate structure. This discourages many companies from changing rate structures to promote conservation.

2.3 WATER FOR AGRICULTURE

Agriculture has been a major user of water in the form of irrigation in most parts of the world. More than 60% of water used in the world each year is diverted for irrigating crops. In Asia alone, 85 per cent of water goes for irrigation. Figure 2.8 provides the World Bank estimates for agricultural use of water for the period from 1980 to 1998 (Down to Earth, 2002). What is not clear in this Figure is the zero agricultural water usage in Germany. Whether agriculture is considered as an industry in Germany needs to be verified. Among the nine countries—excluding Germany, it is observed that the agricultural usage as compared to the industrial and domestic usage is the minimum in UK and the maximum in India. The following sections describe the use of water for agriculture in USA in the developed world and elsewhere.

2.3.1 Agricultural Usage of Water in Arizona, USA

Agriculture has been the predominant user of water in Arizona in the 20th century. Since the 1940s, agriculture has accounted for about 80 to 90 percent of Arizona's water use. In the Tucson area, agriculture's share of water use in the Upper Santa Cruz Basin (which excludes the Avra Valley) was about 84 percent in 1940, shrinking to about 73 percent by 1951.

The general downward trend in agricultural water use has resulted mostly from a reduction in cropped acreage. Cropped acreage in Pima County reached a plateau in 1955 and remained fairly constant until 1975, at about 50,000 to 60,000 acres. Irrigated acreage declined after 1975 as farmland was developed for urban use along the Santa Cruz River

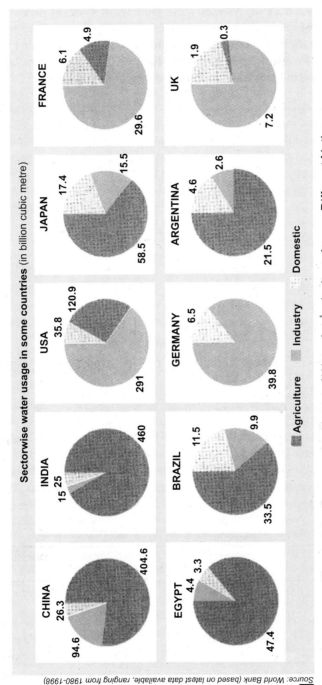

Figure 2.8 Comparative Usage of Water for Agriculture Among Different Nations

floodplain, including the Marana area. The City of Tucson purchased over 20,000 acres of farmland in the Tucson area from the 1950s to the early 1980s, taking that land out of agricultural production in order to use its water rights for municipal purposes. Figure 2.9 provides the historical variation in the irrigated area in Pima County, Arizona, USA.

Figure 2.9 Historical Variation in the Irrigated Area in Pima County, Arizona, USA

[Source: Water in the Tucson Area: Seeking Sustainability: A status report by the Water Resources Research Center, College of Agriculture, The University of Arizona]

After declining through the 1980s and early 1990s, agricultural water use increased to 163.22 million m^3 in 1997. Use in 1997 includes 30.87 million m^3 of Central Arizona Project (CAP) water used in lieu of groundwater under the groundwater savings program set up by the State of Arizona. Table 2.3 gives a summary of use of water for major irrigation in Tucson, Arizona, USA.

Table 2.3 Summary of Water use in Major Irrigation Areas in Tucson AMA

Irrigation Area	Irrigated area (10^4 m^2)	1987–1995 Average Water Use (Million m^3)	Water sources
Cortaro Marana Irrigation District Effluent	4,270	41.13	Groundwater, CAP,
Avra Valley Irrigation District	4,601	32.28	Groundwater, CAP
Farmers' Investment Co.	2,393	34.47	Groundwater
Red Rock Area	1,556	n/a	Groundwater, CAP

Source: Arizona Department of Water Resources, Draft Third Management Plan, Tucson Active Management Area, 1998.

Agricultural water use in TAMA is regulated under the Groundwater Management Act of 1980 (GMA). The GMA regulates agricultural water use in several ways. First, no new agricultural land can be developed for irrigation. Second, farms are given a maximum annual allotment of groundwater to be used for irrigation. Farms using less than their groundwater allowance were given a credit for the difference between their actual water use and the groundwater allowance.

As shown in Figure 2.10, four main groups of farms are clustered in three agricultural areas remaining within TAMA. Two irrigation districts are operating in TAMA: Cortaro Marana Irrigation District (CMID), located north and west of the Town of Marana, and Avra Valley Irrigation District (AVID), located generally just south of CMID, southwest of the Santa Cruz River. Irrigation also is occurring in what is referred to as the Red Rock area, in the Pinal County portion of TAMA. Farms within the Farmers Investment Company (FICO) near Green Valley account for most of the rest of TAMA agricultural land.

Agricultural water use in other areas of TAMA, such as the Tucson area, the Altar Valley and farmland in the Arivaca area, accounts for less than three percent of total agricultural water use in TAMA. The San Xavier (south of Valencia Road) and Schuk Toak districts (western Avra Valley) of the Tohono O'odham Nation have CAP allocations, which may be applied to restore historic farmlands and add additional farmland. Projections show that about 6.15 million m^3 of CAP water may be used on the San Xavier District by the year 2005. The Schuk Toak District currently is developing a farm, which is expected to use 13.28 million m^3 of CAP water per year by 2010. A pipeline to supply CAP water to the San Xavier District is now under construction.

Cotton was reported as the predominant crop grown in TAMA, accounting for about 75 percent of planted acreage. Other crops grown include wheat, barley, sorghum, alfalfa hay, vegetables, nuts, millet and lettuce. Pecans are the predominant crop grown at FICO.

The cost of pumping groundwater for TAMA farmers depends mainly on the depth to groundwater and energy costs. With access to low-cost hydropower generated at Hoover Dam, CMID has about the lowest pumping cost in TAMA. The district controls well pumping and supplies water to farmers at a cost of $30 per acre-foot, plus an annual assessment of $40 per acre for every acre in the district. Individual farmers within AVID have their own wells and control water use decisions. Pumping costs for wells in the district were about $40 to $50 per acre foot in 1995, including

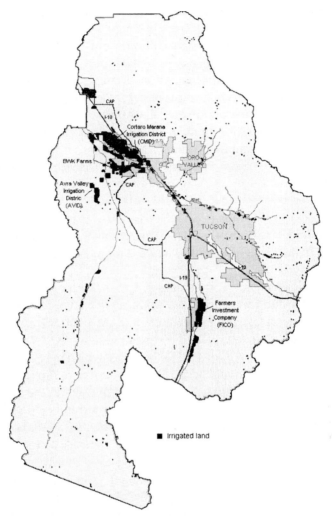

Figure 2.10 Major Irrigation Areas in Tucson, USA

[Source: Water in the Tucson Area: Seeking Sustainability: A status report by the Water Resources Research Center, College of Agriculture, The University of Arizona]

operation, maintenance and repair costs. Average pumping cost for FICO wells was recently reported to be $28 per acre-foot.

TAMA farms, however, are eligible to use CAP water. Under the Groundwater Savings Facilities (GSFs), otherwise known as in-lieu recharge facilities, municipal water providers offer CAP water to farmers at prices below the cost of pumping groundwater. Farms use this CAP water in lieu of groundwater that otherwise would have been pumped. Municipal providers get credits for this "saved" groundwater. The credits then can be used in the future to offset groundwater pumping in efforts to meet state groundwater pumping restrictions.

As is shown in Table 2.4, CAP water use through the groundwater savings program has grown from about 12.3 million m^3 in 1995 to 30.75 million m^3 in 1997. CAP water use through GSFs continues to expand as new facilities are added and existing facilities are permitted to take more water. For example, CMID used almost 12.3 million m^3 of CAP water as a groundwater savings facility in 1997 and has increased its state permit to take up to 24.6 million m^3 per year of CAP water in the future. A groundwater savings facility located within AVID includes several farms. The AVID GSF is permitted to take up to 15.39 million m^3 per year of CAP water. FICO does not use any CAP water currently but is investigating the use of CAP and/or effluent. Use of CAP water at FICO could occur through a GSF with a possible capacity of up to 24.6 million m^3 per year.

Table 2.4 Water Delivered to Groundwater Savings Facilities (million m^3)

Groundwater Savings Facilities	1993	1994	1995	1996	1997
BKW Farms	0.30	2.48	5.21	8.71	10.63
Cortaro Marana Irrigation District	3.26	0	7.26	11.78	11.98
Kai Farm at Picacho	–	–	–	0	8.24
Total In-Lieu Recharge	3.56	2.48	12.47	20.49	30.85

Source: Arizona Department of Water Resources, Draft Third Management Plan, Tucson Active Management Area, 1998.

Kai Farms at Picacho, in the Red Rock area, converted from pecan trees to row crops in 1997, and irrigation with CAP water began under a groundwater savings project arrangement. Total CAP water used at the Kai Farm at Picacho GSF in 1997 was 8.24 million m^3. The facility is permitted to take up to 13.81 million m^3 per year of CAP water. A small amount of treated effluent is used on farms in the Tucson area. CMID purchases an average of about 3,000 acre-feet of effluent per year from Pima County. The effluent is delivered via a ditch from the Ina Road treatment plant and is blended with groundwater for delivery to farms.

2.3.1.1 Improving Agricultural Water Use Efficiency

Water use efficiency ($\eta_{\text{Water Use}}$) is defined as follows:

$$\eta_{\text{Water Use}} = \text{Water Required/Water Applied} \qquad (2.1)$$

Arizona Department of Water Resources (ADWR) has goals for increasing water use efficiency on farms. In 1980, water use efficiency in TAMA averaged about 65 percent of water applied. This means that the average amount of water applied to crops was 35 percent greater than the calculated water need for those crops after accounting for the consumptive use requirement for the crops, the amount of precipitation available for plant growth, any additional water for special needs for crops—such as water needed for germination of lettuce—and a leaching allowance to prevent buildup of salts in the soil. The water use efficiency goal set for farms to reach by the year 2000 was 85 percent.

Flooding irrigation on sloped fields is the most common irrigation method in TAMA. Some farms have saved water by laser-leveling their fields (a method of leveling and slightly sloping fields so that water spreads evenly across the field) or installing systems to pump back water that accumulates at the end of the field. Installation of drip irrigation systems is generally considered too expensive for irrigation in the Tucson area.

ADWR also requires farms to make their distribution systems more efficient. Farms can save water by lining their water distribution canals with concrete or other materials. Under ADWR's Second Management Plan, farms were required to either line all their canals or operate their delivery systems to keep lost or unaccounted water at less than ten percent. The agency reports that most of the largest irrigation districts in TAMA meet this requirement.

Flooding irrigation is very common with a low water use efficiency. Therefore, it is not dealt in this Chapter separately. However, other water efficient irrigation techniques are dealt here. The following sections describe other irrigation techniques, which have evolved with time. These techniques have evolved in various countries in Latin America and the Caribbean nations over the generations.

2.3.2 Small-scale Clay Pot and Porous Capsule Irrigation

This technology consists of using clay pots and porous capsules to improve irrigation practices by increasing storage and improving the distribution of water in the soil. It is not new. The Romans used it for many centuries. This ancient irrigation system has been modernized and reapplied in water-scarce areas in Brazil, India and elsewhere with suitable modifications to take care of the local conditions.

2.3.2.1 Technical Description

This low-volume irrigation technology is based on storing and distributing water to the soil, using clay pots and porous capsules interconnected by plastic piping. A constant-level reservoir is used to maintain a steady hydrostatic pressure. Clay pots are open at the top and are usually fired in home furnaces after being fabricated from locally obtained clay or clay mixed with sand. The pots, usually conical in shape and of 10 to 12 liters capacity, are partially buried in the soil with only the top extending above ground. Distribution is by plastic or Poly Vinyl Chrolide (PVC) piping to ensure a fairly uniform permeability and porosity. Hydrostatic pressure is regulated by maintaining a constant level in the storage reservoir, as shown in Figure 2.11.

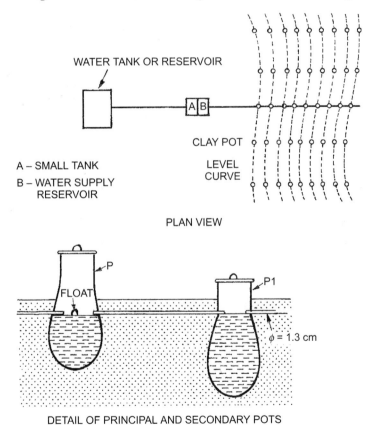

Figure 2.11   Schematic Representation of a Clay Pot Irrigation System

[Source: Aderaldo Silva De Souza, et al. *Irrigación par Potes de Barro: Descripción del Método y Pruebas Preliminares,* Petrolina, PE, Brasil, 1982, (EMBRAPA-CPATSA Boletín de Investigación No. 10)].

P ⇒ Principal Pot, P1 ⇒ Secondary Pot, ---- ⇒ Plastic Piping

A similar system, tested in Mexico and Brazil, uses smaller, closed containers, or porous capsules, completely buried in the soil. These containers distribute the water either by suction and capillary action within the soil, or by external pressure provided by a constant-level reservoir (as in the previous system). Each capsule normally has two openings to permit connection of the plastic (PVC) piping which interconnects the capsules. The capacity of these capsules ranges between 7 and 15 liters, and the storage tanks supplying the system are elevated 1 or 2 m above the soil surface. The capsules are buried in a line 2 meters apart, at least 10 cm under the top layer of the soil.

The number of pots or capsules used is a function of the area of cultivation, soil conditions, climate, and pot size. Up to 800 pots/ha were installed in Brazil; the system there is shown in Figure 2.12.

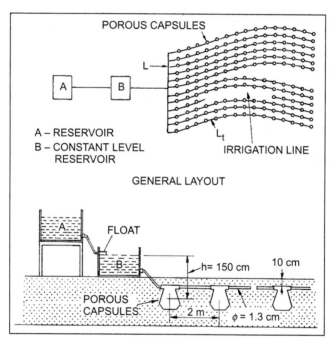

Figure 2.12 Schematic Representation of a Porous Capsule Irrigation System.

[Source: Aderaldo Silva De Souza, et al. *Irrigación par Potes de Barro: Descripción del Método y Pruebas Preliminares*, Petrolina, PE, Brasil, 1982, (EMBRAPA-CPATSA Boletín de Investigación No. 10)].

L : Spacing between irrigation lines

L$_t$: Spacing between porous capsules

ϕ : diameter of plastic piping

2.3.2.2 Extent of Use

This technology is being used for small-scale agricultural irrigation in the arid and semi-arid regions of Argentina, Brazil, Ecuador, Bolivia, and Mexico. It has also been used in tropical countries such as Guatemala, Panama, and the Dominican Republic during drought periods.

2.3.2.3 Operation and Maintenance

The operation is very simple, requiring only the opening of valves to replace the water used from the pots and capsules. However, the installation of the system does require a degree of care since the pots and capsules are made of clay and can be easily broken; also, the gradients must be correct if gravity flows are desired. It is also important to maintain the hydrostatic pressure. If this pressure cannot be maintained, the connections between pots must be checked for possible leaks and/or breakages. Replacement of the pots or capsules is necessary every 3 to 5 years. A soil investigation before the installation is advisable.

2.3.2.4 Level of Involvement

The participation of the community is essential in the implementation of this technology. Further, the support of the government and research institutions is also desirable. In Brazil, the government of the state of Pernambuco built a factory to manufacture porous capsules and developed small areas of bean cultivation for the application of the technology. In Ecuador and Bolivia, universities and government agricultural institutions are testing it.

2.3.2.5 Costs

Costs vary according to the materials and the type of system used. In Brazil, the reported cost was $ 1,300/ha cultivated using clay pots, and $1,800/ha cultivated using porous capsules. A clay pot system in the Dominican Republic reported an annual cost of $ 1,280. Smaller experimental systems in Bolivia and Panama were built for less than $100. Tables 2.5 and 2.6 provide some details of the costs in Brazil around the year 2000.

2.3.2.6 Effectiveness of the Technology

The technology has been shown to improve the stability of the soils. It has allowed agricultural development in areas where climatic conditions and the quality of the soils have prevented the use of conventional irrigation methods. Tests performed in Panama, using fruit trees, show significant

Table 2.5   Installation Costs of a Clay Pot Irrigation System on a 0.2 ha Plot in Brazil in the Year 2000

Item	Quantity	Total cost ($)
Clay pots	166	73.78
½" diameter plastic tubing	800 m	118.52
Tailpiece	0.8 kg	11.85
Float	7	5.19
Labor (digging)	12 person-days	35.56
Other		22.52
Total		267.42

[Source: http://www.oas.org/usde/publications/Unit/oea59e/ch27.htm#TopOfPage]

Table 2.6   Installation Cost of a Porous Capsule Irrigation System on a 1 ha Plot in Brazil in the Year 2000

Item	Quantity	Total cost ($)
Porous pots	2500	745.00
½" diameter plastic tubing	2500 m	815.00
1" diameter Plastic tubing	100 m	23.00
Tailpiece	4 kg	60.00
Labor	50 person-days	150.00
Total		1793.00

[Source: http://www.oas.org/usde/publications/Unit/oea59e/ch27.htm#TopOfPage]

improvements in the size of the stem and the number of fruits per plant; a yield of six fruits per plant was achieved with this system versus two with conventional irrigation. In Bolivia, the use of this technology in the cultivation of potatoes resulted in a yield of 42,000 kg/ha versus 18,000 kg/ha using traditional irrigation methods.

2.3.2.7   Suitability

This technology is suitable for arid and semi-arid regions, and for small-scale agricultural projects in areas affected by periodic drought. Countries like Bolivia, Brazil, Peru, Argentina, and Chile can definitely benefit from the use of this technology in rural areas.

2.3.2.8   Advantages

- This is a low-cost technology.
- Agricultural production is higher with this technology than with other irrigation technologies.

- Agriculture can be undertaken at lower air temperatures.
- Infiltration losses are reduced.
- Weeds can be better controlled, by managing their access to water.
- This system does not cause environmental impacts.
- This technology is very useful in family gardens and in horticulture.
- Water management using this technology allows agricultural development in arid lands and salty soils.
- Vandalism is minimized since most of the equipment is under the soil surface.
- It is easy to operate and maintain.
- It can reduce fertilizer use, by allowing application to defined, cultivated areas.
- Use of this technology can minimize soil erosion.

2.3.2.9 Disadvantages

- The technology is difficult to use in rocky soils.
- Broken pots or capsules can disrupt the irrigation operation and reduce productivity. Some plants with extended root systems are difficult to cultivate using this technology.
- In some areas, it may be difficult to purchase or manufacture the clay pots and/or capsules.
- It is only applicable to small-scale agriculture.

2.3.2.10 Cultural Acceptability

This technology is gaining acceptance among agricultural communities in arid areas. It is well developed as a technology for use in household gardening.

2.3.2.11 Further Development of the Technology

Improvements in the construction of the porous capsules are desirable, perhaps using different materials, which have acceptable levels of porosity but are more robust and can avoid breakages. It is also desirable to develop systems using porous capsules or clay pots that can be used in large-scale or commercial agricultural operations. Educational and informational programming on the benefits of the technology, and training in the manufacture of porous capsules, and pots are required.

This technology was developed and applied in Mexico during the 1970s. It is essentially an intermittent gravity-flow irrigation system. It has been used almost exclusively for small-scale agriculture and domestic gardening.

2.3.3 Raised Beds and *Waru Waru* Cultivation

This technology is based on modification of the soil surface to facilitate water movement and storage, and to increase the organic content of the soil to increase its suitability for cultivation. This system of soil management for irrigation purposes was first developed in the year 300 B.C., before the rise of the Inca Empire. It was later abandoned as more technically advanced irrigation technologies were discovered. Nevertheless, in 1984, in Tiawanaco, Bolivia, and Puno, Peru, the system was re-established. It is known in the region as *Waru Waru*, which is the traditional Latin American Indian (Quechua) name for this technique.

2.3.3.1 *Technical Description of the* Waru Waru *System*

The technology is a combination of rehabilitation of marginal soils, drainage improvement, water storage, optimal utilization of available radiant energy, and attenuation of the effects of frost. The main feature of this system is the construction of a network of embankments and canals. The embankments serve as raised beds for cultivation of crops, while the canals are used for water storage and to irrigate the plants. The soils used for the embankments are compacted to facilitate water retention by reducing porosity, permeability, and infiltration. Infiltration in the clay soils of the region varies from 20% to 30% of the precipitation volume. Thus, clay soils are preferred for this purpose. Sandy soils have too great a porosity to retain the water within the beds.

The cultivation takes place in the "new" soils within the raised bed created by the construction of the embankment. Within the bed, the increased porosity of the new soils results in enhanced infiltration, often increasing infiltration by 80% to 100% of the original soil. This system permits the recycling of nutrients and all the other chemical and biological processes necessary for crop production. Water uptake by the raised beds is through diffusion and capillary movements using water contained within the beds or supplied from the surrounding canals. The soils are kept at an adequate moisture level to facilitate the cultivation of plants such as potatoes and quinoa *(Chenopodium quinoa)*. Thermal energy is captured and retained in the soil as a result of the enhanced moisture levels, which protect the soils of the bed from the effects of frost. The system acts as a thermoregulator of the microclimate within the bed.

The main design considerations for raised bed cultivation include the following:

- Depth of the water table, since a high water table increases the height of the embankment required.

- Soil characteristics, which affect both the dimensions of the embankment and the nature of the cultivation zone.
- Climatic conditions, which include the volume and frequency of rainfall, temperature range, and frost frequency.

An example of a typical raised bed irrigation system is shown in Figure 2.13. Soft fill (e.g., compost or mulch) might be required within the embanked bed to maintain an adequate level of soil moisture.

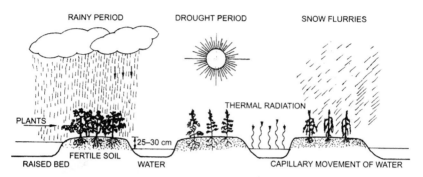

Figure 2.13   Raised Bed Irrigation System in Puno, Peru.
[Source: Alipio C. Murilo and Ludgardo L. Mamani, *Manual Técnico de Waru Waru, Para la Reconstrucción, Producción y Evaluación Económica*, Puno, Peru, Programa Interinstitucional de *Waru Waru*, Convenio PELT/INADE-IC/COTESU, 1992.]

There are three types of raised bed systems, characterized by the source of water:

Rainwater systems, in which rainwater is the primary source of moisture. These systems require small lagoons for storage during dry periods and a system of canals to distribute the water to the beds. They are usually located at the base of a hill or a mountain. Figure 2.14 gives the layout of a rainwater *Waru Waru* system.

Fluvial systems, in which moisture is supplied by water from nearby rivers. These systems require a hydraulic infrastructure, such as canals and dikes, to transport the water. Figure 2.15 provides the layout of a fluvial *Waru Waru* system.

- *Phreatic systems,* in which groundwater is the source of moisture in the beds. These systems are located in areas where the groundwater table is close to the surface of the soil and there is a mechanism for groundwater recharge, such as an infiltration lagoon. Figure 2.16 gives the layout for a typical phreatic *Waru Waru* system.

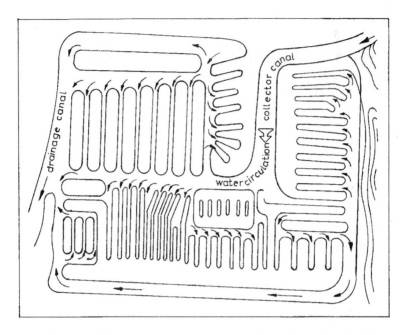

Figure 2.14   Layout of a Rainwater *Waru Waru* System.

[Source: Alipio C. Murilo and Ludgardo L. Mamani, *Manual Técnico de Waru Waru, Para la Reconstrucción, Producción y Evaluación Económica*, Puno, Peru, Programa Interinstitucional de *Waru Waru*, Convenio PELT/INADE-IC/COTESU, 1992]

2.3.3.2   Extent of Use, Operation and Maintenance

Waru Waru technology has been used primarily in the Lake Titicaca region at Puno, Peru, and in the Illpa River basin of Bolivia.

Periodic reconstruction of the embankments or raised beds is necessary to repair damage caused by erosion and water piping. Reconstruction is usually done during the dry season (March to May, in Peru), although in some areas it is done immediately after harvesting because of a lack of available labor at other times of the year. Cultivation of pasture and other grasses of differing heights on the embankments will help to prevent or control erosion caused by torrential rains during the wet season. Cultivation practices can also damage the embankments. Raising animals such as hogs near the embankments should be avoided, since they can damage the cultivation areas in their search for food.

Periodic fertilization of the raised beds is recommended, and the use of insecticides and fungicides may be necessary to limit crop damage. Insecticides are particularly advisable in the cultivation of potatoes.

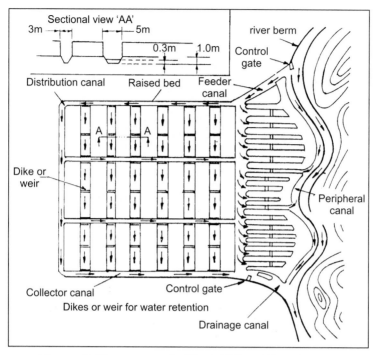

Figure 2.15   Layout of a Fluvial *Waru Waru* System

[Source: Alipio C. Murilo and Ludgardo L. Mamani, *Manual Técnico de Waru Waru, Para la Reconstrucción, Producción y Evaluación Económica*, Puno, Peru, Programa Interinstitucional de *Waru Waru*, Convenio PELT/INADE-IC/COTESU, 1992.]

2.3.3.3   Level of Involvement and Costs

The *Waru Waru* technology has been promoted, and assistance to farmers provided, by several Peruvian governmental organizations, including the Institute Nacional de Investigación Agropecuaria y Agroindustrial (INIAA), the Centre de Investigación Agropecuaria Salcedo (CIAS), the Centro de Proyectos Integrales Andinos (CEPIA), and by a number of NGOs. These organizations intend to reconstruct 500 ha of *Waru Waru* in 72 rural communities in the vicinity of Puno. Such an approach is considered to be representative of the involvement necessary to successfully implement a *Waru Waru* cultivation program in the region. Once established, the operation and maintenance of the systems, like the planting and harvesting of agricultural products, becomes the responsibility of the farmers who benefit from the use of this technology.

Very little information is available on the costs of these systems. The technology is at present largely experimental and limited to portions of the

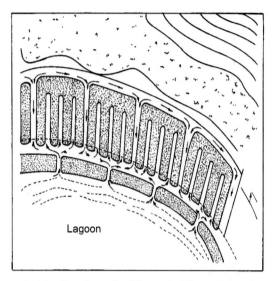

Figure 2.16 Design of a Phreatic *Waru Waru* System.

[Source: Alipio C. Murilo and Ludgardo L. Mamani, *Manual Técnico de Waru Waru, Para la Reconstrucción, Producción y Evaluación Económica*, Puno, Peru, Programa Interinstitucional de *Waru Waru*, Convenio PELT/INADE-IC/COTESU, 1992].

Andean Altiplano in Peru and Bolivia. Nevertheless, the cost per hectare of a phreatic raised-bed system for the cultivation of potatoes is estimated at $1 460 on the basis of the system created in Chatuma, Peru. Of this, 70% is direct cost and 30% is indirect cost. The production cost for 11.2 kg of potatoes using this technology in Chatuma was estimated at $4.80. The technology produces economic benefits during the first 3 years following construction, but, shortly thereafter reconstruction becomes necessary to maintain the productivity of the system.

2.3.3.4 Effectiveness of the Technology and Suitability

In the communities around Puno in Peru, during the seven-year period between 1982 and 1989, 229 ha were converted to this technology with mixed results. Some areas experienced large increases in productivity, particularly in the cultivation of potatoes, while other areas did not. Climatic conditions, such as drought and extremely cold weather, are likely to have contributed to the decrease in productivity in some areas, while poor design and construction of embankments may have led to the decline in productivity recorded in others.

This technology is suitable in areas with extreme climatic conditions, such as mountainous areas that experience heavy rainfalls and periodic droughts, and where temperature fluctuations range from intense heat to frost. It should be very useful in arid and semi-arid areas.

2.3.3.5 Advantages

- This technology can contribute to mitigating the effects of extreme climatic variations.
- The construction cost is relatively low.
- It can increase the production of certain agricultural crops.

2.3.3.6 Disadvantages

- The life span of the technology is relatively short; the systems require reconstruction after about 3 years of operation.
- Testing of soil texture and composition is necessary before implementation.
- *Waru Waru* systems require annual maintenance and periodic repair.

2.3.3.7 Cultural Acceptability and Further Development of the Technology

This is an ancient technology, well accepted in the agricultural communities of Peru and Bolivia. Application of this technology in other areas with different soil and climatic conditions will be a measure of its potential utility outside of the areas where it is traditionally used. Improvements in the design of the raised bed cultivation system are necessary in order to extend the economic life of the technology and to minimize the need for regular reconstruction of the beds to maintain their productivity.

2.3.4 Automatic Surge Flow Irrigation System

This technology was developed and applied in Mexico during the 1970s. It is essentially an intermittent gravity-flow irrigation system. It has been used almost exclusively for small-scale agriculture and domestic gardening.

2.3.4.1 Technical Description

Prior to the development of this technology, electronically controlled valves were used to produce intermittent water flows for irrigation. These valves are expensive and require some technical training to operate. The *diabeto* (from Greek *diabetes* or siphon) was developed for the purpose of replacing these valves with a device that would be more cost-effective and easier to operate and maintain with a minimum consumption of energy. The system consists of a storage tank equipped with one or more siphons, as shown in Figure 2.17. The storage tank must be designed to keep a predetermined head in the system to ensure that the water discharged during the siphoning process does not exceed the water flow into the storage tank, thereby draining the tank.

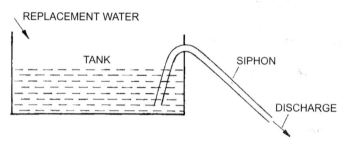

Figure 2.17 Schematic Representation of an Automatic Surge Flow Irrigation System *(Diabeto)*

[Source: P. Martinez Austria and R.A. Aldama, "Dispositive de Control para la Aplicación del Riego Intermitente," *Revista Ingenieria Hidráulica en México*, May-August, 1991]

Another system that produces similar results is the use of a storage tank with a bottom discharge. This system, as shown in Figure 2.18, is equipped with a floater, shown in Figure 2.19, which allows the cyclical opening and closing of a gate at the bottom of the tank. In effect, the operation of the floater is similar to the mechanism in the storage tank of a toilet flushing system.

The materials normally used in the construction of the water storage tanks are gravel, cement, and reinforced concrete. The siphons are usually built of a flexible plastic material; PVC is not recommended.

The design of these systems must consider irrigation water use, available hydraulic load, topographic characteristics in the area of application, physical dimensions of the irrigated land, slope and location of furrows, and soil characteristics. Design manuals, based on laboratory and field experiments, have been developed in Mexico.

2.3.4.2 Extent of Use

This technology has been used primarily in the arid and semi-arid regions of Mexico. The *diabeto* can be used in any gravity irrigation system, but has been particularly useful in the irrigation of 100 to 300 m² fields, using furrow irrigation, and in domestic gardening. This technology is best suited for small-scale (< 4 ha) irrigation in rural areas. At present, it is widely used only in Mexico.

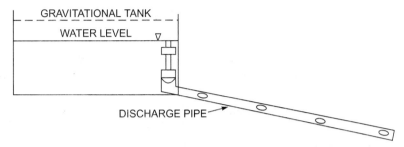

Figure 2.18 Schematic Representation of a Gravitational Tank
Irrigation System

[Source: V.N. García, *Diseño y Aplicación del Riego Intermitente por Gravedad.*
Universidad Nacional Autónoma de México, Facultad de Ingeniería, México D.F., 1995 (Tesis
para obtener el grado de Doctor en Ingeniería Hidráulica)]

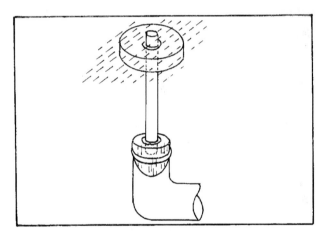

Figure 2.19 Schematic Representation of an Automatic Fluid
Water Control Device used in Gravitational Tanks

[Source: V.N. García, *Diseño y Aplicación del Riego Intermitente por Gravedad.*
Universidad Nacional Autónoma de México, Facultad de Ingeniería, México D.F., 1995 (Tesis
para obtener el grado de Doctor en Ingeniería Hidráulica)]

2.3.4.3 *Operation and Maintenance*

The *diabeto* and the gravitational tanks with bottom discharges function
automatically, based on flow control devices, and do not need outside
energy sources. The water is discharged into a channel that distributes it
into the furrows and to the irrigated crops. Maintenance is very simple,
requiring only periodic cleaning of the tanks, siphons, and/or discharge
pipes.

2.3.4.4 Level of Involvement and Costs

Until around the year 2000, educational institutions, small private agricultural enterprises, and the Mexican Government have promoted this technology. However, it would be desirable if local communities got more involved in implementing it.

A typical surge flow, automatic irrigation device used to cost about $600 in around 2000 AD. This includes an 11.25 m^3 storage tank, feeding system, and siphon. A device of this size can irrigate up to 4 ha. A similar gravitational tank irrigation system, with the same tank capacity, 150 m of piping, and gates, has an estimated cost of $1,500. A smaller system for domestic gardening can cost around $80. The operation and maintenance costs of these systems are practically nil.

2.3.4.5 Effectiveness of the Technology and Suitability

With the surge flow, automatic irrigation systems and the gravitational tank technologies, irrigation efficiencies of over 75% have been achieved in the state of Zacatecas, Mexico. This represents a significant improvement over the 25% rate reported using traditional irrigation technologies. A saving of about 25% in energy consumption costs has also been observed.

The technology is recommended for arid and semi-arid areas where low precipitation and high evaporation rates prevail, and where small storage areas and depleted aquifers exist.

2.3.4.6 Advantages

- This technology can utilize water from small wells of limited capacity, reused wastewater, and small streams.
- Hydraulic energy is used as the driving force; these systems do not require external energy sources.
- The systems are low-pressure.
- Irrigation time and labor force requirements are small, as the systems are automatic.
- The technology is low in cost.
- It is easy to operate and maintain.
- It is applicable to small-scale agricultural systems.
- It is more efficient than traditional irrigation systems.

2.3.4.7 Disadvantages

- The surge flow technology has not been recommended for furrow irrigation in fields with dimensions greater than 200 m long and 25 m wide, as the volume of water required in such applications will require extremely large storage tanks.

- For greater efficiency, the irrigated lands should be leveled.

Apart from these techniques, sprinkler irrigation and drip irrigation are under extensive use in the developed as well as developing nations. The following sections explain both these techniques.

2.3.5 Sprinkler Irrigation

Sprinkler irrigation is a method of applying irrigation water, which is similar to natural rainfall. Water is distributed through a system of pipes usually by pumping. It is then sprayed into the air through sprinklers so that it breaks up into small water drops which fall to the ground. The pump supply system, sprinklers and operating conditions must be designed to enable a uniform application of water.

2.3.5.1 Crops and Slopes Suitable for Sprinkler Irrigation

Sprinkler irrigation is suited for most row, field and tree crops and water can be sprayed over or under the crop canopy. However, large sprinklers are not recommended for irrigation of delicate crops such as lettuce because the large water drops produced by the sprinklers may damage the crop.

Sprinkler irrigation is adaptable to any farmable slope, whether uniform or undulating. The lateral pipes supplying water to the sprinklers should always be laid out along the land contour whenever possible. This will minimize the pressure changes at the sprinklers and provide a uniform irrigation.

2.3.5.2 Suitable Soils and Water for Sprinkler Irrigation

Sprinklers are best suited to sandy soils with high infiltration rates although they are adaptable to most soils. The average application rate from the sprinklers (in mm/hour) is always chosen to be less than the basic infiltration rate of the soil so that surface ponding and runoff can be avoided.

Sprinklers are not suitable for soils which easily form a crust. If sprinkler irrigation is the only method available, then light fine sprays should be used. The larger sprinklers producing larger water droplets are to be avoided.

A good clean supply of water, free of suspended sediments, is required to avoid problems of sprinkler nozzle blockage and spoiling the crop by coating it with sediment.

2.3.5.3 Sprinkler System Layout

A typical sprinkler irrigation system consists of the following components:
Pump unit, mainline, and sometimes submainlines, laterals, sprinklers.

The *pump unit* is usually a centrifugal pump, which takes water from the source and provides adequate pressure for delivery into the pipe system.

The *mainline* and *submainlines* are pipes which deliver water from the pump to the laterals. In some cases these pipelines are permanent and are laid on the soil surface or buried below ground. In other cases they are temporary, and can be moved from field to field. The main pipe materials used include asbestos cement, plastic or aluminum alloy.

The *laterals* deliver water from the mainlines or submainlines to the sprinklers. They can be permanent but more often they are portable and made of aluminum alloy or plastic so that they can be moved easily.

The most common type of sprinkler system layout is shown in Figure 2.20. It consists of a system of lightweight aluminium or plastic pipes, which are moved by hand. The rotary sprinklers are usually spaced 9-24 m apart along the lateral which is normally 5-12.5 cm in diameter. This is how it can be carried easily. The lateral pipe is located in the field until the irrigation is complete. The pump is then switched off and the lateral is disconnected from the mainline and moved to the next location. It is re-assembled and connected to the mainline and the irrigation begins again. The lateral can be moved one to four times a day. It is gradually moved around the field until the whole field is irrigated. This is the simplest of all systems. Some use more than one lateral to irrigate larger areas.

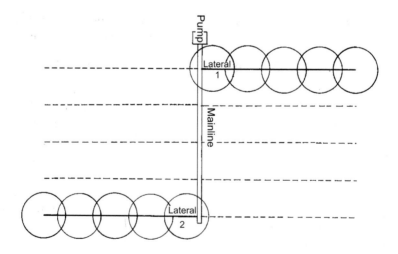

Figure 2.20   Hand-moved Sprinkler System Using Two Laterals

[Source: http://www.fao.org/docrep/S8684E/s8684e06.htm]

A common problem with sprinkler irrigation is the large labor force needed to move the pipes and sprinklers around the field. In some places such labor may not be available and may also be costly. To overcome this problem many mobile systems have been developed such as the hose reel raingun and the center pivot.

Another system that does not need a large labor force is the drag-hose sprinkler system. Main and laterals are buried PVC pipes: one lateral covers three positions. Sprinklers on risers carried by skids are attached to the laterals through hoses (similar to garden sprinklers). Only the skid with the sprinkler has to be moved from one position to another, which is an easy task.

2.3.5.4 Operating the Sprinkler Systems

The main objective of a sprinkler system is to apply water as uniformly as possible to fill the root zone of the crop with water.

Wetting patterns The wetting pattern from a single rotary sprinkler is not very uniform (see Figure 2.21). Normally the area wetted is circular. The heaviest wetting is close to the sprinkler as seen from the cross sectional view. For good uniformity several sprinklers must be operated close together so that their patterns overlap. For good uniformity the overlap should be at

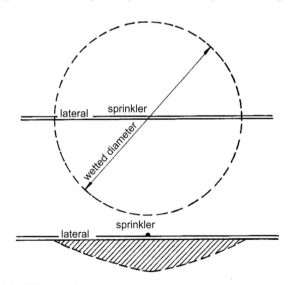

Figure 2.21 Wetting Pattern for a Single Sprinkler (top view and cross-sectional view)

[Source: http://www.fao.org/docrep/S8684E/s8684e06.htm]

least 65% of the wetted diameter. This determines the maximum spacing between sprinklers.

The uniformity of sprinkler applications can be affected by wind and water pressure. Spray from sprinklers is easily blown about by even a gentle breeze and this can seriously reduce uniformity. To reduce the effects of wind the sprinklers can be positioned more closely together.

Sprinklers will only work well at the right operating pressure recommended by the manufacturer. If the pressure is above or below this then the distribution will be affected. The most common problem develops when the pressure is too low. This happens when pumps and pipes wear. Friction increases and so pressure at the sprinkler reduces. The result is that the water jet does not break up and all the water tends to fall in one area towards the outside of the wetted circle. If the pressure is too high then the distribution will also be poor. A fine spray develops which falls close to the sprinkler.

Application rate This is the average rate at which water is sprayed onto the crops and is measured in mm/hour. The application rate depends on the size of sprinkler nozzles, the operating pressure and the distance between sprinklers. When selecting a sprinkler system it is important to make sure that the average application rate is less than the basic infiltration rate of the soil. In this way all the water applied will be readily absorbed by the soil and there should be no runoff.

Sprinkler drop sizes As water sprays from a sprinkler it breaks up into small drops between 0.5 and 4.0 mm in size. The small drops fall close to the sprinkler whereas the larger ones fall close to the edge of the wetted circle. Large drops can damage delicate crops and soils and so in such conditions it is best to use the smaller sprinklers.

Drop size is also controlled by pressure and nozzle size. When the pressure is low, drops tend to be much larger as the water jet does not break up easily. So to avoid crop and soil damage use small diameter nozzles operating at or above the normal recommended operating pressure.

2.3.6 Drip Irrigation

Drip irrigation is sometimes called trickle irrigation and involves dripping water onto the soil at very low rates (2-20 liters/hour) from a system of small diameter plastic pipes fitted with outlets called emitters or drippers. Water is applied close to plants so that only part of the soil in which the roots grow is wetted (see Figure 2.22), unlike surface and sprinkler irrigation, which involves wetting the whole soil profile. With drip irrigation water, applications are more frequent (usually every 1-3 days) than with other

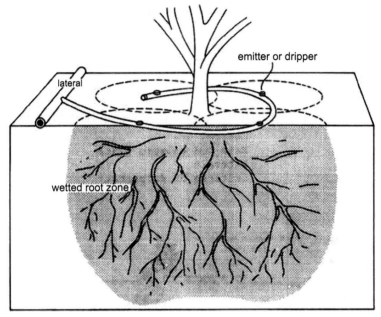

Figure 2.22   Selective Wetting of the Soil Around the Roots in Drip Irrigation

[Source: http://www.fao.org/docrep/S8684E/s8684e07.htm]

methods and this provides a very favorable high moisture level in the soil in which plants can flourish.

2.3.6.1 Suitable Crops and Slopes for Drip Irrigation

Drip irrigation is most suitable for row crops (vegetables, soft fruits), tree and vine crops where one or more emitters can be provided for each plant. Generally only high value crops are considered because of the high capital costs of installing a drip system.

   Drip irrigation is adaptable to any farmable slope. Normally the crop would be planted along contour lines and the water supply pipes (laterals) would be laid along the contour also. This is done to minimize changes in emitter discharge as a result of land elevation changes.

2.3.6.2 Soils and Water Suitable for Drip Irrigation

Drip irrigation is suitable for most soils. On clay soils water must be applied slowly to avoid surface water ponding and runoff. On sandy soils higher emitter discharge rates will be needed to ensure adequate lateral wetting of the soil.

One of the main problems with drip irrigation is blockage of the emitters. All emitters have very small waterways ranging from 0.2-2.0 mm in diameter and these can become blocked if the water is not clean. Thus it is essential for irrigation water to be free of sediments. If this is not so then filtration of the irrigation water will be needed.

Blockage may also occur if the water contains algae, fertilizer deposits and dissolved chemicals, which precipitate such as calcium and iron. Filtration may remove some of the materials but the problem may be complex to solve and requires an experienced engineer or consultation with the equipment dealer.

Drip irrigation is particularly suitable for water of poor quality (saline water). Dripping water to individual plants also means that the method can be very efficient in water use. For this reason it is most suitable when water is scarce.

2.3.6.3 Drip System Layout

A typical drip irrigation system is shown in Figure 2.23 and consists of the following components:

Pump unit, control head, main and sub-main lines, laterals, emitters or drippers.

The *pump unit* takes water from the source and provides the right pressure for delivery into the pipe system. The *control head* consists of valves to control the discharge and pressure in the entire system. It may also have filters to clear the water. Common types of filter include screen filters and graded sand filters which remove fine material suspended in the water. Some control head units contain a fertilizer or nutrient tank. These slowly add a measured dose of fertilizer into the water during irrigation. This is one of the major advantages of drip irrigation over other methods.

Mainlines, *sub-mainlines* and *laterals* supply water from the control head into the fields. They are usually made from PVC or polyethylene hose and should be buried below ground because they easily degrade when exposed to direct solar radiation. Lateral pipes are usually 13-32 mm in diameter.

Emitters or drippers are devices used to control the discharge of water from the lateral to the plants. They are usually spaced more than 1 meter apart with one or more emitters used for a single plant such as a tree. For row crops more closely-spaced emitters may be used to wet a strip of soil. Many different emitter designs have been produced in recent years. The basis of design is to produce an emitter, which will provide a specified constant discharge which does not vary much with pressure changes, and does not block easily. Various types of emitters are shown in Figure 2.24.

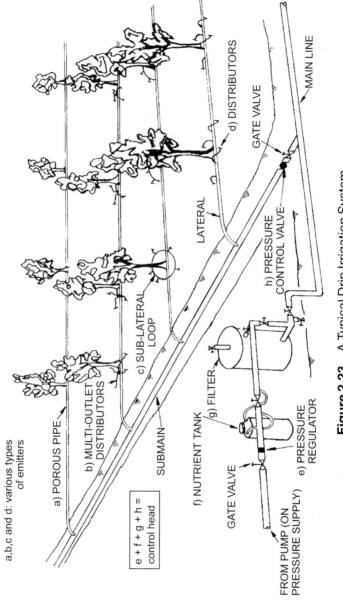

a,b,c and d: various types of emitters

a) POROUS PIPE

b) MULTI-OUTLET DISTRIBUTORS

c) SUB-LATERAL LOOP

d) DISTRIBUTORS

LATERAL

SUBMAIN

MAIN LINE

GATE VALVE

h) PRESSURE CONTROL VALVE

e + f + g + h = control head

f) NUTRIENT TANK

g) FILTER

GATE VALVE

e) PRESSURE REGULATOR

FROM PUMP (ON PRESSURE SUPPLY)

Figure 2.23 A Typical Drip Irrigation System
[Source:http://www.fao.org/docrep/S8684E/s8684e07.htm]

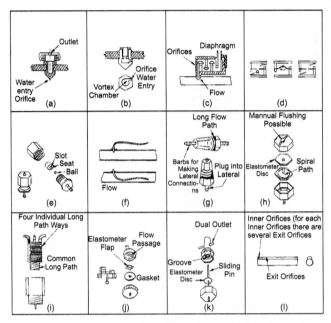

Figure 2.24 Types of Emitters Used in Drip Irrigation

2.3.6.4 *Operating the Drip Systems*

A drip system is usually permanent. When remaining in place during more than one season, a system is considered permanent. Thus it can easily be automated. This is very useful when labor is scarce or expensive to hire. However, automation requires specialist skills and so this approach is unsuitable if such skills are not available. Water can be applied frequently (every day if required) with drip irrigation and this provides very favorable conditions for crop growth. However, if crops are used to being watered each day they may only develop shallow roots and if the system breaks down, the crop may begin to suffer very quickly.

Wetting patterns Unlike surface and sprinkler irrigation, drip irrigation only wets part of the soil root zone. This may be as low as 30% of the volume of soil wetted by the other methods. The wetting patterns, which develop from dripping water onto the soil, depend on discharge and soil type. Figure 2.25 shows the effect of changes in discharge on two different soil types, namely, sand and clay.

Although only part of the root zone is wetted it is still important to meet the full water needs of the crop. It is sometimes thought that drip irrigation saves water by reducing the amount used by the crop. This is not true. Crop

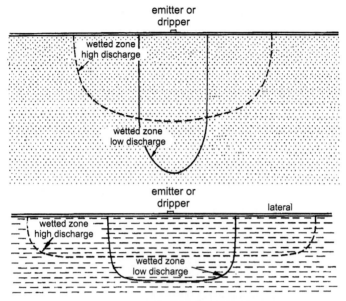

Figure 2.25   Wetting Patterns for Sandy and Clayey Soils with High and Low Discharge Rates (sand and clay)

[Source:http://www.fao.org/docrep/S8684E/s8684e07.htm]

water use is not changed by the method of applying water. Crops just require the right amount for good growth.

The water savings that can be made using drip irrigation are the reductions in deep percolation, in surface runoff, in wetting extra portion of soil and in evaporation from the soil. These savings, it must be remembered, depend as much on the user of the equipment as on the equipment itself. Israel provides the best example of a nation which has overcome its water scarcity problem by successfully implementing drip irrigation to significantly enhance its agricultural production.

2.4   WATER FOR INDUSTRIES

As already explained in section 2.3, many developed nations in Europe and North America use more water for industries when compared to their agricultural usage. UK, France and USA come under this category of nations. In this section, a case study from USA is presented.

2.4.1   Industrial Water Use In Tucson Valley, USA

Industrial water users in the Tucson Active Management Area (TAMA) include metal mines, sand and gravel mining facilities, electric power

112

producers, dairy operations, and other industrial users. As is shown in Figure 2.26, metal mining is the largest water user in the industrial sector, accounting for approximately 70 percent of the total classifies some golf courses and other turf facilities with their own wells as industrial for legal reasons. All turf uses, however, were previously discussed as municipal uses.

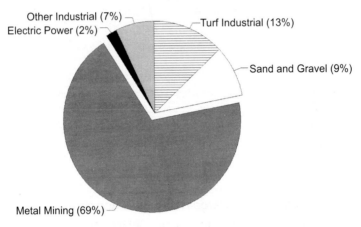

Figure 2.26 Composition of Industrial Water Usage in Tucson, USA
[Source: Water in the Tucson Area: Seeking Sustainability: A status report by the Water Resources Research Center, College of Agriculture, The University of Arizona]

2.4.1.1 *Water for Metal Mining*

Four active metal mines operate in TAMA.

Most mining in the TAMA is open-pit mining followed by milling and flotation. The ore extracted from the pits is crushed and delivered to the mill, where it is crushed again and mixed with water to form a slurry. The slurry is discharged to flotation cells. Here chemicals are added that cause the materials to float to the surface for removal. Waste rock remaining in the flotation cells is sent to thickener tanks. The solids settle in the tanks and the water is recovered. The tailings mixture, which usually is 46 to 55 percent solids by weight, is transported via pipeline to the tailings pond. The tailings slurry is deposited in the tailings impoundment with a spray, leaving standing water in the impoundment, which is skimmed off and recycled back to the mill.

Copper also can be leached from piles of certain types of ore using sulfuric acid. The leachate is then piped to another facility for recovery of copper. Another form of leaching from buried ore, known as "in-situ" leaching, has recently come into use. Both types of leaching use less water than milling and flotation, but "in-situ" leaching has been applied only on a pilot scale in Pinal County, and both forms require the right kind of ore.

Arizona Department of Water Resources (ADWR) funded a study to analyze possibilities for additional water conservation in mining. Increasing the density of the tailings slurry was one of the most effective steps available to decrease groundwater withdrawals. For every one percent increase in the tailings density, 500 to 800 acre-feet of groundwater can be saved per year. In addition, seepage of groundwater underneath the tailings ponds could also be reduced by depositing fine-grained tailings on top of native soils before tailings are delivered to the ponds. Evaporation of water in tailings impoundments can be reduced by installing multiple decant towers to more quickly decant the water.

2.4.1.2 Sand and Gravel Facilities

The approximately 15 sand and gravel facilities operating in TAMA used 6.37 million m^3 of groundwater in 1995. This is projected to increase to 8.61 million m^3 per year by 2025. Most sand and gravel facilities recycle much of the water used for washing mined stream deposits. Facilities can save additional water by reducing water use for dust control and other clean-up related activities. Sand and gravel facilities could theoretically use CAP water, effluent or poor quality groundwater. Secondary effluent could be an inexpensive and feasible alternate supply. Because sand and gravel operations are able to pump groundwater at relatively low cost, switching to CAP water would not be economical without a subsidy.

2.5 WATER FOR HYDROPOWER GENERATION

Water is required for the generation of hydropower. However, the water used for hydropower generation can be used again for other purposes. The energy in falling water can be converted into electrical energy or into mechanical energy to pump water or grind grain. The amount of energy that can be captured is a function of the vertical distance by which the water drops (i.e., the *head*, h) and the *volumetric flow rate (i.e., the discharge,* Q) of the water. The theoretical power (P_{theo}) that can be developed in a hydro power scheme is given by:

$$P_{theo} = \gamma hQ \qquad (2.2)$$

Here γ = specific weight of water. For most hydro projects, water is supplied to the turbine from some type of storage reservoir, usually created by a dam or weir. The reservoir allows water to be stored and electricity to be generated at more economically desirable times—during periods of peak electrical demand, for example—when the electricity can be sold for a higher price. In these systems the amount of electrical power that can be generated is determined by the amount of water that is stored in the reservoir and the rate at which it is released. The most environmentally sound hydro system does not impact the amount or pattern of water flow that normally exists in the river or stream. Such *run-of-river systems* may use a special turbine placed directly in the river to capture the energy in the water flow. On the other hand, *pumped storage hydropower schemes* pump water from a lower reservoir to a higher reservoir during periods of low demand to create sufficient head and discharge during peak demand periods. Description about pumped storage projects may be found elsewhere (Dandekar and Sharma, 1979).

2.5.1 Small Hydropower

In the past, hydropower stations were often built as a part of large dam projects. Due to the size, cost, and environmental impacts of these dams (and the reservoirs they create), hydro developments today are increasingly focused on smaller-scale projects. Although the definition of small-scale varies, only projects that have less than 10 megawatts (MW) of generating capacity are considered here. Table 2.7 provides the definition criteria for different types of hydropower, which are in use in different parts of the world.

Table 2.7 Definition Criteria for Types of Hydropower in Various Regions

Country	micro (kilowatts)	mini (kilowatts)	small (megawatts)
United States	<100	100–1000	1–30
United States	<100	100–1000	–
China	–	<500	0.5–25
USSR	<100	–	0.1–30
France	5–5000	–	–
India	<100	101–1000	1–15
Brazil	<100	100–1000	1–30
Other Countries	<100	<1000	<10

[Source: http://www.iash.info/worldpotential.htm]

The main components of a small-scale hydro (SSH) system are the turbine and the generator. Other components include the physical structures to direct and control the flow of water, mechanical and/or electronic controllers, and structures to house the associated equipment. Different types of turbines are available and the optimum choice depends strongly on the head and the water flow rate. Generally, a high head site will require smaller, less expensive turbines and equipment.

2.5.1.1 Planning, Developing and Operating Small Hydro

Although most small hydro projects are different, the following steps provide a good outline of the main stages in the development and operation of a project:

1. *Reconnaissance surveys and hydraulic studies*
This first phase of work frequently covers numerous sites and includes map studies; delineation of the drainage basins; preliminary estimates of flow and floods; and a one day site visit to each site (by a design engineer and geologist or geotechnical engineer); preliminary layout; cost estimates (based on formulae or computer data); a final ranking of sites based on power potential; and an index of cost.

2. *Pre-feasibility study*
Work on the selected site or sites would include: site mapping and geological investigations (with drilling confined to areas where foundation uncertainty would have a major effect on costs); a reconnaissance for suitable borrow areas (e.g. for sand and gravel); a preliminary layout based on materials known to be available; preliminary selection of the main project characteristics (installed capacity, type of development, etc.); a cost estimate based on major quantities; the identification of possible environmental impacts; and production of a single volume report on each site.

3. *Feasibility study*
Work would continue on the selected site with a major foundation investigation program; delineation and testing of all borrow areas; estimation of diversion, design and probable maximum floods; determination of power potential for a range of dam heights and installed capacities for project optimization; determination of the project design earthquake and the maximum credible earthquake; design of all structures in sufficient detail to obtain quantities for all items contributing more than about 10 percent to the cost of individual structures; determination of the dewatering sequence and project schedule; optimization of the project layout, water levels and components; production of a detailed cost estimate; and finally, an economic and financial evaluation of the project including an assessment of the impact on the existing electrical grid along with a multi-volume comprehensive feasibility report.

4. *System planning and project engineering*

This work would include studies and final design of the transmission system; integration of the transmission system; integration of the project into the power network to determine precise operating mode; production of tender drawings and specifications; analysis of bids and detailed design of the project; production of detailed construction drawings and review of manufacturer's equipment drawings. However, the scope of this phase would not include site supervision nor project management, since this work would form part of the project execution costs.

5. *Financing*

The process of arranging financing for small-hydro projects is often difficult. The developer has to complete two steps to realize their development plans. The first is to obtain a contract with a utility or organization, which will purchase the produced electricity. With this contract in place the next step is to negotiate a bank loan or other source of financing. However, many banks lack knowledge of small hydro projects and have no experience with this type of loan. In recent years, there are some banks which specialize in these kinds of loans.

2.5.1.2 *World's Small Hydropower Potential*

Hydropower technology has been extensively deployed throughout the world at both large and small scale for electricity generation, and at small scale for mechanical power. This generally places hydropower at an advantage over other renewable technologies for new deployment, since operational or abandoned schemes are often available within the target country. Operational, design and construction experience may well also be available there. The technology is technically and commercially mature. Small scale hydro schemes can make a useful contribution to rural electrification strategies, presenting a suitable alternative to decentralized diesel generation, particularly where fuel supply is a problem. Some countries are encouraging deployment through subsidies and incentives.

About 680 GW of hydropower have been developed globally, 47 GW of which is small hydropower (<10 MW/site). The remaining non-utilized global hydropower potential is estimated at 3,000 GW, 180 GW of which is at small hydro sites. 70% of this small hydro potential (126 GW) is in developing countries. Table 2.8 provides the details about small hydro capacity and production.

Table 2.8 Estimates of Small Hydro Capacity and Production by Region in 1995

Region	Capacity (MW)	Production (GWh)
North America	4,400	18,000
Latin America	1,000	3,500
Western Europe	9,740	40,000
CEE and FSU	2,070	8,500
Mid East and Mediterranean	180	700
Africa	400	1,600
Pacific	160	700
Asia	10,000	42,000
Total	27,950	115,000

[Source: http://www.iash.info/worldpotential.htm]

2.5.1.3 Small Hydropower in India

While there has been a continuous increase in the installed capacity of hydro power stations in India, which was reported as 22,439 MW in 2000, the share of hydropower has been reduced to only 25% in the total installed for power generation from 50.62% in 1963. Ministry of Power in the Government of India is responsible for the development of large hydropower projects in India. Ministry of Non-conventional Energy Sources (MNES) has been responsible for small and mini hydro projects up to 3 MW station capacity since 1989. The subject of small hydro between 3-25 MW has been assigned to MNES w.e.f. 29th November, 1999. In order to maintain the balance between hydropower and thermal power, Ministry of Power has announced a policy for accelerated development of hydropower in the country. Development of small hydropower at an accelerated pace is one of the tasks in the policy.

Development of Small Hydro Power (SHP) projects and thrust areas Small hydropower development is one of the thrust areas of power generation from renewables in the Ministry of Non-conventional Energy Sources. The ministry is encouraging development of small hydro projects in the State sector as well as through private sector participation in various States. In the 1990's, the capacity of Small hydro projects up to 3 MW has increased 4-fold from 63 MW to 240 MW. From 1994 onwards, the thrust of SHP programme is to encourage private sector for setting up of commercial SHP projects. 13 potential States have announced their policies for private sector participation in SHP sector. The main thrust areas are nation-wise resource assessment, setting up of commercial SHP projects, renovation and modernization of old SHP projects, development and upgradation of water mills and industry-based research and development.

Planning small hydro projects require many stages of technical and financial study to determine if a site is technically and economically feasible. The viability of each potential project is very site-specific. Power output depends on flow of water and the height of the drop of the available water. The amount of energy that can be generated depends on the quantity of water available and the variability of flow throughout the year. The economics of a site depends on the power (capacity) and energy a project can produce, if the power can be sold, and the price paid for the power sold. In a remotely located community the value of power generated for consumption is generally significantly more than for systems that are connected to a central grid. However, remotely located communities may not be able to use all the available energy from the small hydro plant or, may be unable to use the energy when it is available because of seasonal variations in water flow and energy consumption.

2.5.1.4 Small Hydropower in China

China's small hydro is making a significant contribution to sustainable energy policy and development. Over 26 GW of small hydropower capacity has already been installed, the sector employs 1.2 million people, and over 43,027 SHP stations had been built with a total installed capacity of 26,262 MW and an annual output of 87.1 billion kWh by the end of year 2001. SHP stations mainly supply about 300 million people in 760 counties. Many of the benefits of small hydro are social and environmental.

About 300 million people across 760 counties have the majority of their electricity supplied by SHP stations, and development of the technology is encouraged by the worldwide call for reductions in greenhouse gas (GHG) emissions. With the global power industry's focus shifting towards the development of clean, cost-effective renewables, large hydropower (LHP) remains one of the most controversial issues. As this has a great impact on the surrounding environment, including soil, water, forest and climate, increasingly severe conditions are applied to the construction of new LHP projects. For instance, it is now generally accepted in the European Union that projects with installed capacity over 15 MW are not recognized as renewable energy. In such a context, there is a worldwide trend towards development of small hydropower.

Since 1980's, China has witnessed remarkable development in this field. By the end of 2001, over 43,027 SHP stations had been built there, with a total installed capacity of 26,262 MW and an annual output of 87.1 billion kWh. The Chinese leaders, from the Premier down, have all shown deep commitment to rural electrification, and SHP is considered as one of the most important ways of achieving this. The Chinese government has launched a series of programmes to promote SHP development, including 'Sending Electricity to Villages', 'Replacing Firewood by Electricity' and '400 Rural Electrification Counties'. As a result of this, China is becoming a significant arena for SHP development, accounting for 39% of the worldwide capacity under construction. However, in China today, the deregulation of the power industry is in progress, and a new market framework will take shape in the future. Meanwhile, new mechanisms are shaping up for SHP development and management, with the rising number of independent power producers (IPPs). China's SHP is facing both the challenges and opportunities within the changing energy infrastructure.

The Status of SHP in China by 2001 China has abundant SHP resources, with 100 GW of potential capacity that it would be economically feasible to develop. This potential is spread over 1500 of the almost 2300 counties in China. The country has always given great importance to SHP development. According to statistics, by the end of 2001, the number of SHP stations operating in the country was 43,027, with an total installed capacity of 26,262 MW—accounting for over a quarter of the total potential that could be developed. The annual output has reached 87.1 GWh. Table 2.9 shows the list of provinces in China that have over 300 MW of installed SHP capacity.

Table 2.9   At Least 300 MW of SHP Development in the Provinces of China by Year 2001

Province	Number of stations	Installed capacity (MW)	Annual output (GWh)	Proportion of total installed SHP capacity in China (%)
Guangdong	6525	3576	11,928	13.6
Sichuan	4395	3529	13,933	13.4
Fujian	5096	3057	11,540	11.6
Yunnan	1904	2250	9614	8.6
Hunan	4615	2033	7198	7.7
Zhejiang	2738	1873	4693	7.1
Hubei	2292	1466	3670	5.6
Guangxi	2547	1407	5061	5.4
Jiangxi	3965	1180	3301	4.5
Guizhou	1154	958	3657	3.6
Chongqing	1061	839	2851	3.2
Xinjiang	525	662	2114	2.5
Shaanxi	2158	453	1037	1.7
Gansu	467	363	1299	1.4
Henan	735	318	552	1.2
Hebei	195	317	304	1.2

[Source: Renewable Energy World, Jan-Feb. 2003]

Apart from the rapid development in scale, a unique management mechanism has also been created, as discussed below. SHP stations themselves can generally be characterized as being one of the following:

- Focused on economic, social and environmental benefits, with some input coming from governments at different levels, or
- Owned by independent power producers (IPPs), focusing on economic returns with input from the private sector.

Tables 2.10, 2.11 and 2.12 categorize these by installed capacity, and by management and operation modules.

Table 2.10 SHP Stations by Installed Capacity in China by Year 2001

Type		Micro	Mini	Small	Total
Station	Number	18,944	19,606	4427	43,027
	Percentage (%)	44.0	45.6	10.4	100
Installed capacity	MW	687	7171	18,404	26,262
	Percentage (%)	2.6	27.3	70.1	100
Annual output	GWh	1860	20,245	65,036	87,141
	Percentage (%)	2.1	23.2	74.6	100

[Source: Renewable Energy World, Jan-Feb. 2003]

Table 2.11 SHP Stations by Operation Module in China by Year 2001

Module		Connected to national grid	Local grid	Isolated operation	Total
Station	Quantity	4722	20,465	17,840	43,027
	Percentage(%)	10.9	47.6	41.5	100
Installed capacity	MW	6412	17,869	1981	26,262
	Percentage(%)	24.5	68.0	7.5	100
Annual output	GWh	20,097	60,792	6252	87,141
	Percentage(%)	23.1	69.8	7.2	100

[Source: Renewable Energy World, Jan-Feb. 2003]

Table 2.12 SHP Stations by Ownership in China by Year 2001

Ownership		State	Others and Private	Total
Station	Quantity	8244	34,783	43,027
	Percentage (%)	19.2	80.8	100
Installed capacity	MW	17,500	8762	26,262
	Percentage (%)	66.6	33.4	100
Annual output	GWh	62,954	24,187	87,141
	Percentage (%)	72.2	27.8	100

[Source: Renewable Energy World, Jan-Feb. 2003]

Furthermore, the development of SHP expands local grids and changes the composition of power consumption in the area supplied. It also ensures that the increase in power consumption by the Town and Village Enterprises (TVEs) and households is sustainable, Indeed, electrification rates have risen between 1985 and 2001 as follows:

- Village: 91.8% to 97.3%
- Town: 78.1% to 97.5%
- Household: 65.3% to 96.1%.

The fact that 300 million people in rural areas of China have access to electricity shows that SHP has been playing an increasingly important role in promoting social and economic development and environmental protection in such areas as well. SHP is not only a solution to rural energy shortage, but has become the leading industry in rural development in China.

SHP practice and experience in China Since 1990's, in line with the principle of self-reliance and other characteristics of rural development in China, a unique SHP development mechanism has come into being—covering planning, design, construction, operation and management. It is summarized below.

Decentralized development and management mechanism focusing on local stakeholders Unlike the centralized development model in other developing countries, or commercially oriented development in some industrialized countries, most SHP development in China is focused on decentralized management. SHP development strategies, objectives, standards and policies are decided by the central government, while planning, construction, design, operation, management, and so on are undertaken by the local governments. This decentralized management mechanism results in the complementary coexistence of national grid development, local grid development and isolated SHP stations.

Policy stimulates development For a long time, the Chinese government has made a series of preferential policies, to support and encourage local governments and local people to develop their nearby, rich small hydro power resources. 'Self-construction, self-management, and self-consumption' has been a well known policy, guiding SHP development in China since the early 1960s. This determines that local authorities or local people who invest in and construct the SHP stations have the right to manage the stations, and to utilize the electricity thus generated. In addition, 'self-consumption' also implies that there should be an SHP market—that is, SHP stations should have their own distribution/supply area, instead of selling all the energy to the national grid for distribution. As for taxation policy, value added tax on electricity generated by SHP systems has since 1994 stood at 6%—much more favorable than the 17% tax for large hydro power stations.

Funding sources

Joint investment Joint investment, whether undertaken by individuals or community or private enterprise, is mostly encouraged with investment from outside, including overseas markets. Under the policy of 'who invests, will own the project, and receive benefits', anyone is welcome to invest. As a result of this, there are different types of ownership for SHP enterprises.

Loans and grants Moreover, the developer can get financial support from different levels of government and from banks, in the form of different soft loans and grants. For example, the central government gives grants of ¥300 million RMB (US$36 million) to the 400 Rural Electrification Counties annually, and the soft loans account for 38% of the total investment for this. This offers local developers different sources for funds, as Table 2.13 shows.

Table 2.13   Investment Sources for Chinese SHP Station Construction in Year 2001

Sources	State input	Bank's loan	Foreign investment	Local input	Others	Total
All nation input	30.17	110.93	2.36	49.75	20.35	213.56
Percentage (%)	14.1	52.0	1.1	23.3	9.5	100

[Source: Renewable Energy World, Jan-Feb. 2003]

Note: The investment sources are the construction fund composition for the project owner. State Input is composed of State Investment, Special Hydropower infrastructure Fund and Agriculture & Water Allowance.

Close relationship with rural electrification programme Electricity availability plays a more and more important role, a role which it was essential to develop for the rural areas. However, in a vast country like China, economic development and resources are unevenly distributed, making it unrealistic to realize rural electrification by relying solely on State support or by depending on the national grid. So in 1982, China's State Council decided to establish 100 counties with rich hydro resources as 'Rural Electrification Counties', on a trial basis. Through this measure, it became clear that SHP should provide the basis for rural electrification. This significant strategy paved the way for large-scale SHP construction, uncommon elsewhere in the world. Within 15 years, 653 Rural Electrification Counties across China have been established; another 400 more counties are expected to be electrified through SHP before the end of 2005.

Emphasis on cost-effective SHP technology The rapid development of SHP in rural areas also resulted from the emphasis on cost-effective application of an appropriate technology, which is in line with rural economic development. This involved the standardization of a series of technical criteria for planning, design, construction, installation, testing and acceptance, operation etc., to ensure quality, improve efficiency, and help popularize new technology and innovative products. Progress has included the rolling concrete arch dam, the plate rock-filling dam, automatic control systems, optimum operation and trans-basin hydro power development, efficient runners and simplified turbine operators. The application of these not only saves money but also shortens the construction period and increases the power reliability.

Formed local grid development and SHP own supply area China has always been focusing on developing SHP's own supply area, where an integrated power generating, supplying and distribution system has been established in those areas with rich hydro resources. The local grid can be interconnected to the national grid at a certain point to make full use of the seasonal power and to get more benefits.

Future tasks for SHP development in China

Government strategy and rural electrification The Chinese Government has always attached great importance to the significant role SHP plays in promoting the sustainable development of rural areas and for the remote hilly areas in the middle and western parts of the country to implement programmes such as 'Sending Electricity to Villages', 'Replacing Firewood by Electricity' and 'Reducing Cropland for Forests' (or Afforestation) using SHP as part of a sustainable development mechanism. To this end, the State Council has, based on the successful experience of the 653 SHP-electrified counties already established, approved the allocation of special funding to help establish 400 more electrified counties during the Tenth Five-Year Plan period. This will solve the issue of non-electrified villages and electricity shortage at the household level.

Rural electrification and improvement Access to electricity in villages should be ensured, as this will help alleviate poor conditions in remote areas. Most of the population of China which has no access to electricity is scattered across Sichuan, Qinghai, Xinjiang and Tibet. At present, 7.06 million households in 16,509 villages—a total of 28 million people—are without electricity in China. In some rural areas, where electricity for cooking and heating cannot be guaranteed as might be expected, the forests around villages are deteriorating.

It is difficult to electrify the villages using conventional approaches, as they are remote and isolated, with insufficient transportation facilities. They are also a long distance from the national grid, have a low intensity of electricity load, and are decentralized. Therefore, the Chinese government is arranging special funding to solve this issue locally by developing SHP.

The 'Programme of Replacing Firewood with Electricity' The 'Programme of Replacing Firewood with Electricity' aims to protect the environment and promote sustainable energy development in remote areas. According to investigation, one household of four people consumes a total of 2000–3500 kg of firewood annually for cooking and heating. For example, in the Litang County of Sichuan Province, annual consumption of 320,000 m³ of firewood and 350 million kg of dung causes serious environmental problems. On the other hand, if cost-effective electricity from SHP (the electricity price is set below ¥0.15/kWh, or 18 cents/kWh) can be used for cooking and heating, 4 tons of firewood can be saved annually per kW of installed SHP capacity. This will greatly improve the forestry coverage rate, preserve water resources and alleviate soil erosion, and the Programme will be implemented to promote local social and economic development in areas in the fringes of forests.

Independent development SHP is a clean and renewable energy source, which promotes sustainable social development. In line with the strategy of global hydro development, independent power producers (IPPs) will be established with different ownership patterns. Because of the richness and scattered locations of SHP resources, local economies can be developed through the construction of IPPs, which are mainly characterized by self-consumption and the export of surplus electricity to the grid, so as to upgrade efficiency and ensure optimum distribution of resources.

In line with the strategy of 'SHP Going Out', SHP should take an active part in global competition. After several years of development, there are 1.2 million employees in the SHP industry in China, of which 300,000 are design and construction technologists, qualified for SHP engineering projects of various types. There are over 100 major equipment manufacturers, with the capability of supplying all types of SHP equipment. Therefore, Chinese SHP enterprises are well qualified to take up all overseas SHP projects.

Power deregulation, development and management SHP should promote and participate in the reforms that power deregulation brings about. It also needs to devise a new development and management mechanism, and look at how the CDM could work for SHP. In China, both the large power plants with the national grid and the small plants with local grids have for a long time had their own supply areas. Such a monopoly situation is ripe for change during the establishment of a competitive market. To achieve development, co-ordination is needed between the power sector and society, with a new mechanism established.

Developing the International Network on Small Hydropower (IN-SHP) The establishment of IN-SHP with its headquarters in Hangzhou, China represents one of the achievements of China's policy of opening to the outside world (i.e. being more accessible to international markets). It also shows that the whole world needs SHP, reflecting China's achievements in SHP and the international status it enjoys. IN-SHP has created an innovative Triangular South–South Co-operation Model among developing countries, developed countries and international organizations. It has established bases and Asian sub-centers in India. Meanwhile, it promotes international technical and economic co-operation by various channels. Further development of IN-SHP will help serve the global SHP industry.

SHP—innovation and its role in sustainable development Although SHP is the most practical and beneficial renewable energy in China, there remain problems and difficulties. Over 28 million people are still without access to electricity, and appropriate SHP technology is not as well developed as it might be. The rural power market deregulation is also facing new challenges.

Through technology innovation, improvement of mechanisms and international co-operation enhancement, however, SHP can become a more competitive player in China. SHP offers substantial social and environmental benefits in solving rural electricity shortages, integrated development of mountain areas, water, forest and road networks, protection of the rural environment, alleviation of poverty, and promotion of the local social sustainability and economic development. These benefits cannot be achieved by other, solely commercial, power supply practice. As the immense social and environmental benefits of SHP have not been paid back, up to now, the nature of SHP enterprises need to be recognized as a special, non-profit entity in the power market through legislation. In line with the strategic shift of hydro power development, new innovation will promote SHP as a sustainable energy policy, incorporating the sustainable development of SHP itself and local society.

2.5.1.5 Small Hydropower in Japan

Japan is almost destitute of natural resources. Water has been regarded as the one and only resource granted to the people across the country. Throughout Japan's history, the utilization of river flows has been of great importance. In fact, people have built the environment to sustain their own life by using the fertile land fed by plentiful water. About 110 years ago, hydropower was first put to practical use to generate electricity, and during the first half of this century it expanded at a good pace throughout the country. Hydro electricity has been an important user of Japan's water resources since that time. Local people accepted it with comparatively little resistance, because it generally did not consume the water, and it did not significantly impede other users, especially during the early stages of development.

During the early stages, most hydropower stations were of the run-of-river type, because of the low cost of generation and the ease of supplying to scattered communities. Gradually, pondage and reservoir type installations came into being. Subsequently, a number of reservoir type hydropower stations with large dams were planned and completed, aiming at a more effective exploitation of water resources and at the regulation and stabilization of the electricity supply system. Hydropower was positioned as the supplier of base energy in the overall electricity supply structure.

In more recent years, the growth of the national economy and the accompanying rapid increase in demand, together with the global expansion of fossil fuel trading, have led to hydropower being assigned a secondary role. The primary energy supply moved to an oil-oriented system. At present, further hydropower development is not thought to hold much potential, except for pumped-storage type installations, which are indispensable to cope with fluctuating peak demand. This shift in the role of hydropower is also due to depletion of suitable sites and the resulting higher cost of generation in comparison with thermal power.

2.5.1.6 Small Hydropower in Nepal

Nepal is rich in hydro-resources, with access to one of the highest per capita hydropower potentials in the world. The estimated theoretical power potential is approximately 83000 MW. However, the economically feasible potential has been evaluated at approximately 43000 MW, of which only 262 MW or 0.2% of this potential had been harnessed. As stated earlier, hydropower accounts for 75% of the energy supply in Nepal.

Micro-hydropower development has a long history in Nepal. In 1988 approximately 615 micro-hydro turbines were in operation and were being used to grind grain, hull rice, and expel oil from oilseeds, as well as generating electricity in the hills of Nepal. More than 90% of these installations were used exclusively for agro-processing works. Estimates in 1998 indicate that almost 1000 micro plants were in operation.

The Remote Area Development Committee (RADC)—a non-government organization (NGO)—is helping to promote micro-hydro projects in 22 remote districts in Nepal. The RADC works with international NGOs such as CARE-Nepal, Canadian Cooperation Office, German Development Service, etc. Future plans for the RADC include improving efficiency of traditional projects for agro-processing works and electricity generation.

2.5.1.7 Small Hydropower in Macedonia, Greece

Best Practice in Rehabilitation of SHP Plants: Transferable Solution This project is considered to be the one with the best practice because it established a detailed model for maximizing the efficiency of electricity generation in Macedonia. It established a technical, environmental and financial framework for rehabilitating older, hydropower plants. The project methodology can be transferred and implemented by other public or private organizations that involve hydropower production. Transferability is enhanced by (1) determining the energy production regime in which the rehabilitated plants work (base, variable or peak—typically, small hydropower plants perform as peak energy plants); and, (2) determining the price of energy that will be used as a baseline for financial/profitability analyses.

Project Summary The system of electrical energy production and distribution in Macedonia is centralized and operated by the Electric Power Company of Macedonia. Macedonia produces 6500 GWh of electrical energy per year. Hydropower is used to generate approximately 20% (i.e., 1300 GWh) of the total electrical energy produced in Macedonia. Of this 20%, 170 GWh are produced by small hydropower plants (SHPPs). Seven of the 18 SHPPs have been in operation for more than 40 years. To rehabilitate the seven older plants, more information was needed on the technical requirements, costs, financing options, and environmental impacts of rehabilitation. The goal of the EcoLinks project was to analyze the basic technical, financial and environmental aspects of rehabilitating the seven small-scale hydropower plants. The project was led by the Electric Power Company of Macedonia in cooperation with Elektroprojekt, a Croatian consulting firm. Additional assistance was provided by the faculty of mechanical engineering at the Hydraulic Engineering and Automation Institute in Macedonia. A Rehabilitation Study was prepared and is to be used to obtain rehabilitation financing.

Based on the project findings, the Electric Power Company of Macedonia decided to finance the rehabilitation of the analyzed SHPPs through a rehabilitate-operate-transfer model. The procedure for the selection of an investor/co-owner was initiated during the project. The EcoLinks Challenge Grant ultimately led to $21 million of private funding for the rehabilitation, operation and transfer of the seven, small, hydroelectric power plants. The Electric Power Company of Macedonia signed a concession contract with Hydropol Management, a consortium of British and Czech investors, to establish a joint venture that will modernize all seven plants over a period of eleven years and extend their useful lives well into the future.

Each plant was visited providing a visual review of each facility and its current condition. The powerhouses, mechanical and electrical equipment, and intake facilities were reviewed in detail. Also, the civil works that need to be undertaken were noted. All existing data and documentation from the project leader's database on SHPPs were collected. This activity was crucial because it set up the basis for all of the following activities. Documents collected for each plant included development plans, tariff information, plant location maps, monthly generation figures, information on traffic connections, coverage of water management issues, technical data and documentation, maps of the electric power system, and single line diagrams of the plant's grid connections.

The collected data and documentation were reviewed and analyzed to determine the present condition of each plant. The present condition of each plant's status was evaluated to determine the need for rehabilitation and the possibilities for enlargement of the installed capacity. Product(s): Rehabilitation needs for each plant.

Three possible alternatives (i. e., case studies) were formulated regarding the rehabilitation of the plants. Case study 1 involved rehabilitation of the existing equipment without enlargement of its basic technical and operational performances. Case study 2 provided rehabilitation and upgrading of the equipment with possible increases in the installed flow within existing dimensions. Case study 3 required a full replacement of main and ancillary equipment with the ability to increase the installed flow.

Based on the findings during plants inspections and the analyses completed during the previous activities, recommendations for the reconstruction of civil engineering works as well as hydro-mechanical and electrical equipment rehabilitation were presented for each plant.

The costs and benefits of each solution alternative were computed for the rehabilitation of all seven plants. The best option for rehabilitation was retained for each plant.

Two financing options were considered:

a) Foreign commercial bank loan to owner using standard credit terms (10 tenor, 3-5 % spread over 6 –12 months grace period) and

b) ROT (Rehabilitate, Operate and Transfer).

Based on the company's present financial status, the Company decided to use a Rehabilitate-Operate-Transfer (ROT) financing model for financing the rehabilitation of all seven SHPPs.

Information and results regarding technical solutions, economic and financial assessments, proposed strategies for project implementation, and the assessment of project financing schemes were discussed within the team. Recommendations for rehabilitation were developed and included the following:

1. All seven SHPPs are to be maintained in a condition that justifies their rehabilitation;

2. Rehabilitation should be completed in a manner that allows the plants to operate as variable or peak (energy production) plants. This results in the construction of daily inflow compensation reservoirs for the plants where construction is possible;

3. The existing hydro-mechanical equipment should be partially replaced and upgraded in all seven SHPPs. The extent of rehabilitation depends on the existing condition of the equipment and is based on the feasibility of rehabilitation and/or upgrading for each plant; and

4. The existing electrical equipment should be rehabilitated on all seven SHPPs.

Project Benefits The project generated several benefits. It built the capacity of the Electric Power Company of Macedonia to collaborate and improve operations. This project improves energy production producing both environmental and economic benefits.

Capacity Building Benefits This project strengthened the Electric Company's capacity to structure a public/private partnership that addressed both its energy production needs and its financial situation. With more experience working with other firms, the Electric Power Company of Macedonia further strengthened its capacity to work collaboratively to improve efficiency and prevent further environmental impacts of energy production. The Electric Power Company could now implement other similar feasibility studies as needed to improve their operations.

Environmental Benefits The rehabilitation of the seven SHPPs will result in avoided emissions of 136.3 tons of CO_2 by avoiding alternative energy production methods and resources (e.g., coal and fossil fuel combustion). In addition, since all seven SHPPs are an integral part of the water management system in Macedonia, the rehabilitation and life extension of the hydropower plants will result in sustained water management and reduced soil deterioration.

Economic Benefits Plant rehabilitation will increase the generation of electric energy by 17% from 86.4 GWh to 101.4 GWh annually. With this increase in production, an additional income of $320,000 to $640,000 per year is expected. The total investment outlays for the rehabilitation amount to $14.5 million. For all seven plants, the Internal Rate of Return (IRR) ranges from 45.8% to 16.9%. IRR for one of the analyzed plants has a negative value, –6%, but this plant was also planned to be rehabilitated using the increased income of the other six plants to cover costs.

Lessons Learned The following lessons were learned during this project:

- Rehabilitating existing hydropower plants can lead to an increase in electricity generation as well as positive environmental effects, and should be considered a favorable alternative to developing new facilities.
- A critical factor for the successful implementation of similar projects is the scope and reliability of existing data on flow measurements, electrical energy production records, and technical documentation for the plants.
- The Rehabilitate-Operate-Transfer (ROT) financing model is a viable option for many enterprises in the region that have limited financial resources.

2.5.2   Medium and Large Hydropower

After the discussion on small hydropower so far, the following section describes case studies of some countries in medium and large hydropower. While Norway represents an almost complete development of medium and large hydropower, UK and Turkey represent nations, which depend on hydropower and other energy sources. Hydropower development in these countries is discussed here.

2.5.2.1 Hydropower in Norway

More than 99 percent of electricity in Norway comes from hydropower. Representing the fifth largest hydropower system in the world and regarded by many as the most efficient and modern, the Norwegian hydropower system generates an average of 11,500 GWh of electricity a year and has a total installed capacity of 29,932 MW. Co-sponsored by the Norwegian government and private companies, a nationwide research and development programme has been implemented to develop ways to reduce costs and increase output of hydro structures. The entire project is aimed at making the existing power system more valuable in the marketplace. Through refurbishment and R&D, the country plans to increase its own hydro generating capacity. The hope is that as demand for power increases while budgets remain limited, hydro owners can rehabilitate existing plants to increase capacity for a relatively modest investment. Norwegian hydro leaders hope that technology developed through these programs can be useful throughout the world and help to keep Norway at the leading edge of the industry.

2.5.2.2 Hydropower in the United Kingdom

Hydropower provides only about 2% of the total electricity consumption of the United Kingdom, Most of the generation takes place in Scotland, which is both mountainous and wet, with Wales also making a contribution. England and Northern Ireland have many quite small hydroelectric projects but their total contribution is not large. Total installed capacity (excluding pumped storage sites) is 1349 MW.

The great period of hydropower construction occurred in the years after the Second World War—from 1948 to 1965—when most of the Scottish schemes were built by the North of Scotland Hydroelectric Board. Since that time there has been little further construction due to increasing opposition to reservoir construction, on environmental grounds. Today, there are over 50 main hydropower stations in Scotland with an installed capacity of around 1050 MW. In addition there are three pumped storage power stations with a combined capacity of 830 MW. In Wales the Dinorwig pumped storage scheme has an installed capacity of 1200 MW and there are two other main hydro stations providing 16 MW.

Various estimates have been made about how much potential hydropower is available for development in the UK. There is probably a further 750 MW of capacity in Scotland, but it is difficult to say how much of this would be economically viable and equally importantly, environmentally acceptable.

So far as small hydro is concerned, there is a potential for about 300 MW of plants less than 10 MW, but probably not more than one third of this would be economically exploitable. Water resources in England and Wales are controlled by the Environment Agency (EA), a recently established quasi-governmental body which has subsumed the duties of the previous National Rivers Authority. The EA issues licences for all "abstractions" from water courses. In Scotland this duty is fulfilled by the River Purification Boards and in Northern Ireland by the Environment Department.

Total hydroelectric capacity in the UK has remained approximately the same over the last ten years, at about 1,400 MW, but because of the gradual increase in total generating capacity, currently about 60,000 MW, its share is gradually decreasing. There is however a definite increase in small hydro, but this is too small to compensate for the declining market share.

2.5.2.3 Hydropower in Turkey

The gross theoretical hydropower potential of Turkey is 433,000 GWh/year and the technically feasible potential is 215,000 GWh/year. The economically feasible potential has been re-evaluated in 1997 as 123,385 GWh/year, equivalent to 34 862 MW. So far, about 17 percent of the technically feasible potential has been developed. It has been planned that by 2010, 60 percent of the economically feasible hydropower potential will be developed and the installed capacity will reach 22,509 MW.

The total installed capacity of all thermal and hydro plants is 21 247 MW. There is about 10108 MW of hydro capacity in operation, generating on average 36 866 GWh/year. Actual generation in 1996 was 40,475 GWh.

It is estimated that the hydro potential in operation could be uprated between 10 and 15 per cent. About 57 of the 493 hydro projects are part of multipurpose schemes. A further 3938 MW of hydro capacity is under construction, and 19 433 MW is planned by 2010.

The largest hydro plants under construction include: Birecik (672 MW), Deriner (670 MW), Berke (510 MW), Obruk (203 MW), Batman(198 MW), Karkamis (189 MW), Ozluce (170 MW) and Alpaslan (160 MW). The mean cost per kW of the hydro capacity under construction in 2000 was US$ 1350/kW (excluding transmission costs).

The Southeast Anatolia Project (GAP), a large water resources development project involving 22 dams, 19 hydro plants (7,474 MW, 27 TWh/year), which will provide irrigarion for agricultural production over an area of 1.7 million ha in the region, is important for the sustainable development for the whole of Turkey. So far, about 60 per cent of the hydro capacity is in operation, including the Keban (1330 MW), Karakaya (1,800 MW) and Ataturk (2,400 MW) plants, and 15 percent is under construction, which includes Birecik and Karkamis.

About 40 percent of the total US$ 32 billion cost of the GAP has been expended so far. Among the small hydro there were 57 small, mini or micro hydro plants in operation, with a total capacity of 126.5 MW. Two plants were under construction (3.6 MW) and 98 were planned (493 MW).

Among the individual large hydropower projects the Itaipu Project with a capacity of 12,600 MW jointly developed by Brazil and Paraguay in South America by year 2000 is worth mentioning here. China is building an 18,600 MW capacity hydropower project at Three Gorges Dam on Yangtze River. The project is expected to be completed by 2009. It has 26 units of 700 MW capacities each.

2.6   WATER FOR NAVIGATION

Navigation has a long history, navigable canals date back to the age of Roman empire more than 2000 years ago. Due to their increasing orders of the coefficients of friction, the load carrying capacities for water, rail and road transport are in the ratio of 80:10:3 . As early as the sixth century, China had an extensive system of waterways that eventually led to the development of the Grand Canal, which links the Yellow and Yangtze Rivers. Substantial freight was carried on navigable rivers in Europe, some of which were connected with canals. River Amazon in South America is navigable by steamers for a length of 3,200 km from its mouth to Iquitos in Peru, where it has a width of 5 km. Various types of locks and devices were used to elevate boats from one level to another.

Waterways offered comparatively easy routes through unmapped wilderness for the exploration of new lands. As these lands were developed, boats were often the only feasible means of moving cargo too heavy for pack animals or wagon trains.

In several countries water transport plays an important role because of its inherent advantages. In USA, in 1965, traffic handled on the inland waterways was as much as 147 billion ton miles.

Navigable waterways on rivers and canals in India are nearly 15000 km in length but most of them are not in use due to silting and poor maintenance. About half of the navigable waterways are on rivers and the rest are on canals and backwaters. These are fit for country boat traffic only. The main waterways exist in Ganga and Brahmaputra basins, which account for more than 60% of the traffic.

2.6.1 Requirements of Navigable Waterways

There is no absolute criterion of navigability and, in the final analysis, economic criteria control. The physical factors that affect the cost of waterborne transport are depth of channel, width and alignment of channel, locking time, current velocity, and the terminal facilities. Commercial inland water transport is mostly accomplished by barge tows consisting of 1 to 10 or more barges pushed by a shallow-draft river tug. The cost for a trip between any two terminals is the sum of the fuel cost and wages, fixed charges, and other operating expenses dependent on the time of transit.

2.6.2 Methods for Improving River Navigability

There are three basic methods for improving a river for navigation—open channel, lock and dam, and canalization.

Open Channel Methods

Open channel methods seek to improve the existing channel to the point where navigation is feasible. The requirements of a stream for open channel improvement are as follows:

- Sufficient flow to permit navigation a reasonable portion of the year
- A channel cross-section sufficient for modern barge tows.
- A satisfactory alignment without excessively sharp bends.
- Channel slope sufficiently flat so that velocities are not excessive.
- Bed and bank materials permitting satisfactory treatment by one or more of the open channel methods.

Lock and Dam

Dams create a series of slackwater pools through which the traffic can move, with locks to lift the vessels from one pool to the next. Lock and dam construction may be indicated where conditions are unfavorable for open-channel methods. A series of locks and dams can be maintained if flow is sufficient to provide water for lockages, sanitary releases as may be required, and evaporation losses from the pools. The slackwater pools behind the dams submerge rapids and channel bends and because of their relatively large cross section have low flow velocities.

The general features of a stream suitable for lock and dam construction for navigation improvements are:

- Conditions unsatisfactory for open channel methods

- Low sediment transport
- Suitable dam sites

Canalization

Canalization provides a totally new channel cut by artificial means around an otherwise impassable obstruction or between two navigable water bodies. Canalization is usually feasible only when a short length of channel opens a large length of navigable waterway.

The development of navigation in the United States is explained in the following section.

2.6.3 Navigation Development in the USA

The Mississippi ranks third in length and fifth in volume among the world's rivers. It ranks third also in area drained. It is 5,609 km long and has played a great part in economic development of USA. Together with its tributaries, a network of inland navigation waterways form a gigantic waterway system of 19,884 km. The water borne traffic on this river has reached 250 million tonnes a year. It has become one of the nation's greatest industrial attractions and cities along the river are rapidly providing port and terminal facilities to handle the expanding river commerce.

Mississippi is navigable upto Minneapolis for commercial purposes and, for another 80 km, upto Lake Itasca, small craft navigation is possible. In this reach there are 14 dams without locks.

Between Minneapolis and St. Louis, there are 27 locks and barrages. There are no locks or barrages downstream of St. Louis. The use of the locks is free to all types of crafts. Between Cairo and Head of Passes, the navigational channel is 2.7 m deep and 78 m wide. In the Upper Mississippi, the navigation depth is only 1.58 m and to keep pace with the growing traffic the depth is being increased to 2.7 m. Between Baton Rouge and the Gulf of Mexico, the navigation depth and widths are, on an average, 12 m and 180 m. The navigation depths in the lower Mississippi are maintained by dredging, which is required between Cairo and Baton Rouge to maintain 2.7 m depth channel and the quantity dredged total varying from 22 million cu.m to 50 million cu.m in a year.

Along the tributaries navigation depth of 2.7 m is maintained, as for example along the Ohio from Cairo to Pittsburgh. Similarly, by the Illinois canal to Chicago, access to the ports of the Great Lakes is provided. Thus Mississippi has more than justified the expectations of the people who named it as the father of the waters.

2.7 WATER FOR PISCICULTURE

The use of water for pisciculture has been explained with an example of the wetlands of Calcutta (renamed as Kolkata) in the following article.

2.7.1 Wetlands of Calcutta, India

Calcutta is sustained by a unique and friendly water regime. To its west flows the river Hooghly, along the levee of which the city has grown. About 30 km eastwards flows the river Kulti-Bidyadhari that carries the drainage to the Bay of Bengal. Underneath the city lies a copious reserve of groundwater. Finally, and central to this regime is the vast wetland area beyond the eastern edge of the city that has been transformed to use city wastewater in fisheries, vegetable gardens and paddy fields in successive tracts of land. These combine to what has been described as the 'waste recycling region' (Ghosh, 1985). This wetland and waste recycling region covers about 12,500 hectares, and is the largest of its type in the world. This remarkable work of transformation and wise use of wetlands was innovated, developed and upgraded by local fish producers and farmers through a large part of this century.

Since 1980, a good amount a data has been collected and scientific studies carried out to understand this system better. Seldom does one find an ecosystem of such size to be equally significant for bringing economic gains and ensuring ecological balance using a reliable traditional technology that has changed the grammar of conventional wastewater treatment. The system that grew in the early decades of this century peaked around mid century and is now losing ground to urban expansion, poses one of the most formidable challenges to conservation activists in this part of the world. The twin challenges are to save this city and provide a least cost alternative to wastewater utilization.

An introductory note describing the unique ecosystem, the practice and the emerging conflicts is presented here.

The east Calcutta wetlands area is an urban facility that treats the city's huge wastewater and utilizes the treated water for pisciculture and agriculture, through recovery of wastewater nutrients in an efficient manner. Here, wastewater has been used in fisheries and agriculture covering about 12,500 hectares that has been designated as conservation area by an order of the Calcutta High Court. The conservation area, also described as the waste recycling region, has three major sub-regions of economic activity—fishponds (*bheris*), garbage farms and paddy lands.

The smallest recycling sub region on the edge of the city covers the vegetable fields that grow vegetables on a garbage substrate and are uniquely planned with alternate bands of garbage filled lands and elongated trench-like ponds locally known as 'jheels'. In these jheels, sewage is detained for sometime, after which the treated effluent is used for irrigating the garbage fields for growing vegetables. In the fishponds, the city's wastewater is made to flow through a network of drainage channels. The wastewater fishponds act as solar reactors and complete most of their biochemical reactions with the help of solar energy. Reduction of BOD (biochemical oxygen demand) takes place due to the unique phenomenon of algae-bacteria symbiosis where energy is drawn from algal photosynthesis. In this way, requirement and consumption of energy remains the minimum. Unlike conventional mechanical sewage treatment plants, wastewater ponds can ensure efficient removal of coliforms that are prone to be pathogenic. The fishponds drain out the used water to irrigate paddy fields.

The fishpond ecosystem of east Calcutta is one of the rare examples of environmental protection and development management where a complex ecological process has been adopted by the fish producers and farmers for mastering the resource recovery activities. What is remarkable is that the fish yield rate attained is among the best in any freshwater pisciculture in the country. The east Calcutta wetland has the largest number of sewage fed fishponds in the world that are located in one place. The knowledge that has emerged based on traditional skill, enterprise and innovation provides an alternative to the conventional option of wastewater treatment by an ecologically sustainable and wise use of wetlands. Here the task of reusing nutrients is linked with the enhancement of food security and development of livelihood of the local community using nutrient rich effluent in fisheries and agriculture. The conventional sewage treatment plant is considered an externality in the basic social and economic culture of Calcutta and its fringe. Interestingly, a large part of this folk technology is still retained in an oral tradition. New generation environment friendly engineering has been quick to incorporate the advantages of natural biological processes and principles of ecological regulation. In this context, using wetland functions for reducing wastewater pollution and reuse of nutrients is an example of an effort that has opened new areas of research and application in other parts of the world too. In 1980, on an initiative of the Government of West Bengal, the wetland area and its reuse practices were assessed. By 1983, the first scientific document on this ecosystem was published which enabled the rest of the world to know about the ecological significance of this outstanding wetland area. In 1985, the map of the waste recycling region that forms the basis of all planning and development activities on this wetland area was prepared. In the same year, the state government put forward a proposal to introduce a resource efficient stabilization tank (REST) system for the treatment and reuse of city sewage. This was accepted by the Ganga Project Directorate as an alternative to conventional energy expensive and capital-intensive mechanical treatment plants. A number of such projects under the Ganga Action Plan have now been completed and are working.

Towards the end of the 1980's real estate interest reached a new high and there was a strong tendency to convert water bodies and wetlands into housing complexes. To combat this, the Government of West Bengal initiated a number of development control measures. For a comprehensive planning and development of the entire region, a baseline document for management action plan using Ramsar guidelines has been completed.

The east Calcutta wetlands are distributed nearly equally on the two sides of the Dry Weather Flow (DWF) channel reaching the Kulti Gong to the east. The region is suitable for using solar radiation and improving wastewater quality. Each hectare of a shallow water body can remove about 237 kg of BOD per day. In winter, the clearness of the sky is satisfactory (about 90 percent) for carrying out biochemical activities in water purification. The wetlands area falls within the south Bengal ecological zone where the process of land formation is largely influenced by the Ganga river system that continues through centuries to create land by raising it through upland and tidal deposition. This pattern of delta building has also undergone significant changes due to natural and manmade reasons. Calcutta grew along the Hooghly levee. While riverbanks are always raised to first allow human settlement and productive activities, the spill areas also gradually become higher up to the tidal level for subsequent reclamation. Unfortunately, with the death of the Bidyadhari, the tidal channel that used to deposit silt in this area, the process of natural deposition and raising of level of the spill area has completely stopped since the end of the last century. This incomplete process of delta building did not allow the low-lying areas behind the Hooghly levee, to the east of Calcutta, to rise higher. Since the early 15th century when the Ganga changed its main course from the Bhagirathi to the Padma, this eastward change of the course of the main flow of the river Ganga brought metamorphic changes in the process of delta building in central and south Bengal. A number of distributaries and redistributaries were cut off from any upland flow that signalled the end of those channels. However, human interference in the region further reduced the spill area and the channel beds heaved up to quicken the process of decay. This way, the Bidyadhari became defunct and the opportunity of silt deposition in its spill areas ceased. Prior to wastewater fish farming, this region was a brackish water fishery area gradually rendered derelict on account of the receding Bidyadhari spill channel. Interestingly, these fisheries were responsible for reducing the spill area and compounded with the dwindling upland flow from the river Hooghly to the spill channel caused the death of the tidal creek. During the thirties, the Bidyadhari carried only the city sewage and in the process got choked further due to the high silt content of the sewage. Figure 2.27 provides a sketch of the wetlands of East Calcutta.

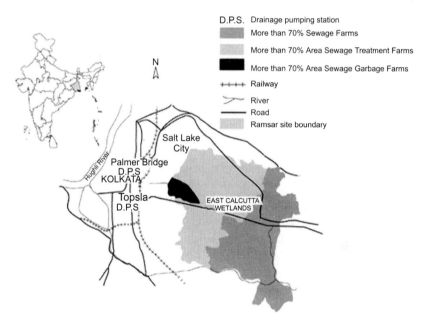

Coordinates—22°25'—22°40' N latitude and 88°20'—88° 35' E longitude
Area—12500 ha
Elevation—2 m GTS (Global Telecommunications System)

Figure 2.27   Wetlands of East Calcutta (i.e., Kolkata)
[Source: East Calcutta Wetlands, Identification and designation of potential Indian Ramasar sites, World Wide Fund for Nature (WWF)—India 2002]

In such a situation a leading fish producer of that time was successful in growing fish in large water area using city sewage in the same process sequence as it is done today. The basic feature of this mono pond wastewater fishery was that it combined the process of sewage treatment and recovery of its nutrients through aquaculture within the same pond. Application of sewage was sequenced skillfully on the basis of detention time needed to improve the water quality appropriate for growing fish.

New varieties of fish are being constantly introduced in these ponds. Fish producers in this area have shown keen interest to design the optimum crop group. In these fisheries, we find one of the most efficient yield rates from a multiple species polyculture.

In spite of the early success, lack of wastewater supply restricted the spread of such fisheries in this area. Things improved with the construction of the Dry Weather Flow (DWF) channel in 1943-44 to carry city sewage to the Kulti Gong as a new outfall receptacle with provision to channelize sewage for fishery feeding. Sewage-fed fishery spread rapidly since then. However, the scenario changed as the spectre of Salt Lake loomed larger and larger. The wetlands began to lose stability since 1956. There were instances of forcible attempts to take over the right of land by violent farmers. Such action was long overdue. For more than 50 years owners of small parcels of land were systematically forced to give up their property rights in favor of a few large landowners. But the unfortunate fallout of such forcible occupation of land of big owners by landless farmers lay elsewhere. Fisheries were drained and cultivation of paddy was attempted. It needed much persuasion to impress upon the landless farmers that it was the fishery owners and not the fisheries that would be affected. Fishery was the most efficient ecosystem for the farmers who were natural growers of fish rather than paddy, apart from the multiple benefit that fishery provided. At present, most village people realize this and forcible takeover has largely disappeared. Organized movements by fish farm labourers have succeeded in bargaining for higher terms of labour from the employers.

The resources recovered from city sewage are being used in three kinds of economic activity—wastewater fisheries, vegetable farming on a garbage substrate and paddy cultivation using pond effluent—in that order. The fishponds are central to the entire waste recycling region and have therefore been described in great detail.

There are 264 fish farms operating on a commercial basis. They cover a total area of about 2858.65 hectares. The fish farms consist of units of various sizes from large holdings locally called *bheris* and relatively smaller ones called *jheels* due to their trench-like elongated shapes. But all these fish farms generally have similar types of produce, farming practice and distribution system.

2.7.1.1 *Types and Amount of Fish Produced*

A variety of sweetwater fishes are produced in the *bheris*. The main varieties in the existing practice of polyculture include:

- Indian Major Carp—Rahu (*Labeo rohita*), Catla (*Catla catla*), Mrigal (*Cirrihinus mrigala*)
- Indian Minor Carp—Bata (*Labeo bata*)
- Exotic Variety—Silver Carp (*Hypophthalmichthys molitrix*), Common Carp (*Cyprinus carpio*), Grass Carp (*Tenopharyngodon idella*)
- Tilapia Nilotica (*Oreochromis nilotica*), Mosambica (*Tilapia mosambica*)

Apart from these cultured varieties (except Tilapia), some other varieties including forage fishes are also available occasionally in the *bheris*. They are Punti (*Puntius japonica*), Sole (*Channa striatus*), Lata (*Channa punctatus*), Chyang (*Channa gachua*), Singi (*Heteropneustes fossillis*), Magur (*Clarias batrachus*), Fouli (*Notopterus notopterus*), Pungus (*Pangasius pangasius*), etc. The average marketable sizes of fishes is shown in Table 2.14.

Table 2.14 The Average Marketable Sizes of Fishes in East Calcutta Wetlands

Fish variety	Average marketable size (in gm)	Fish variety	Average marketable Size (in gm)
(I) Indian Major Carp		(III) Exotic Variety	
Rahu	75 – 100	Silver carp	200 – 300
Catla	125 – 150	Common carp	250 – 350
Mrigal	75 – 100	Grass carp	500 – 750
(II) Indian Minor Carp		(IV) Tilapia	
Bata	50 – 75	Nilotica	70 – 80/100
		Mosambica	50 – 75

[Source: Baseline Document 1997]

It has been observed that the average size of marketable fishes are not of the optimum weight (in grams) because fishes are normally netted/harvested much before they attain mature marketable sizes that can command better prices. The main reasons for this are that the management/owners are compelled to create the maximum number of man-days possible in a year to provide employment to direct labourers viz. harvesters, carriers, etc. and the impending threat of poaching.

Production and yield per hectare of fish varies among the *bheris* depending upon the conditions of production. It has not been possible to study all possible variables that affect the production and yield of fish in this region.

However, enough data is available to give reasonably sensible indication in this matter. Table 2.15 gives an idea of the yield in relation to the size of the *bheris*.

Table 2.15 Yield in Relation to the Size of the Fish Farms in East Calcutta Wetlands

Sl. No.	Area-range (In ha.)	No. of fish farms	Gross effective area (ha.)	Total production (in MT)	Yield / ha. (in MT)
1.	Up to 2	76	80.83	318.72	3.94
2.	Above 2 to 10	125	608.57	2443.50	4.01
3.	Above 10 to 20	34	449.53	1850.99	4.12
4.	Above 20 to 30	13	315.94	1100.25	3.48
5.	Above 30 to 40	3	93.59	270.95	2.89
6.	Above 40 to 50	1	46.13	180.00	3.90
7.	Above 50 to 60	3	170.33	760.00	4.46
8.	Above 60 to 70	5	331.57	1496.85	4.51
9.	Above 70	4	384.66	2493.89	6.48
	Total	264	2481.15	10915.15	4.40

[Source: Baseline Document 1997]

The maximum number of *bheris* (around 125) fall within the area range of above 2 and below 10 hectares, followed by 76 *bheris* within the range up to 2 hectares and 34 *bheris* fall within the range above 10 hectares to 20 hectares. More than 89 per cent of *bheris* fall within the area range of up to 30 hectares. The maximum average yield (i.e., 6.48 MT per ha) has been achieved by the *bheris* of more than 70 hectares. Initially the yield is seen to increase with the increase in the size of *bheris*; thereafter, it tends to decrease as the size gets bigger and is lowest (i.e., 2.89 MT where *bheri* size is between 30 and 40 hectares). Yield per hectare again increases steadily with the increase in the *bheri* size and is the highest at 6.48 MT in fish farms where the effective area of water body is above 70 hectares. Larger size *bheris*, though few, are more organized in terms of operations, planning through efficient management of production schedules, utilization of manpower, sewage, better procurement planning, monitoring water quality and fish health and efficient personnel management, etc. All these combine to provide the right synergy for achieving better production performance and per hectare yield.

In terms of annual production, 45 operating *bheris* reported 'increasing trend', 61 *bheris* reported 'decreasing trend', while in 76 *bheris* production was reported to be at an even level during the past three years. No proper indication of production trend was available in the remaining 82 *bheris*.

For a *bheri* to be operated efficiently, the following conditions are critical:

a) Maintenance of the required depth of water at all the three stages of the production process e.g., at nursery pond, rearing pond and stocking pond with proper inlet-outlet management of sewage.

b) Availability of quality spawn/fry/fingerlings at required time and quantity.

c) Proper and efficient deployment of working personnel, ensuring satisfactory labour productivity and congenial labor relations.

The most important requisite to run a fishpond efficiently is adequate and safe supply of wastewater. This is related to a number of other factors and deserves more elaborate discussion. Poor quality of sewage brings in lower quantity of nutrients and higher toxic load for the fish to feed upon. Low quality of sewage borne nutrients requires supplementing with nutrients from outside. This entails more expenditure, and increased operational costs that affect viability. In such situations the *bheri* owners, with additional input costs, seek to add value to their produce (fish) to get higher returns and recover the additional expenditure incurred on fish nutrients/feed. One way of countering this situation is by allowing the fishes to grow bigger. As this means lesser number of netting (harvesting) days resulting in the loss of man-days, the workers' union does not allow this to happen. This, in turn, gives birth to a situation of conflicting interests.

In wastewater pisciculture, labor cost constitutes the major part of expenditure. Apart from supervisory staff and skilled or unskilled labourers who live in the *'alas'* or field offices located within the *bheri* area engaged by the management/owners, all other categories of laborers are temporary workers in the strict sense of the term and the costs pertaining to them are 'variable' in nature. The labor union has a pervading dominance over supply of labor in the east Calcutta wetlands. The labor union periodically settles with the *bheri* management/owners, the terms and conditions of labor including the number of workers to be engaged, daily wage rates and benefits, job specification, total minimum days of employment in a year for each category of worker, leave and other conditions of employment. Once settled, no *bheri* can take unilateral decision on any of the terms of employment including deployment of working personnel. Thus, the number of labor engaged that has been fixed for a particular *bheri* remains so for the whole of the agreement period irrespective of the volume of production attained. While the Union affirms the labor cost to be fixed in nature, the *bheri* owners consider the category of workers as 'temporary' and the costs pertaining thereto as 'variable costs'. In spite of the conceptual nicety involved under cost accounting procedures, most of the *bheri* owners agree that this has assumed the nature of 'fixed' costs. This will continue till the status of the workers in this predominantly agricultural vocation is defined by the concerned authorities (Baseline Document, 1997). Employer-employee relationship is a sensitive area in the fish farms. The issues that often cloud owner-worker relationship in a *bheri* are daily wage rate, leave with pay, bonus and non-cash incentives, fixation of yearly working days, deployment of labour per bheri, days of rearing and harvesting, daily harvest quantity, distribution and sale of marketable surplus, job rotation etc.

The owners and the workers addressed this problem through a dispute settlement exercise. In 1992 a code of conduct was adopted at the meeting of the district level labour union officials and the 24 Parganas District Fish Producers' Association, a representative body of private *bheri* owners. In March 1996, the agreement was signed and ratified by both the parties, after minor modifications.

Although being identified as an ecosystem that provides copious ecological subsidy to sustain the city of Calcutta commensurate action to conserve the same has always been uncertain. Major conservation issues and causes of concern and uncertainty in the region can be grouped as follows:

1. Institutional and regulatory inadequacies
   - Information asymmetry
   - Disappearing traditional skill and wisdom
   - Absence of conservation policy guidelines and legislation
2. Physical factors
   - Siltation in the canals and fishponds.
   - Inadequate availability of waste water in fishponds
3. Human interference
   - Information asymmetry
   - Theft, poaching and pilferage
   - Volatile and sensitive employer-employee relationship
   - Non-cooperation among *bheri*-owners themselves for sharing sewage
   - Improper inlet-outlet management of sewage for operating *bheris* and cultivating paddy.
   - Tenurial instability in the garbage farms owned by the Kolkata Municipal Corporation (KMC).
   - Contamination of wastewater due to untreated industrial effluent.

- Tenurial instability in the workers' managed cooperative fish farms.
- General level of insecurity and law and order problems.
- Fragmentation and conversion of land.
- Threats of real estate takeovers
- Coordination among the various government departments/agencies viz. irrigation and waterways, fisheries, agriculture, urban development, labor, health, environment, KMC, Kolkata Metropolitan Development Authority (KMDA), Kolkata Metropolitan Water and Sewerage Authority (KMW&SA).

4. Factors arising out of legislation, tradition and rights
   - Lack of institutional credit from rural banks
   - Absence of life/health insurance facilities
   - Absence of crop insurance facility

To remedy this state of distress a set of desirable actions have been listed to initiate confidence. Within the wetland conservation area, wastewater is used by all the three major recycling practices (viz. solid waste farming, fisheries and paddy cultivation). However, perennial fisheries, which need waste water supply throughout the year, are the most important users of wastewater and are also most critically dependent upon the availability of requisite supply of the same. Variations in availability put the operational viability to strain. Effluents released by most of the fishponds are ideally suitable for paddy cultivation and are profusely practiced within the designated wetland area. Inappropriate understanding of the significance of these wetland practices has led to gradual loss of system efficiency. The actions listed below can restore stability of these recycling practices and ensure wiser use of wetlands.

a) Entire flow through the dry weather flow (DWF) channel should be distributed to the fishponds as an obligatory function of the agencies responsible for such work. Existing practice of releasing a part of the wastewater flow to Storm Weather Flow channel should be stopped.

b) All drainage channels and distributaries within East Calcutta Wetland Region should be brought under a comprehensive action plan for its maintenance. This will be implemented by a consortium of concerned local self government (known as *gram panchayats*) facilitated by necessary technical support from appropriate departments.

c) Restoration of the vital drainage structures lying defunct within the wetland area.

d) Dredging should be initiated in the Dry Weather Flow channel and the fishery feeding channels to augment uninterrupted supply of wastewater to the *bheris*.

The existing practice of growing vegetables on locally compostable garbage substrate using waste water stored in shallow ponds (locally known as *jheels*) suffers from institutional indifference and a few more years of apathy will lead to the collapse of this unique urban facility. Immediate action to save this waste recycling practice would include the following steps:

a) There are 49 water bodies (*jheels*) under the Kolkata Municipal Corporation (KMC) area designated for solid waste dumping. There should be no further filling up of any of these water bodies by garbage disposal.

b) All interconnecting drainage network linking these water bodies will have to be restored.

c) All formal KMC plans for solid waste recycling must include the provision for the agricultural uses of solid waste. (At present no such provisions are made in any of the KMC plans for solid waste recycling).

d) Farmers should be provided with a list of safe species that can be grown in the garbage farms around Dhapa and will have to be persuaded to discontinue the growing of relatively unsafe species. Local *Krishak Sabhas* (farmers' councils) and non-governmental organizations (NGOs) may take active role in this upgrading initiative and introduce modified guidelines.

e) Elaborate plantation should be taken up in this urban agricultural area taking the help of the local farmers, NGOs and the concerned agencies in the Government.

f) A shadow of uncertainty looms largely over the farmers/occupiers of the garbage farms over title/tenurial rights, as the entire land is owned by the KMC. Removal of land use uncertainty over the entire KMC area is imperative. The KMC should collect rent from the occupiers of these farming plots. A list of farmers who are presently tilling the land be prepared and appropriate rent depending upon the size of the plot be introduced. Similarly a reasonable license fee should be introduced for pisciculturists for use of KMC water bodies in the area in consonance with ground realities.

g) Extension of health care facilities for the dangerously exposed workforce engaged in rag picking in the municipal dumping grounds at Khanaberia (West) and Durgapur villages.

h) Ensuring supply of vegetable seeds particularly of the right quality to ensure and optimize cropping intensity.

Developing a technical information base for better management of the resource recovery systems and wise use of wetlands. Major components of the information base will include:

a) Relative relief of the wetlands area on the basis of a detailed topographical survey, including pond bathymetry and longitudinal profiles of all drainage channels and distributaries within the area.

b) Wastewater (both qualitative and quantitative) data along wastewater outfall channels (at least six sampling points).

c) Biological indicators (both aquatic and terrestrial) of environmental stress within the wetlands area.

d) Water dependent population including their habitats and food habit.

e) Types of species grown (on land and water) using city waste, their origin, spread, threats and risk to the users.

A large number of small and medium scale industries are availing of the drainage system laid by KMW&SA since last couple of decades to release their untreated effluent. These drainage channels are linked with the main outfall channels leading to the river Kulti. This industrial wastewater can cause undesirable impact on the fish and vegetables grown using the same. It is imperative to identify such industrial units which discharge contaminated effluent to bring them under the purview of pollution control regulations. Common in-situ effluent treatment plants for the polluting industries are considered to be the likely solution to this emerging problem.

Phenomenal deposition of silt has raised the Kulti-Bidyadhari river-bed to critical limits. Immediate programmes will have to be taken up to restore the conservancy of this outfall receptacle to save Calcutta from the disastrous problem of waterlogging.

The work of restoring biodiversity will include the following activities:

a) Plantation of trees (specially along the dykes)

b) Introduction of reed-beds along the existing water hyacinth buffer margins in fishponds

c) Awareness campaigns among the local people for carrying out the programme stated above;

d) Promotion of scientific research through existing institutions and agencies.

Educational institutions and schools in particular should be encouraged to include the east Calcutta wetland model as a case study on urban waste management and demonstrative field trips organized where possible. These schools can be local, regional and even from other parts of the world. Almost every year, students from a few European countries have been visiting this unique ecosystem.

A state-level wetland conservation and management authority should be set up for conservation, development, monitoring and control of inappropriate use and conversions. Research and public awareness programmes should be initiated involving national and state level institutions, universities and NGOs on wetland values, wise use functioning and sustainable development of resources and their utilization. This could help address the systemic drawbacks like public ignorance and indifference about these wetland practices.

The Ramsar Bureau has selected 17 case study sites from all over the world to demonstrate and understand wetland wise use. In that list the East Calcutta Wetlands is the only entry from India and also the only one that is by the side of a city and is largely acclaimed as an urban facility for using the city sewage in traditional practices of fisheries and agriculture. No wonder that core Calcutta has not been provided with any fund for constructing sewage treatment plant under the Ganga Action Plan. Historically, and where we tend to become forgetful, waste water management system of Calcutta, since pre-independence period, undertook all the three steps that form the basis of today's Ganga Action Plan, viz. interception, diversion and resource recovery. In course of this not only the local people, the farmers, the managers and even political leaders remained engrossed with the research team but there was, more importantly, a commonality of purpose which was shared by all alike. Such participatory initiatives build the bedrock for any conservation programme on such sensitive tracks. By this time, much more has been known than before that conservation research is primarily a grass-roots exercise.

2.8 WATER FOR RECREATION

Navigation rivers are increasingly used for recreation and sporting purposes. In USA and other advanced countries, thousands of small boats owners cruise round the rivers. Rafts, canoes and small boats are used for recreation. The demand for this type of recreation will increase with the improving standard of living and increase of urban population.

The popularity of water sports is on the increase in all countries, but the diversified activities such as camping, boating, fishing and hunting will develop on an extensive scale only when people have their minimum needs met comfortably and their growing tendency to come closer to nature becomes more pronounced. Acquisition of frequently inundated flood plains, and reservation of open space for recreational activities will minimize pollution.

Limited in area, the 0.32 percent of Arizona's surface covered by lakes, rivers and streams includes some of the state's most popular recreational spots. Two out of three Arizonans visit water-based recreation areas at least once per year. Destinations range from large lakes along the Colorado River to small mountain and canyon streams. Closer to home, water is enjoyed in some 190,000-backyard pools and numerous water parks.

Water-based recreation also attracts visitors to Arizona. Of the 25 million tourists visiting the state annually, more than half flock to places along the Colorado River, such as Lake Mead, Lake Powell, and Lake Havasu, helping to create a statewide tourism industry that employs 100,000.

These lakes, actually reservoirs behind dams, are used by the state's 150,000 boat owners, 325,000 anglers, and countless swimmers, water skiers and windsurfers. Because these reservoirs are operated for multiple purposes, including flood control, water supply and power generation, conflicts increasingly arise over the way some dams are operated.

The following paragraphs give an account of the water based recreational activities in Sri Lanka.

2.8.1 Water Based Recreation in Sri Lanka

Sri Lanka provides the holidaymaker with a wide range of recreational activities at its tourist resorts. Water sports, beach games, golf, tennis, squash; group events conducted by trained animators are among the many activities provided at the resorts. Wind surfing, water skiing, boating, swimming, diving and snorkeling are popular events at the west and south coast resorts. Diving schools run by experienced divers help the uninitiated. These resorts are popular with almost all holidaymakers coming to Sri Lanka. Australian surfers have a special affinity to surfing at Hikkaduwa.

Submerged wrecks in Sri Lanka's coastal waters are an added attraction to the skilled divers. Negombo, Bentota, Ambalangoda, Hikkaduwa, Galle, Unawatuna, Weligama along the west and south coastal belt are ideal locations for water-based activities. Wind surfing and water skiing are popular events at coastal hotels. So are boating and boat trips. Inland boating facilities are available at the Bolgoda Lake in Colombo. Boating along rivers has become a favourite pastime with the tourists.

2.9 REFERENCES

Aderaldo Silva De Souza, et al. (1982), *Irrigación par Potes de Barro: Descripción del Método y Pruebas Preliminares,* (EMBRAPA-CPATSA Boletín de Investigación No. 10, (English translation), Petrolina, PE, Brazil

Alan Guttmacher Institute, (1995), *Hopes and Realities: Closing the Gap Between Women's Aspirations and Their Reproductive Experiences,* New York, USA

Arizona Department of Water Resources, (1998), Draft Third Management Plan, Tucson Active Management Area, Phoenix, Arizona, USA

Asian Development Bank, (1991), *Human Resource Policy and Economic Development.* Manila, Philippines

Baseline Document for East Calcutta Wetlands and Waste Recycling Region, (1997), Calcutta Metropolitan Water and Sanitation Authority, Calcutta, India

Basic Manual, Integrated Wetland System for Wastewater Treatment and Recycling for the Poorer Parts of the World with Ample Sunshine, (1995), (Prepared by Dr. D. Ghosh for RHUDO/ USAID, Calcutta, India

Bouwer, H., (1985). "Renovation of wastewater with rapid infiltration land treatment systems", Asano, T. (ed.), *Artificial recharge of groundwater.* Butterworth Publishers, Boston, Massachusetts, USA.

Dandekar M.M. and Sharma, K.N. (1979), *Water Power Engineering,* Vikas Publishing, New Delhi, India.

David, A., (1959), "Effects of Calcutta Sewage upon the fisheries of the Kulti Estuary and the connected cultivable fisheries", *Journal of Asiatic Society (Bengal).* Vol. 1 No. 4, pp. 339-363, Calcutta, India

Down to Earth, (2002), "Nurturing land or depleting water resources?", Apr. 15, 56, pp. Centre for Science and Environment, New Delhi, India

EIP Working Report 5, (1996), *Sustaining Calcutta : Present Status Report of the Urban Peoples Environment* (Calcutta, CMW & SA), Calcutta, India

García, V.N. (1995), *Diseño y Aplicación del Riego Intermitente por Gravedad.* Universidad Nacional Autónoma de México, Facultad de Ingeniería, México D.F., (Tesis para obtener el grado de Doctor en Ingeniería Hidráulica) (English translation), Mexico

Gelt, J., Henderson J., Seasholes, K., Tellman, B., Woodard, G., Carpenter K., Hudson C., Sherif S., (1998) *Water in the Tucson Area: Seeking Sustainability,* A status report by the Water Resources Research Center, College of Agriculture, The University of Arizona, Tuscon, USA

Ghosh D., (1985), *Cleaner Rivers: The Least Cost Approach. State Planning Board,* Government of West Bengal, Kolkata, India.

Ghosh D., (1991), "Ecosystem Approach to Low-cost Sanitation in India, Where the People Know Better". *Proceedings of the International Conference on Ecological Engineering for Wastewater Treatment,* Sweden

Ghosh D., (1994), "Ecosystem Approach to Wastewater Management in Urban Areas— Lessons for Poorer Countries with Ample Sunshine", *International Workshop on Integrated Water Resources Management in Urban and Surrounding Areas,* International Hydrological Programme of UNESCO, Essen, France

Ghosh D., (1997), "Environmental Agenda for Calcutta: Conserving the Water Regime for an Ecologically Subsidized City", *Banabithi—Environment Special Issue,* Department of Environment and Forests, Government of West Bengal, Kolkata, India

Ghosh, D., (1983), *Sewage Treatment Fisheries in East Calcutta Wetlands*. Report to the Department of Fisheries, Government of West Bengal, Calcutta, India

Ghosh, D., (1993), "Towards Sustainable Development of the Calcutta Wetlands", *Towards the Wise Use of Wetlands*, T.S. Davis, Ed., Ramsar Convention.

Ghosh, D. and C. Furedy, (1984), "Resource Conserving Traditions and Waste Disposal: The Garbage Farms and Sewage-fed Fisheries of Calcutta". *Conservation and Recycling*, Vol. 7, No. 2-4.

Ghosh, D. and Sen, S. (1992), "Developing Waterlogged Areas for Urban Fishery and Waterfront Recreation Project". AMBIO, *Journal of the Royal Swedish Academy of Sciences*, Vol. 21, No. 2., Sweden

Ghosh, D. and Sen, S. (1987), "Ecological History of Calcutta's Wetland Conservation", *Environmental Conservation*, Vol. 14, No. 3

Ghosh, D. (1999), *Wastewater Utilization in East Calcutta Wetlands*, Urban Waste Expertise Programme (UWEP) Occasional paper, July, CW Gouga, The Netherlands gopher:// gopher.un.org:70/ga/docs/S-19/plenary/ES5.TXT

Gupta, S.K., (1993), "Water for recharging of aquifers in Ahmedabad, India", *Indian Journal of Earth Sciences*. Vol. 20, No. 1, pp. 28-36, India

Gupta, S.K. and Sharma, P. (1995), "An approach to tackling of fluoride problem in drinking water", *Current Science*, Vol. 68, No. 8, pp. 774, Bangalore, India

Gupta, S.K. and Sharma P. (1996), "Rejuvenating our rivers—Akshaydhara concept", *Current Science*, Vol. 70, No. 8, pp. 694-696, Banglore, India

http://www.oieau.fr/ciedd/contributions/at3/contribution/suresh.html

http://www.oieau.fr/ciedd/contributions/at3/contribution/gupta.html

http://www.oieau.fr/ciedd/contributions/at2/contribution/2nepal.htm

http://ag.arizona.edu/AZWATER/publications/sustainability/report_html/

http://www.earthsummitwatch.org/cleanwater.html

http://www.fao.org/docrep/S8684E/s8684e06.htm

http://www.fao.org/docrep/S8684E/s8684e07.htm

http://www.iash.info/worldpotential.htm

http://www.waste.nl/docpdf/OP_calc.pdf

http://www.wwfindia.org/programs/fresh-wet/calcutta.jsp?prm=104

http://www.jxj.com/magsandj/rew/2003_01/small_hydro.html#author

http://www.water.org/ *(1997)*

http://www.inshp.org

Jiandong, T., (2003), "Small Hydro on a Large Scale: Challenges and Opportunities in China", *Renewable Energy World*, Jan-Feb., James & James, London, UK

Martinez Austria, P., and Aldama, R.A. (1991), "Dispositive de Control para la Aplicación del Riego Intermitente," (English translation), *Revista Ingenieria Hidráulica en México*, May-August, Mexico City, Mexico

Murillo, A.C. and Ludgardo L. Mamani, (1992), *Manual Técnico de Waru Waru, Para la Reconstrucción, Producción y Evaluación Económica*, Programa Interinstitucional de *Waru Waru*, Convenio PELT/INADE-IC/COTESU, (English translation), Puno, Peru

Nema, P., (1996), Soil-aquifer-treatment—an emerging technology. *The Times of India*, June 4, Ahmedabad, India

Rao, K.L. (1979), *India's Water Wealth*, Orient Longman, Hyderabad, India

Scherr, J., and Bohart, B. (1998) *Clearing The Water: A New Paradigm for Providing the World's Growing Population With Safe Drinking Water,* Earth Summit Watch, Washington, USA

Suresh, V. (1998), Strategies for Sustainable Water Supply for All: Indian Experience, Earth Summit Watch, Washington, USA

UNDP-World Bank, (1996), Water and Sanitation Program. Annual Report. July, 1994-June, 1995

UNEP, (2003), *Source Book of Alternative Technologies for Freshwater Augmentation in Latin America and the Caribbean,* Chapters 4.1, 4.2, 5.7, Unit of Sustainable Development and Environment of the General Secretariat of the Organization of American States (OAS), Joint United Nations Environment Programme (UNEP) Water Branch and International Environmental Technology Centre (IETC)

UNFPA, (1991) United Nations Population Fund, *Population, Resources and the Environment: The Critical Challenges,* London, UK

UNFPA, (1997), United Nations Population Fund, *The State of World Population,* New York, USA

UNGASS, (1997), United Nations General Assembly Special Session, 23-27 June 1997, *Programme for the Further Implementation of Agenda 21* (UN PCSD, Advance Unedited Text, New York, USA

UNCSD, (1997), Trends, United Nations Commission on Sustainable Development, *Global Change and Sustainable Development: Critical Trends: Report of the Secretary General* (Department of Policy Coordination and Sustainable Development), Fifth Session, 7-25 April, New York, USA

UNCSD, (1997), Assessment, United Nations Commission on Sustainable Development, *Comprehensive Assessment of the Freshwater Resources of the World: Report of the Secretary General* (Fifth Session, 5-25 April, E/CN.17/1997/9), New York, USA

Wilhelm, S.R., Schiff S.L. and Cherry, J.A. (1994a), "Biogeochemical evolution of domestic waste in septic systems: 1. conceptual model", *Ground Water.* Vol. 32, No. 6, pp. 905-916.

Wilhelm, S.R., Schiff S.L. and Robertson W.D. (1994b), "Chemical fate and transport in a domestic septic system: unsaturated and saturated zone geochemistry", *Environmental Toxicology and Chemistry.* Vol. 13, No. 2, pp. 193-203

World Bank, (1997), *World Development Indicators 1997* (International Bank for Reconstruction and Development/The World Bank, Washington D.C., USA

World Wide Fund for Nature (WWF)-India, (2002), *East Calcutta Wetlands, Identification and Designation of Potential Indian Ramasar Sites,* New Delhi, India

Chapter 3

Hydraulic Principles and Eco-friendly Design Approach

3.1 CONSERVATION PRINCIPLES IN GENERAL

In this Chapter, we deal with the basic principles that govern any branch of mechanics in general and hydraulics in particular. These principles are known as conservation principles. With the advancement of science into the new domains of relativity and nuclear science, we know very well that mass can be converted into energy and vice versa. In such a case also, if we express mass as well as energy into a common physical parameter, that physical parameter must be a constant for any particular system. In other words, any physical system (either nuclear or non-nuclear) is essentially a conservative system as far as the preservation of the quantum of the parameters of that system is concerned.

In fluid mechanics or hydraulics, all the processes are essentially non-nuclear in nature. It means that physical parameters such as mass and energy cannot be converted into each other. The quantity of these parameters gets conserved or maintained in a *hydraulic system*. Here a *hydraulic system* implies that portion of fluid specifically identified for analysis. Thus the conservation of mass, conservation of energy and conservation of momentum constitute the three fundamental conservation principles in hydraulics. The principle of conservation of angular momentum will be applicable only in the case of rotation of a fluid in a hydraulic system. While the mass conservation equation is also known as *continuity equation* or *mass balance equation*, the energy conservation equation is known as the *energy equation* or *Bernoulli's Equation*. The *general statement for conservation of either the mass or the energy* can be stated as follows:

Mass or energy can neither be created nor be destroyed in a hydraulic system and only their form can change from one form to another.

For simplicity as well as for the convenience of identification, a hydraulic system is generally considered to be limited in its spatial extent and is commonly referred to as a *control volume*. Essentially a control volume is the hydraulics equivalent of free body diagram, used very commonly in solid mechanics. A *control volume* is either real or fictitious volume bounded by a control surface, which may be either pervious or impervious to mass or heat. If the flow is unidirectional, the portion of the control surface through which the fluid flows is commonly known as the *cross section of flow*.

When the flow is in only one direction as shown in Figure 3.1, the control volume will be consisting of a bunch of flow lines. Each of these lines will have more or less the same flow direction. Such flow lines are called *streamlines*. A streamline is a line whose tangent at any point provides us with the flow direction at that point. A collection of adjacent streamlines is called a *streamtube*. Similar to a streamline, a streamtube will also have flow in only a particular direction. However unlike a streamline, a streamtube has certain cross-sectional area.

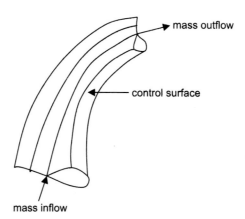

Figure 3.1   A Typical Control Volume With a 1-D Flow (i.e., a Streamtube)

In real life situations flows may be either 1-dimensional or 2-dimensional or 3-dimensional and the cross sections as well as the flow directions may overlap with one another. In such cases, the flow may be resolved into different components along convenient directions for simplicity.

The momentum conservation principle is generally known as either the *impulse-momentum equation* or *the equation of motion*. It is *generally stated as* follows:

The impulse of the resultant force on a fluid (i.e., the product of force and the time increment during which it acts) will be equal to the change in the linear momentum of the body in the direction of force.

The angular momentum conservation principle is also known as the *moment of momentum equation* or the *angular momentum equation*. Its *general statement* is as follows:

The resultant torque (i.e., twisting moment) acting on a fluid is equal to the rate of change of angular momentum or moment of momentum.

The following pages describe these principles in detail. Typical applications are also included to ensure easy comprehension.

3.2 MASS CONSERVATION PRINCIPLE AND APPLICATIONS

As already mentioned earlier, another name for the mass conservation principle is the *continuity equation*. It implies the continuity or the maintenance of the constant amount of mass of a fluid flowing through a hydraulic system. In other words, the quantity of fluid entering a hydraulic system will be exactly same as the quantity of the fluid leaving that system. Figure 3.2 shows this principle diagrammatically.

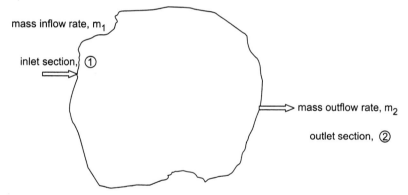

Figure 3.2   Continuity Equation for the Flow through a Hydraulic System

The continuity equation for a *steady flow* (i.e., flow having no variations with time) can be expressed in terms of the mass rates of flow as follows:

$$m = \rho_1 \cdot A_1 \cdot V_1 = \rho_2 \cdot A_2 \cdot V_2 \tag{3.1}$$

where, ρ_1 : mass density of fluid at inlet section;
A_1 : cross-sectional area at the inlet;
V_1 : average velocity at the inlet;
ρ_2 : mass density of fluid at outlet section;
A_2 : cross-sectional area at the outlet;

V_2 : average velocity at the outlet.

The continuity equation for the steady flow of an incompressible fluid (i.e., having a constant mass density always) will be in terms of volume rate of flow (i.e., discharge) as follows:

$$Q = A_1 \cdot V_1 = A_2 \cdot V_2 \tag{3.2}$$

A number of examples can be cited to illustrate this principle. Some of the most common examples observed in nature are oxygen cycle, water or hydrologic cycle and nitrogen cycle. One of the famous Sanskrit sloka says, "water falling from sky must reach the ocean" to enunciate this principle through the example of hydrologic cycle described in Section 1.2 of Chapter 1.

3.2.1 Typical Applications of Mass Conservation Principle

The continuity equation can be applied in many practical situations. If there is flow occurring in a particular hydraulic system, essentially the continuity equation has to be satisfied in the form of equality between mass inflow and mass outflow. Otherwise, it implies that there is no complete fluid flow. Thus the continuity equation can be applied in all cases where there is flow. The majority of the flows can be any of the following types:

a) Types of Flows

Flow can be classified as either *steady flow* or *unsteady flow* depending on whether the flow parameters are constant with time or not. On the other hand the flow can be classified as either *uniform flow* or *non-uniform flow (i.e., varied flow)* depending on whether the flow parameters are constant with respect to space or not. Another criterion for classification of flow is the layered nature of flow without any intermixing. Based on whether the flow satisfies this criterion or not, it can be classified as either *laminar flow* or *turbulent flow*. Other types of flows such as *1-Dimensional flow* or *2-Dimensional flow* or *3-Dimensional flow* mean whether the velocity has components predominantly in only one direction or two mutually perpendicular directions or three mutually perpendicular directions in that flow.

Flows can also be classified as either *irrotational flow* or *rotational flow* depending upon whether the *vorticity* (i.e., the flow along a closed curve per unit area or the line integral of the tangential velocity about a closed curve per unit area or the *circulation* per unit area) is either zero or non-zero. In simple terms, rotational flows have angular deformations of fluid particles caused due to shear stresses acting on them. All the flows near the boundary are driven by viscosity or resistance and hence are rotational flows. Flows around the centers in pipes can be approximated as *irrotational flows* or *potential flows*.

Figure 3.3 illustrates these types of flows. Among these types of flows, uniform and non-uniform flows are generally described in open channels in terms of the uniformity in flow depth everywhere. Steady and unsteady flows, laminar and turbulent flows, 1-D, 2-D and 3-D flows as well as rotational and irrotational flows occur both in pipe flows and in open channel flows.

Another major application of mass conservation principle is the concept of *flow net*. *It consists of streamlines and equipotential lines, which are two orthogonal sets of lines.* Flow net is found to be very useful in analyzing 2-Dimensional irrotational flows (i.e., potential flows). They are discussed in the following paragraphs.

b) Flow Net

To study the flow net, we need to know about *velocity potential* (ϕ) and *stream function* (ψ). *Velocity potential* (i.e., *velocity potential function* or *potential function*) is defined as the scalar function of space and time whose negative partial derivative with respect to any direction gives the velocity component in that direction.

As already mentioned, flow net is the orthogonal grid obtained by drawing equipotential lines and streamlines. The flow net construction is possible only when the flow is steady and irrotational. A typical sketch of a 2-Dimensional flow net is given here in Figure 3.4.

3.3 ENERGY CONSERVATION PRINCIPLE AND APPLICATIONS

The energy conservation principle (i.e., Energy Equation or Bernoulli's Equation) relates to the maintenance of the total energy in a hydraulic system. The predominant components of energy in a hydraulic system are:

i) *Flow Energy* (i.e., *Pressure Energy*, PrE) due to the pressure of the liquid. It can be expressed as the work done, which is product of either force (F) and length (l) or the product of pressure (p) and volume (Ψ). Volume can further be expressed as the ratio of weight (W) and specific weight (γ). Equation 3.3 gives this expression:

$$PrE = p\,W/\gamma \qquad (3.3)$$

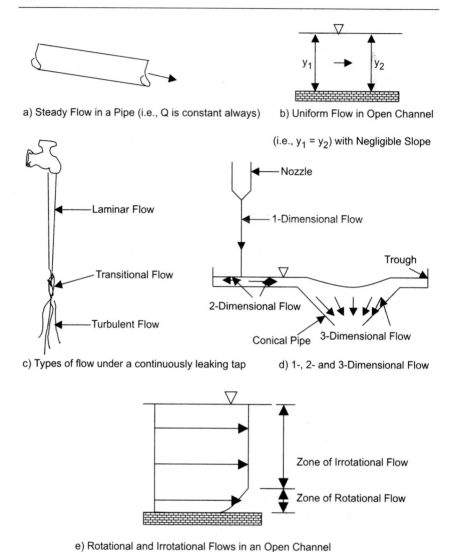

a) Steady Flow in a Pipe (i.e., Q is constant always)

b) Uniform Flow in Open Channel (i.e., $y_1 = y_2$) with Negligible Slope

c) Types of flow under a continuously leaking tap

d) 1-, 2- and 3-Dimensional Flow

e) Rotational and Irrotational Flows in an Open Channel

Figure 3.3 Various Types of Flows through Examples in Real Life

ii) *Kinetic Energy (KE)* due to the motion of the liquid. It can be expressed in terms of weight (W), gravitational acceleration (g) and the average velocity of flow (V). Equation 3.4 gives this expression:

$$KE = \frac{WV^2}{2g} \qquad (3.4)$$

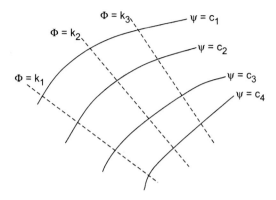

Figure 3.4 A Typical 2-Dimensional Flow Net.

iii) Potential Energy (PE) due to the position of the liquid. It can be expressed by Eqn. 3.5 as the product of weight (W) and the elevation of the liquid (z) w.r.t. a certain datum or reference level.

$$PE = W \cdot z \tag{3.5}$$

The total energy (TE) of the liquid will be the sum of these three energy components. The total energy per unit weight will have the dimensions of length and is called the *total head* (H). It is given by Eqn. 3.6.

$$H = \frac{TE}{W} = \frac{p}{\gamma} + \frac{V^2}{2g} + z \tag{3.6}$$

Each of the three terms on the RHS of Eqn. 3.6 also have the dimensions of length. The first term among these (i.e., p/γ) is called the *pressure head*, which is the flow energy or pressure energy of the liquid per unit weight. The middle term (i.e., $V^2/2g$) is called *velocity head* or *kinetic head*, which is the kinetic energy per unit weight of the liquid. The last term (i.e., z) is called the *datum head* or *potential head*, which is the potential energy w.r.t. a datum per unit weight of the liquid.

Let us consider an element of an incompressible fluid under steady irrotational flow moving between sections 1 and 2 of a conduit. By summing up the three major energy components at both these sections, we get the expressions for total energy at each of the sections as under.

$$TE_1 = \frac{p_1 W}{\gamma} + \frac{WV_1^2}{2g} + Wz_1 \tag{3.7}$$

$$TE_2 = \frac{p_2 W}{\gamma} + \frac{WV_2^2}{2g} + Wz_2 \tag{3.8}$$

In an ideal case, energy is neither added nor removed. Therefore by the energy conservation principle, the total energy values can be equated. Since the weight (W) of this element is common in both these expressions, we can eliminate it. Thus we get the expression equating total heads at both the sections (H_1 and H_2). The resulting equation is popularly known as the *Bernoulli's Equation*. Figure 3.5 illustrates Bernoulli's Equation for an ideal case.

$$\frac{P_1}{\gamma} + \frac{V_1^2}{2g} + z_1 = \frac{P_2}{\gamma} + \frac{V_2^2}{2g} + z_2 \tag{3.9}$$

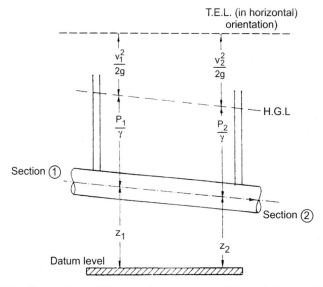

Figure 3.5 Illustration of Energy Conservation Principle for an Ideal Case

There are certain *restrictions for applying the Bernoulli's Equation* to various flow problems. These restrictions are:

1. The fluid should be incompressible;
2. The fluid has to be inviscid (i.e., the flow is irrotational);
3. The flow has to be steady;
4. The Bernoulli Equation is applicable only along a streamline and
5. There should not be any energy conversion devices (i.e., pumps, turbines, etc.).

In real life situations, it may not be possible to satisfy all these restrictions in general and it may not be possible to satisfy the inviscid condition in particular. Therefore, the Bernoulli's Equation is modified, taking into consideration the energy input per unit weight (H_E) and energy lost due to friction and other energy loss phenomena (H_L) as follows:

$$\frac{p_1}{\gamma} + \frac{V_1^2}{2g} + z_1 + H_E - H_L = \frac{p_2}{\gamma} + \frac{V_2^2}{2g} + z_2 \qquad (3.10)$$

In case of pumps, there will be a positive energy input and in case of turbines, there will be a negative input or extraction of the energy from the system. Figure 3.6 clearly shows such a real life situation.

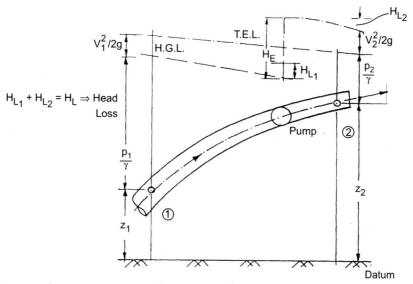

Figure 3.6 Energy Conservation Principle Applied to a Real Life Situation

In Figures 3.5 and 3.6, if we connect all the *piezometric heads* [i.e., head denoted by a gage pressure measuring device = $(p/\gamma + z)$] by a line, we get what is known as *hydraulic grade line (HGL)*. On the other hand, if all the points denoting the total head are connected by a line, we get the *total energy line (TEL)*. TEL is always above the HGL by a distance equal to the velocity head ($V^2/2g$). For the ideal case TEL should be horizontal. In real life situations the total energy line will show ups or downs, which may be either sudden or gradual as indicated in Figure 3.6.

Whenever there is a viscous fluid, there will be resistance at the fluid boundary. This results in a non-uniform velocity throughout the flow cross section. Therefore the actual velocity head will be different from the velocity head ($V^2/2g$) based on the cross sectionally averaged velocity (V). The actual velocity head in such cases is taken as $(\alpha V^2/2g)$, where the coefficient (α) is known as the *kinetic energy correction factor*. The kinetic energy correction factor (α) is evaluated in terms of the area of flow cross section (A), variable velocity (v) over an elemental area (dA) as follows:

$$\alpha = \frac{1}{A}\int\left(\frac{v}{V}\right)^3 dA \qquad (3.11)$$

When there is uniform velocity throughout the cross-section of flow, $\alpha = 1$. For laminar flow through pipes, $\alpha = 2$ and for turbulent flow through pipes, α might vary from 1.01 to 1.20. Whenever there is no data available in regard to α, it is taken as 1.

3.3.1 Typical Applications of Energy Conservation Principle

Energy Equation has many applications. Some of the typical applications of the energy equation are in the form of devices for velocity and discharge measurement such as Pitot Tube and Venturi Meter. The following paragraphs describe these applications of Bernoulli's Equation.

a) Pitot Tube

A Pitot Tube is a velocity measuring device for open channels as well as pipes. A Pitot Tube consists of a right angled (i.e., L-shaped) tube which is sufficiently large to overcome the effects of capillary action due to surface tension. It is introduced with one of its open end held against the liquid flow direction and the other end is open to atmosphere. Due to the right angled bend in the tube, the liquid velocity head just upstream of the Pitot Tube gets converted into pressure head completely. This is known as the *stagnation phenomenon*, wherein the liquid velocity gets reduced to zero. The point where the stagnation phenomenon takes place is called the *stagnation point*. Thus additional pressure head is created, which will appear in the vertical limb of the L-shaped tube. It represents the total head at the stagnation point. It is also called the *total pressure head*, which includes *static pressure head* (i.e., the sum of pressure head as well as the datum or potential head) and velocity head.

To measure the velocity of the fluid, we need to measure the velocity head. It is simply the difference between total pressure and the static pressure. In case of open channels, the water surface itself represents the static pressure head as indicated in Figure 3.7a. Whereas, in case of pipe or conduit flow, we need a *piezometer* (i.e., a simple tube emerging out of the pipe body, normal to the pipe axis) as shown in Figure 3.7b.

Many a time, both these tubes are combined into two concentric tubes to constitute what is known as the *Pitot Static Tube* (consisting of two concentric l-shaped tubes) shown in Figure 3.7c. In the body of the Pitot Static Tube's outer pipe, there will be radially symmetrical holes to measure the average static pressure head. At one end of the inner concentric pipe, there will be a single hole corresponding to the pipe diameter to measure the total pressure head. To get the actual velocity (V_a), we need to consider friction and use velocity coefficient (C_v) as given in Equation 3.12. Generally, the value of C_v is around 0.98. Here 'h' is the head and 'g' is the gravitational acceleration.

$$V_a = C_v \sqrt{2gh} \qquad (3.12)$$

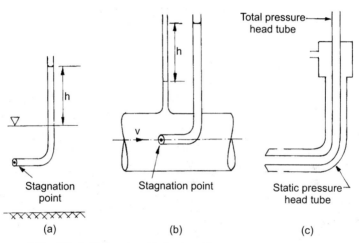

Figure 3.7 Pitot Tube for Velocity Measurement in a) Open Channels; b) Pipes; c) Pitot Static Tube

b) Venturi Meter

A Venturi Meter is a flow measurement device for pipe flow as shown in Figure 3.8. In case of open channels, a Venturi Flume is used. It is also based on the same principle as the Venturi Meter. The Venturi Meter consists of two tapering pipes known as converging cone and a diverging cone and a short cylindrical pipe known as the throat fixed in between both the cones. The bigger diameter of each of the cones corresponds to the diameter of the pipe where the flow needs to be measured. The converging cone is shorter in

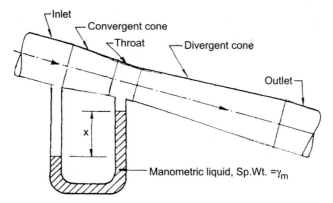

Figure 3.8 Venturi Meter for Flow Measurement in Pipes.

length and is situated upstream w. r. t. the diverging cone. There is also a U-tube manometer, generally with mercury as its manometric liquid, connecting the pipe just upstream of the converging cone with the throat, to measure the pressure difference.

Venturi Meter can be in either a horizontal or a vertical or an inclined orientation. At the throat, the velocity gets increased due to its smaller cross sectional area. Correspondingly, the pressure head at throat will be smaller so as to maintain the total head. Generally the throat diameter will be about 60% of the pipe diameter. The diverging cone is of a larger length, so as to avoid flow separation. By applying the 1-Dimensional continuity equation (i.e., Eqn. 3.3), the Bernoulli's Equation (i.e., Eqn. 3.9) and the manometric equation for pressure (i.e., Eqn. 3.13) between the inlet and throat sections, considered as sections 1 and 2, we get the following expression (i. e., Eqn. 3.14) for theoretical discharge (Q_{th}) after simplification:

$$\frac{p_1}{\gamma} - \frac{p_2}{\gamma} = h = x\left(\frac{\gamma_m}{\gamma} - 1\right) \tag{3.13}$$

$$Q_{th} = \frac{A_1 A_2 \sqrt{2gh}}{\sqrt{A_1^2 - A_2^2}} \tag{3.14}$$

Here, γ_m is the specific weight of the manometric fluid, p_1 is the pressure at inlet section, p_2 is the pressure at throat section, γ is the specific weight of fluid flowing, h is the difference in pressure heads between inlet-throat, A_1 is the inlet cross-sectional area, A_2 is the throat cross sectional area and x is the vertical difference between the levels of the manometric fluid in both the limbs of the manometer. In deriving the Eqn. 3.14, friction and other losses are neglected. However, due to friction and other losses, the actual discharge or flow rate (Q_{act}) will be smaller than Q_{th} and is expressed in terms of coefficient of discharge (C_d) as under:

$$Q_{act} = C_d \cdot Q_{th} \tag{3.15}$$

Energy equation can also be applied for various other cases of fluid flow such as nozzles, compressible fluid flow examples, different hydraulic machines and so on. Some of these applications will be discussed in different context in this book and a variety of references are available on almost all the applications.

3.4 MOMENTUM CONSERVATION PRINCIPLE AND APPLICATIONS

Momentum Conservation is based on the Newton's Second Law of Motion. According to Newton's Second Law, the time rate of change of momentum in any particular direction will be equal to the external force acting on the system. If we choose any arbitrary direction x, we can mathematically write as follows:

$$\sum F_x = \frac{dM_x}{dt} \tag{3.16}$$

Here F_x represents the external force and M_x represents the momentum in the arbitrary direction x. If we multiply the force (F_x) by the time increment (dt), we get impulse which will be given by the change of momentum $[d(M_x)]$. This is the *impulse momentum equation*. As already explained, it is also known as *momentum equation* or *equation of motion* or *momentum conservation equation*.

Figure 3.9 Provides us with a sketch for the momentum conservation for an incompressible fluid under steady flow occupying a control volume. Here the mass of the fluid in the region 1 2 3 4 gets shifted to a new position 1' 2' 3' 4' after a short time interval due to the external forces acting on the stream. Due to a gradual increase in the cross sectional area of flow, the velocity as well as the momentum is gradually increasing in the direction of flow. Since the area 1' 2' 3 4 is common to both the regions 1 2 3 4 and 1' 2' 3' 4' it will not experience any change in momentum. By applying impulse-momentum principle, it can be shown that the sum of all external forces acting on the system in the arbitrary x-direction (ΣF_x) is given by the expression as under:

$$\sum F_x = \rho Q [V_{2x} - V_{1x}] \tag{3.17}$$

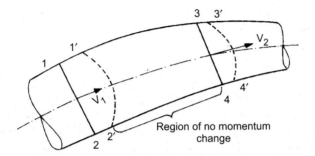

Figure 3.9 Momentum Equation for Steady Flow of an Incompressible Fluid.

Analogous expressions in the y- and z-directions can be written as follows:

$$\sum F_y = \rho \, Q \, [V_{2y} - V_{1y}] \qquad (3.18)$$

$$\sum F_z = \rho \, Q \, [V_{2z} - V_{1z}] \qquad (3.19)$$

In Eqns 3.17 to 3.19, only 1-Dimensional flow is considered. However, in situations where there is a cross-sectional variation in velocity, we need to multiply the momentum term by the *momentum correction factor* (β), which accounts for this change in momentum. The momentum correction factor (β) is given by the following expression:

$$\beta = \frac{1}{A} \int \left(\frac{v}{V}\right)^2 dA \qquad (3.20)$$

Here v = variable flow velocity across an elemental area 'dA', A = cross-sectional area of flow and V = cross-sectionally averaged velocity. For uniform velocity distribution, $\beta = 1$. For laminar flow through circular pipes, β value is generally about 1.33. For turbulent flow through circular pipes, β value is around 1.05. In case of non-availability of data, it is assumed to be 1.

For a general case, the momentum principle has to be derived from the basic concept of Newton's Second Law of motion considering various forces that might possibly act on fluids. These forces may be due to gravity, pressure, viscosity, turbulence, surface tension and compressibility or elasticity. Among these forces, the forces due to surface tension and compressibility are generally not significant in most of the fluid flow problems. Considering only the forces due to gravity, pressure, viscosity and turbulence, we obtain what are known as *Reynolds' Equations*. Considering only the forces due to gravity, pressure and viscosity, we get what are known as the *Navier-Stokes' Equations*. Considering only the forces due to gravity and pressure, we get what are known as *Euler's Equations*.

a) Force on a Pipe Bend

Let us consider a 2-Dimensional pipe bend carrying an incompressible fluid under steady flow. For simplicity, let us assume uniform velocity throughout the inlet and outlet cross sections of the pipe bend. The only forces to be considered in this case are due to gravity and pressure, as shown in Figure 3.10. The fluid will exert a force on the pipe bend, which can be resolved into two components (F_x and F_y) along the two directions. By applying momentum equation in both x and y directions, we get the following expressions:

$$p_1 \cdot A_1 \cdot \cos \theta_1 - p_2 \cdot A_2 \cdot \cos \theta_2 - F_x = \rho \, Q \, [V_2 \cdot \cos \theta_2 - V_1 \cdot \cos \theta_1] \qquad (3.21)$$

$$p_1 \cdot A_1 \cdot \sin \theta_1 - p_2 \cdot A_2 \cdot \sin \theta_2 - F_y - W = \rho \, Q \, [\, V_2 \cdot \sin \theta_2 - V_1 \cdot \sin \theta_1] \qquad (3.22)$$

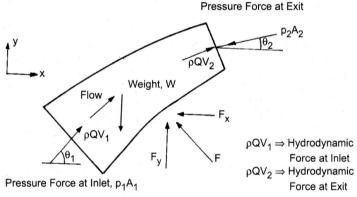

Figure 3.10   Force on a Pipe Bend due to Fluid Flow.

Using the Eqns. 3.21 and 3.22 as well as the 1-D continuity equation, the total force on the pipe bend can be easily computed. The momentum equation can also be applied to analyze problems such as flow through a suddenly enlarged pipe. It is discussed here.

b) Flow through a Sudden Enlargement of a Circular Pipe

Let us consider a fluid to be flowing through a pipe, which is suddenly enlarging into a bigger size. For simplicity, let us assume the velocity at the inlet and the outlet to be uniform as shown in Figure 3.11.

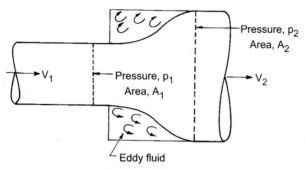

Figure 3.11   Momentum Conservation for Circular Pipe Flow with Sudden Enlargement

Here it is required to determine the loss in head (H_L) due to the sudden enlargement of pipe. Considering all the forces acting on the control volume in the direction of flow and applying momentum equation, we get the following expression:

$$p_1 \cdot A_1 + p_e (A_2 - A_1) - p_2 \cdot A_2 = \rho\, Q [V_2 - V_1] \qquad (3.23)$$

Here p_e is the mean pressure on the annular area connecting both the pipes due to eddy flow of fluid. It has been established experimentally that the mean pressure due to eddies is nearly same as the pressure at the inlet of the suddenly enlarging pipe (i.e., $p_e \approx p_1$). Also by applying the Energy Equation (i.e., Eqn. 3.10) with $H_E = 0$ and simplifying, we get the following expression for the head loss:

$$H_L = \frac{(V_1 - V_2)^2}{2g} \qquad (3.24)$$

Momentum conservation principle can be applied to a variety of flow problems. Some of these are hydraulic jump, force on a sluice gate, force due to jet impact on plates and moving vanes, force on a nozzle etc. These will be dealt with later as per their importance and relevance.

3.5   ANGULAR MOMENTUM CONSERVATION PRINCIPLES AND APPLICATIONS

As already stated, the time rate of change of angular momentum will be equal to the torque (T) on the rotating fluid having a density (ρ) and a steady flow rate (Q). Mathematically, this can be written as follows:

$$T = \rho\, Q\, (V_2\, r_2 - V_1\, r_1) \qquad (3.25)$$

where V_2 and V_1 are the absolute velocities of the fluid as it exits and enters the control volume; r_2 and r_1 are the corresponding radial distances of the fluid from axis about which the torque is acting. Thus the angular momentum (i.e., moment of momentum) is conserved and any change in its value about an axis will appear as torque (i.e., twisting moment) acting on the fluid about the same axis.

Angular momentum equation has many applications especially for the hydraulic machines, which will be discussed in another chapter. As a typical application of this equation, a simple lawn sprinkler is described here.

a) Lawn Sprinkler

Consider a lawn sprinkler with two rotating arms as shown in Figure 3.12a. In Figure 3.12b, the velocity triangle for one of the rotating arms of the lawn sprinkler is shown.

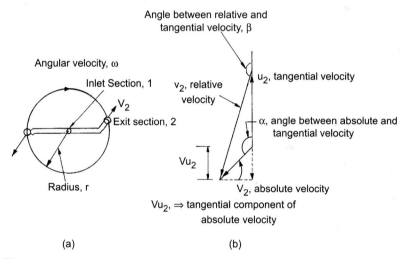

(a)   (b)

Figure 3.12 a) Angular Momentum Principle Applied to a Lawn Sprinkler; b) Velocity Triangle for One of the Sprinkler Arms.

The retarding torque acting on the lawn sprinkler can be expressed as in Eqn. 3.26. From these expressions, we can determine the maximum angular speed (i.e., runaway speed, ω_{max}) of the occurring when the torque on the sprinkler, $T = 0$ and also the maximum torque (T_{max}) occurring when the sprinkler's angular speed, $\omega = 0$. Equations 3.27 and 3.28 provide us with the necessary expressions

$$T = -\rho \cdot Q \cdot V_{u_2} \cdot r \qquad (3.26)$$

$$\omega_{max} = \frac{v_2 \cos \beta}{r} \tag{3.27}$$

and $$T_{max} = \rho \cdot Q \cdot r \cdot v_2 \cdot \cos \beta \tag{3.28}$$

So far we have discussed in this Chapter the fundamental principles of hydraulics and some of their direct applications. In the next section, we shall deal with an integrated approach whose guidelines can help us to ensure eco-friendly water systems for all living beings.

3.6 FLOW MEASUREMENT DEVICES IN PIPES AND OPEN CHANNELS

Flow measurement is very important in pipes as well as open channels. It involves the measurement of discharge i.e., the volumetric flow rate per unit time. The flow measurement devices will either measure the flow rate directly or will measure the flow rate as a product of flow velocity and area. The selection of an appropriate method or device for flow measurement will generally depend upon the cost, type of the conduit or channel, accessibility of the conduit or the channel, hydraulic head available, type of the liquid stream and characteristics of the liquid stream. Generally, flow-measuring devices are different for pipe conduits and open channels. However, some measuring systems are commonly applicable for both pipes and open channels.

For flow measurement, a system in which discharge is related to one easily measurable variable is adopted. The discharge values can be easily obtained by calibration curves developed for the flow measuring system. Table 3.1 provides a list of different types of flow measuring devices.

Table 3.1 Types of Flow Measurement Devices Commonly Used

Flow measurement devices	Flow measurement principle
I. For Pressure Pipes or Conduits	
a. Venturi meter	Differential pressure is measured.
b. Flow nozzle meter	Differential pressure is measured.
c. Orifice meter	Differential pressure is measured.
d. Pitot tube	Differential pressure is measured.
e. Electromagnetic meter	Magnetic field is induced and voltage is measured.
f. Rotameter	Rise of float in a tapered tube is measured.
g. Turbine meter	Uses a velocity driven turbine vane/wheel.
h. Acoustic meter	Sound waves are used to measure the velocity.

Contd.

Table 3.1 *Contd.*

Flow measurement devices	Flow measurement principle
i. Elbow/Bend meter	Differential pressure is measured around an elbow/a bend.
II. For Open Channels	
a. Flumes (Parshall, Palmer-Bowlus)	Critical depth is measured at the flume.
b. Weirs	Head is measured over the crest of a weir.
c. Current meter	A rotational element is used to measure velocity.
d. Pitot tube	Differential pressure is measured.
e. Acoustic meter	Uses sound waves to measure velocity and depth.
f. Floats (surface floats, rod floats and Canister floats)	Travel time for a floating object is measured to compute flow velocity.
III. For Freely Discharging Pipes Flowing Full	
a. Nozzles and orifices	Water jet data is recorded.
b. Vertical open-end flow	Vertical height of water jet is recorded.
IV. For Freely Discharging Pipes Flowing Partially Full	
a. Mildly sloping open-ended pipe	Dimensions of free falling water jet are recorded.
b. Open flow nozzle (Kennison nozzle or California pipe method)	Flow depth at free falling jet is determined.
V. For Other Cases	
a. Dilution method	Either a sudden injection of a dye tracer or a constant rate injection of a dye tracer is adopted.
b. Collecting tank/bucket method	a collecting tank/bucket is used and the collection time is recorded.
c. Emptying tank method	Water level difference in an emptying tank and the time of emptying is noted.

Among the above-mentioned flow measuring devices, Pitot tube and Venturi meter have already been described in Section 3.3.1. Flow nozzle meter and orifice meter are very much similar to venturi meter in their operation. The following paragraphs describe some of the flow measuring devices listed in Table 3.1.

3.6.1 Electromagnetic Meter

It is based on Faraday's Law which states that if a conductor (the flowing water in this case) is passed through a magnetic field, voltage will be produced proportional to the velocity of the conductor. The method involves sophisticated and expensive instrumentation and is found to be accurate to ±3% for conduits as well as open channels with a maximum width of 100m. The minimum detectable velocity is 0.005 m/s. Due to its extensive instrumentation, it requires trained personnel to handle routine operation and maintenance. Figure 3.13 gives a schematic representation of an electromagnetic meter.

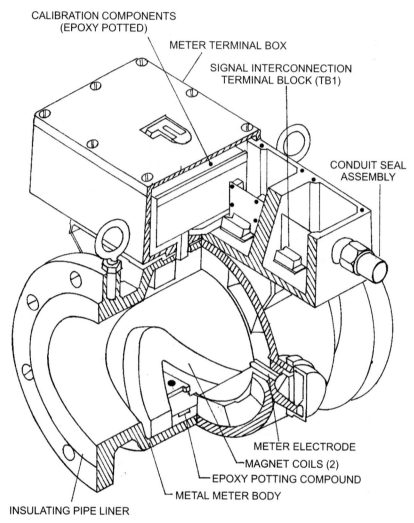

CALIBRATION COMPONENTS
(EPOXY POTTED)

METER TERMINAL BOX

SIGNAL INTERCONNECTION
TERMINAL BLOCK (TB1)

CONDUIT SEAL
ASSEMBLY

METER ELECTRODE

MAGNET COILS (2)

EPOXY POTTING COMPOUND

METAL METER BODY

INSULATING PIPE LINER

Figure 3.13 An Electromagnetic Meter for Conduit Flow Measurement

3.6.2 Turbine Meter/Current Meter

This device utilizes a rotating element of a turbine such as the vane or the blade or a propeller or an assembly of cups/cones, which are radially attached to an axis. Velocity of the rotating element will be proportional to the flow velocity of water. Turbine meters are generally used for flow measurement in pipes running full, while current meters are generally used in flow measurement for open channels. Figure 3.14 gives a sketch of a horizontal axis propeller type current meter. Current meters/turbine meters are quite accurate over a wide range of flows.

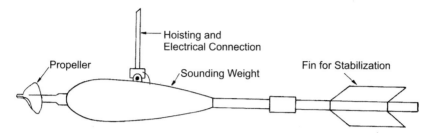

Figure 3.14 A Horizontal Axis Propeller Type Current Meter [Courtesy: Subramanya, 1996]

3.6.3 Acoustic Meter

Acoustic meters utilize sound waves to measure the flow. Depending upon whether the sound waves are in or above the audible range, they are classified as sonic meters or ultrasonic meters. They can determine water level, flow area and velocity using the travel time for a sound wave in the upstream and downstream directions. From the flow area and velocity, discharge is calculated. Figure 3.15 provides a schematic of an acoustic meter for flow measurement. Acoustic meters have a low head loss and are quite accurate. They can be used in pipes and open channels of any size over a wide range of flows. They require skilled manpower for operation and maintenance.

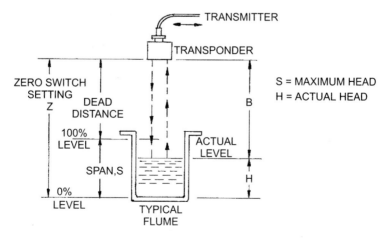

Figure 3.15 Acoustic Meter for Flow Measurement

3.6.4 Palmer-Bowlus Flume

This type of flume creates a change in the flow depth by decreasing the channel width and maintaining the slope. A head-discharge calibration chart obtains the discharge. Palmer-Bowlus flume has a lower head loss than the Parshall flume but is less accurate. Figure 3.16 gives a simple sketch of this type of flume.

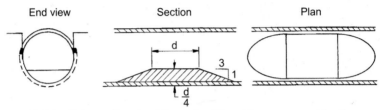

Figure 3.16 Palmer-Bowlus Flume for Flow Measurement in Open Channels

3.6.5 Open Flow Nozzle

Open flow nozzles are very simple devices used for flow measurement at the freely discharging ends of pipes. They require a flat pipe for a length of at least 6 times the pipe diameter upstream of the open flow nozzle. Kennison Nozzle and California Pipe are the most commonly used types of open flow nozzles. They result in huge head loss due to their freely discharging property. Figure 3.17 provides a typical sketch of a Kennison Nozzle.

The description about Parshall Flume, different types of weirs and other flow measuring devices listed in Table 3.1 can be obtained in any standard text book of Hydraulics or Fluid Mechanics. Hence they are not described here.

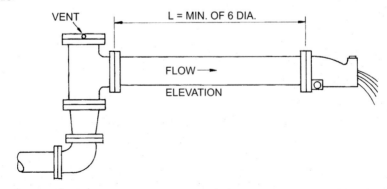

Figure 3.17 Kennison Nozzle for Flow Measurement at a Freely Discharging Pipe.

3.7 BASIC CONSIDERATIONS FOR ECO-FRIENDLY DESIGN OF WATER SYSTEMS

In recent times much has been said about eco-friendliness and sustainable development. Eco-friendliness means being aware of maintaining ecological balance. Sustainable development means the development that ensures humanity that it meets the present needs without compromising the ability of the future generations to meet their own needs. Eco-friendly design principles are considered as basic as physical principles, like hydraulic principles. Therefore they are addressed in this chapter along with the hydraulic principles. Among human needs, water comes second only to air, even before other requirements such as food, clothing, shelter and energy in that order. Hence it is extremely necessary to ensure proper quality and quantity of all these basic requirements including water, for all living beings in addition to human beings.

Therefore it is essential for human beings as well as other living things to live in harmony, so that everyone's basic needs are fulfilled. This is basically the principle of "live and let others live" which is very well said in an ancient Indian concept of governance. As per this concept, there is neither a ruler nor the ruled but every one performs their own duty and thereby protects one another. In the next few paragraphs, we shall discuss the basic considerations that need to be addressed in the eco-friendly design of water systems. These water systems can be mainly grouped into the following categories:

a) Drinking Water Systems,
b) Other Water Supply and Irrigation Systems,
c) Domestic and Industrial Wastewater Systems,
d) Water Power Systems, and
e) Other Water Systems.

The reason behind such a categorization is explained as below. The quantity of water required exclusively for drinking and cooking per person per day is within 13-15 liters. However, the general trend in the design of centralized water supply systems is to ensure the drinking water quality standards for water used for all purposes. This has led to enormous cost of water treatment, which may not be affordable in many cases in the developing world. Wherever law enforcement is not stringent enough, it has led to slackness in the quality of water, which is supposed to be used for drinking and cooking also. This slackness in the water quality is either due to inadequate treatment process or due to the corrosion/contamination in the water distribution system. Therefore, depending on the cost constraints drinking and other water supply systems need to be either combined or separated. Keeping all these issues in mind, in this chapter, the basic considerations for eco-friendly design of drinking water systems and other water systems are grouped separately so as to ensure their applicability in both the situations.

3.7.1   Basic Considerations for Eco-friendly Designs for Drinking Water Systems

These considerations can be listed as follows:

1. Drinking water systems for communities need to strictly meet the international water quality standards set up by agencies such as World Health Organization (WHO).

2. Drinking water systems need to be cost effective so that they can satisfy the entire human as well as animal population even in remote areas. The example of Thailand's National Jar Program comprising 8 million jars of 6000-liter capacity in the predominantly rural North-eastern Thailand, commissioned by 1992, is worth mentioning here. In this case, the water quality was found to be actually improving with the storage due to death of pathogens.

3. Drinking water systems should promote soil as well as water conservation practices among the users by adopting/enforcing appropriate policies such as community involvement as compared to centralized supply, rainwater harvesting as compared to groundwater exploitation, proper pricing as compared to huge government subsidies, use of mechanical pumping devices as compared to electric pumps, which may result in wastage of precious quantity of treated water. The revival of River Arvari in Rajasthan, India by community participation is worth mentioning here.

4. Drinking water systems need to be free from contamination due to either the natural reasons such as presence of arsenic on account of geological activities or artificial reasons such as corrosion of the water pipelines, discharge of untreated municipal/industrial wastewater into their sources, etc. It is worth noting here that all the water supply pipelines are replaced once in every 5 years in Singapore. The construction of 'sustainable house' in Sydney, Australia, the construction of 'Resource Efficient TERI Retreat for Environmental Awareness and Training (RETREAT)' developed by the Tata Energy Research Institute (TERI) in Gual Pahari, near New Delhi, India are some of the relevant examples worth mentioning here. Another

example is the construction of the Confederation of Indian Industry (CII)—Godrej Green Business Centre in Hyderabad, India in 2003. It received the upgraded rating of Version 2 Platinum—the highest possible award for sustainable design in November 2003 and was the only building in the world with that rating. In all these cases, drinking water is obtained by rainwater harvesting.

5. Drinking water systems need to have adequate stand-by units to overcome even the worst scenario such as droughts, sudden and uncontrolled ruptures in pipelines, contamination generally occurring during floods/cyclones due to dead bodies of animals, etc.

3.7.2 Basic Considerations for Eco-friendly Design of other Water Supply and Irrigation Systems

These considerations can be listed as under:

1. If water supply systems are designed separately for purposes other than drinking, they need to be cost effective by ensuring only the required water quality standards, which are less stringent than drinking water standards. The earlier mentioned designs of 'Sustainable House' and 'RETREAT' can be cited here again, wherein the water for other purposes is obtained by biological purification of wastewater.

2. Irrigation systems need to be designed to promote water conservation by adopting appropriate policies/techniques such as proper pricing of water and electricity instead of huge subsidies, sprinkler or drip irrigation instead of flooding or other water application techniques requiring excessive quantities of water and so on. As an example, the innovative research on sprinkler irrigation for rice cultivation in Arkansas State of USA can be mentioned here.

3. Irrigation systems should promote the use of bio-fertilizers and insecticides/weed-killers, which are easily bio-degradable and hence hardly cause any environmental pollution, instead of artificial chemicals which may be hazardous to human as well as animal health. This may be achieved by law-enforcement and/or proper education/training. The success story of organic cultivation in a small farm in a village in Haryana State in northern India and the dissemination of this know-how through the farmer organization 'Haryana Kisan Welfare Trust' is worth mentioning here. The complete ban on insecticides like DDT in many developed countries can be cited here.

4. Other water supply and irrigation systems should be designed with provision for proper drainage so that problems like soil salinity due to water-logging, spreading of diseases such as malaria due to stagnant water can be avoided.

5. Irrigation systems need to be designed such that losses due to evaporation and deep percolation are minimized and soil moisture is retained to the maximum possible extent even during severely dry conditions. The application of certain non-hazardous liquid chemicals like hexadecanol and octadecanol, which form a mono-molecular film on the surface of water bodies, can be mentioned here. It has been reported that such chemical films were able to reduce evaporation losses by 64% in certain cases. Similarly, the design of concrete-lined irrigation canals to minimize percolation losses can also be cited here.

3.7.3 Basic Considerations for Eco-friendly Design of Domestic and Industrial Wastewater Systems

These considerations are listed as under:

1. Domestic and industrial wastewater systems need to be properly treated to avoid bad odor in the vicinity. The discharge of untreated wastewater into sources of drinking water systems such as rivers, tanks and wells should be avoided at any cost. Proper law enforcement as well as general public awareness plays a very important role in preventing such cases of pollution.

2. Wastewater systems should be designed such that they consume minimum quantity of water and they can be operated at an optimum cost. The use of flushing cisterns of smaller volumetric capacity is worth mentioning here.

3. Wastewater system designs need to promote the *three R's (reduce, recycle and reuse) of environmental conservation*. In this connection, the use of treated domestic wastewater for sewage farming as well as for many purposes other than drinking and cooking can be mentioned here. This practice is being followed in various parts around the world.

4. Wastewater systems need to be designed such that the storm water does not get into the treatment system during floods and result in an unnecessary increase in the volume, detention time as well as the treatment cost. Many instances of storm water being segregated and used for artificial recharge of groundwater are relevant to mention here.

5. Proper precaution needs to be taken in ensuring that the sewage pipelines are always below the water supply pipelines. This step can ensure that even accidental contamination of wastewater especially into the drinking water systems is avoided as far as possible. In this case also, general public awareness as well as proper law enforcement plays a vital role.

3.7.4 Basic Considerations in the Eco-friendly Design of Water Power Systems

These considerations are listed as under:

1. Water power systems need to be designed such that plant and animal species are not endangered due to submergence as well as there is minimum distress of rehabilitation to the human population inhabiting the catchment. Proper rehabilitation of all the affected persons as well as animals is essential in all the water power projects. The rehabilitation programmes proposed in the Three Gorges Project in China and the Sardar Sarovar Project in India are worth remembering here.

2. Water power systems need to be designed in such a way that proper head as well as flow rate of water, within the acceptable range of values, is ensured during high as well as low flows. Especially during high flows, excessive flow of silt and floating impurities need to be prevented. The diversion canal water power projects are precisely meant for serving these purposes.

3. Water power systems need to ensure safe as well as cost-effective power by taking adequate safety measures in all the operations as well as minimizing loss of power in the transmission system. The adoption of high voltage water power transmission lines has been found to be capable of reducing the power transmission losses significantly to about 8% in many developed countries.

4. Water power systems must ensure as many other benefits as possible such as flood control/management, municipal/industrial water supply, irrigation/aquaculture, recreation/navigation etc. All the existing multi-purpose water resources development projects, engaged in hydropower production can be cited here.

5. Water power systems need to promote poverty alleviation especially in the remote rural areas by implementing the policy of "self-construction, self-management and self-consumption". The small hydro projects being implemented throughout the world in general and in China in particular are worth mentioning here.

3.7.5 Basic Considerations for the Eco-friendly Design of other Water Systems

Other water systems include systems such as ponds/reservoirs for raising fish and other aquatic animals such as prawn, recreational facilities such as swimming pools, channels for navigation etc. Some of the basic considerations for their design are enlisted here:

1. The water systems for raising fishes need to have devices such as fish ladder for their smooth movement upstream as well as downstream. Many water resources projects have fish ladders these days.

2. The water systems for raising prawn need to ensure proper care of co-existing fish also. This can be done by properly regulating the movement of trawlers so that they do not scare/kill the fish unnecessarily.

3. Recreational water systems such as swimming pools need to be designed/maintained to always meet the required water quality standards so that they do not cause hazards to human health such as skin irritation, spreading of contagious and water borne diseases affecting eyes, ears, etc. Well-maintained swimming pools with periodic recirculation and treatment of water are worth mentioning here.

4. Navigational water systems need to be designed for safe transport of the human beings as well as other commodities. The channel banks need to be properly stabilized at all times. The canals of Venice, Italy, the floating market of Bangkok, Thailand and the inland waterways of Yangtze River in China and Mekong River in Vietnam can be cited here.

5. All these other water systems should have adequate provisions for regular maintenance to enhance their utility value as well as their working life span. The Grand Anicut built in 1st Century, A. D. in Tamil Nadu, India for irrigation and other purposes, which is still functioning on account of regular maintenance, is worth remembering here.

3.8 REFERENCES

Agarwal, A. and Narain, S. (Ed.), (1997), *Dying Wisdom, State of India's Environment-4, A Citizen's Report*, Centre for Science and Environment, New Delhi, India.

American Water Works Association (AWWA), (2000), 'Water Conservation Tips for Home', *Water Wiser*, Denver, USA.

Bertin, J., J., (1984), *Engineering Fluid Mechanics*, Prentice-Hall, Englewood Cliffs, New Jersey, USA.

Dandekar, M.M. and Sharma, K.N., (1979), *Water Power Engineering*, Vikas Publishing, New Delhi, India.

Daugherty, R.L., and Franzini, J.B., (1965), *Fluid Mechanics with Engineering Applications*, 6th Ed., McGraw-Hill, New York, USA.

Down to Earth, (1999), 'The Arvari—Coming Back to Life', Vol. 7, No. 19, Feb. 28, Centre for Science and Environment, New Delhi, India.

Gerston, Jan, *'Rainwater Harvesting: A New Water Source'*, http://twri.tamu.edu/twri.pubs/WtrSavrs/v3n2/article-1.html

Gilbrech, D.A., (1965), *Fluid Mechanics*, Wadswoth Publishing, Belmont, California, USA.

http://www.greenbusinesscentre.com/, (2003).

Jamwal, N., (2004), 'Small Farms can be Profitable', *Down to Earth*, Vol. 12, No. 23, April 30, Centre for Science and Environment, New Delhi, India.

Janna, W., S., (1983), *Introduction to Fluid Mechanics*, Brooks/Cole Engineering Division, Monterey, California, USA.

Jiandong, T., (1997), *Small Hydropower: China's Practice*, Hangzhou International Center on Small Power, Hangzhou, China.

Kumar, A., (1998), "The Value of a Raindrop", *Down to Earth*, Vol. 7, No. 10, Oct. 15, Society for Environmental Communications, New Delhi, India.

Linsley, R.K., Franzini, J.B., Freyberg, D.L. and Tchobanoglous, G., (1992), *Water Resources Engineering*, 4th Ed.., McGraw-Hill, New York, USA

Mariusson, J.M., (1997), *Study of the Importance of Harnessing the Hydropower Resources of the World*, The National Power Company, Reykjavik, Iceland.

Mobbs, M., (1998), *Sustainable House*, Choice Books, Sydney, Australia

Modi, P.N. and Seth, S.M., (1985), *Hydraulics and Fluid Mechanics*, 7th Ed., Standard Book House, Delhi, India.

Mott, R.L., (1972), *Applied Fluid Mechanics*, Charles Merrill Publishing, Ohio, USA.

Munson, B.R., Young, D.F. and Okiishi, T.H., (1994), *Fundamentals of Fluid Mechanics*, 2nd Ed., John Wiley & Sons, New York, USA.

Mutreja, K.N., (1986), *Applied Hydrology*, Tata McGraw-Hill, New Delhi, India.

Nigam, A., Gujja, B., Bandyopadhyay, J. and Talbot, R., (1997), *Freshwater for India's Children and Nature*, UNICEF and WWF Report, New Delhi, India.

Novak, P., (Ed.), (1988), *Developments in Hydraulic Engineering-5*, Elsevier Applied Science, London, England.

Rajput, R.K., (1998), *Fluid Mechanics and Hydraulic Machines*, S. Chand, New Delhi, India.

Rao, K.L., (1979), *India's Water Wealth*, Orient Longman, Hyderabad, India.

RETREAT, (2000), 'Resource Efficient TERI Retreat for Environmental Awareness and Training', http://www.teriin.org/retreat/index.htm

Rouse, H., (1946), *Elementary Mechanics of Fluids*, John Wiley, New York, USA.

Sardar Sarovar Project, (1999), *Meeting the Challenges of Development*, Sardar Sarovar Narmada Nigam, Gandhinagar, Gujarat, India.

Shames, I.H., (1962), *Mechanics of Fluids*, McGraw-Hill, New York, USA.

Sivanappan, R.K., (2000), "Strategies in Surface and Ground Water Management", *Eighth National Water Convention*, organized by National Water Development Agency, Ooty, India.

Subramanya, K., (1996), *Engineering Hydrology*, 2nd Ed., Tata Mc-Graw Hill, New Delhi, India.

Subramanya, K., (1993), *Fluid Mechanics*, Tata McGraw Hill Publishing, New Delhi, India.

Three Gorges Project, (1999), China Yangtze Three Gorges Project Development Corporation, Yichang, Hubei, China.

Todd, D.K., (1995), *Groundwater Hydrology*, 2nd Ed., John Wiley (SEA), Singapore.

Verghese, B. G., (1999), *Waters of Hope*, 2nd Ed., Oxford & IBH Publishing, New Delhi, India.

World Bank, (1998), *Pollution Prevention and Abatement Handbook*, Washington, DC, USA.

World Health Organization (WHO), (1999), *Fact Sheet No. 210, 'Arsenic in Drinking Water'*, Feb., Geneva, Switzerland.

Chapter 4

Water Hazards and their Management

4.1 HAZARDS IN GENERAL: AN OVERVIEW

Natural hazards have been occurring in different parts of our planet earth from the day of its origin. Their impacts were being felt due to global increase in size of population as a result of which more and more areas and people living therein are coming under the risk of natural hazards. The fast growing population and increasing developmental activities are causing considerable damage to the environment and ecology thereby further aggravating the rise of natural hazards and increased vulnerability to disasters.

Most of the natural disasters can be attributed to earthquakes, floods, landslides, volcanoes, droughts, avalanches and cyclones. Technological and industrial developments have opened up areas of man-made disasters such as famine and food shortages, accidents, civil strife and fire. During the year 1990 alone about 283 major disasters have been reported over the world. Whereas it is obvious that man-made disasters can be averted, natural disasters cannot be prevented. On the other hand, it can be said that with increasing degradation of environment and ecology the frequency and magnitude of natural disasters have increased.

In the present chapter we shall be discussing the hazards arising out of water-related activities and some aspects of their management. Under this emphasis will be given to pollution of water bodies, damages due to floods generated as a result of heavy precipitation as a result of cyclone depression, occurrence of landslides, blocking water courses in hills due to landslides and subsequent collapse resulting in widespread damage downstream. Similar catastrophes due to earthquakes effect on water retaining structures such as dams and their failure may result in widespread damages.

4.2   WATER AND THE NATURE OF ITS POLLUTION

Due to increase in population in developing and developed countries these regions face serious water problems. It is thus important to know the methodologies adopted in assessment of water resources and determining the quality of water after the risks posed to it from man-made pollution.

Fresh water resources of a country play a crucial role in providing basic amenities to common man. Modern methods of agriculture, development of industries and provision of safe drinking water require fresh water. Apart from the exploitation of surface water resources, groundwater is also utilized for the purposes of irrigation, industry and drinking. Groundwater is available at site and its other advantages in comparison with surface water is the common belief that it is free from pathogenic bacteria, does not contain harmful constituents and is free from suspended matter which however is not really true in actual circumstances. Thus it is necessary to study both groundwater availability and assessment of its pollution.

All naturally occurring water contains some impurities. Water is considered polluted when the presence of impurities is sufficient to limit its use for a given domestic, agricultural or industrial purpose. Pollution results in a modification of the physical, chemical and biological properties of water. Further all solids introduced into the hydrological environment due to human activities contaminate the water. It is however termed polluted when the concentration exceeds some accepted permissible limits. The primary sources of pollution are due to the presence of both solid and liquid wastes generated from domestic use, commercial use, and industrial establishments. Untreated wastewater if allowed to accumulate starts decomposition of organic matter, which will lead to production of obnoxious gases, which pollute the atmosphere. Additionally it contains numerous pathogens or disease carrying micro-organisms. It also contains nutrients, which stimulate growth of aquatic plants which may contain toxic elements.

Processing of solid and liquid wastes are not generally carried out in most countries in the world especially the underdeveloped ones, and in majority of the cases waste is disposed of to streams or in low lying areas, where it often comes into contact with surface or groundwater thus contaminating it. In the case of industrial units the effluents are mostly discharged into pits, open ground or open unlined drains near factories to move to low lying depressions resulting in groundwater pollution. Indiscriminate use of fertilizers in certain areas in quantities far in excess of optimum requirements has resulted in very high concentrations of some of the constituents in ground water. The problem of both surface and ground water pollution has become quite acute in many parts of the world and unless steps are taken for its detailed analysis and abatement there will occur extensive damage to the precious water resources.

4.2.1 Classification of Sources of Pollution

The pollution source whether natural or man-made can be classified as either point or non-point that means diffused source of pollution. Point sources enter the pollution transport routes as discrete, identifiable locations. They can be measured directly for or their quantification is possible. Its impact can be evaluated. Point sources include effluent from sewage or industrial plants and effluents from buildings or disposal sites for solid wastes.

Pollution from diffused sources can be related to weathering of minerals, erosion of virgin lands, forests, residues of natural vegetation or artificial sources, which means application of fertilizers, pesticides, herbicides for agricultural farming. Others are due to animal feeds, construction sites, mining, transportation etc. In this connection the difference between non-point source and background pollution may be distinguished the latter being the result of contact of water with rocks, undisturbed soils, geological formations, natural erosion and elutriation of chemical and biochemical components from forest liter, migration of salt water into estuaries etc.

It can thus be seen that the quality of stream water is directly affected by the amount of waste discharged into the stream and this quality in turn decides the extent up to which the water can be used for purposes like water supply and recreation. Similarly groundwater gets contaminated due to untreated city sewage being discharged into agricultural lands and also due to solid wastes being disposed of in land fills. Apart from domestic wastes the various industries discharge their effluents into pits or passed to nearby depressions. This results in toxic substances percolating down into the groundwater systems.

4.2.2 Water Quality Criteria

The principal water pollution and its associated consequences are mentioned in Table 4.1.

The analysis of water quality problems thus boils down to the following components:

  I. Inputs that is discharge of residues into the environment from natural and man-made activities.

Table 4.1 Water Pollution and its Consequences

Manifestation of the problem	Water quality problem	Water quality parameters
Bad odors, death of fishes, radical change in ecosystem	Low content of dissolved oxygen	BOD, NH_3 organic solids, DO, phytoplankton
Disease transmission, gastrointestinal disturbance, eye irritation	Bacteria level is very high	Total coliform bacteria, fecal streptococci, viruses
Tastes and odors, blue-green algae, Beach nuisances-algal unbalanced ecosystem	Excessive growth of plants (Eutrophication)	Nitrogen, Phophorous, Phytoplankton
Carcinogens in water supply, fishery abondoned, unsafe toxic levels of mortality, impairment of reproductive system	High toxic chemical levels	Metals, pesticides, herbicides, toxic chemicals, radioactive substances

II. The reactions and physical transport that is the chemical and biological transformations and movement by water result in different levels of water quality in the aquatic system.

III. The output is the resulting concentration of substance such as dissolved oxygen or nutrients at a particular location in the water body during a particular time of the day or year.

Normally the inputs are discharged into an ecological system such as river, estuary, lake or ocean and due to chemical, physical and biological phenomena the inputs result in specific concentration of the substance in the water body system. Consequently for a desirable water use by the people certain standards of water quality parameters have been agreed upon by National and International agencies. Such standards are then compared to the concentration of the substance resulting from the discharge of the residue and such a comparison may result in the need for an environmental engineering control. Environmental engineering controls are then instituted on the inputs to provide the appropriate reductions of the concentration levels so as to reach the desired level. The analysis and execution of various engineering control measures to reach the objective forms the main criterion in the decision making process of water quality management.

4.2.3 Water Quality Monitoring

The word monitoring in true sense means watching any matter closely to ensure that, no rules or standards are violated. In the case of water the monitoring refers to sampling, measurement and predictions of water quality variables at different times and locations. Precisely it means the effort to obtain quantitative information on the physical, chemical and biological characteristics of water via sampling.

The basic questions to be answered while planning monitoring of a river basin are as follows:

I What are the objectives of the programme?
II From where the samples are to be collected?
III Which determinants are of interest?
IV When and how often the samples are to be taken?
V What is to be done with the results?

Once the objectives of the program are decided the most important issue is the selection of sampling sites.

4.2.4 Selection of Water Quality Sampling Sites

I. Based on the Number of Contributing Tributaries

Sharp (1970) has indicated procedures for specifying the sampling station locations for monitoring purposes in a river basin. According to it, the basin in which the river is located is systematically subdivided into portions, which are relatively equal in terms of the contributing tributaries. Each link contributing to the main stream of a river is assigned a magnitude of one. An exterior tributary is considered to be a stream having a specified minimum discharge. A stream formed by the intersection of two exterior tributaries becomes a second order tributary. Continuing downstream in this manner a section of river formed by intersection of two upstream tributaries could have a magnitude equal to the sum of the magnitude of intersection streams. At the mouth of the system the magnitude of the final river section will be equal to the number of contributing exterior tributaries. As regards the hierarchy of sampling station the centroid of the basin is fixed by dividing the stretch by two and new centroid could be found for each portion, which is designated as first hierarchy.

Division of the entire network into quarters is defined as the second hierarchy sampling reaches. Successive subdivision defines increasing level of hierarchy. To identify the downstream sampling location for second and third hierarchy the basin divided for first hierarchy is renumbered considering the main stream as a tributary and new centroid is obtained using Sharp's approach. The same procedure is applied for identifying the third hierarchy. To isolate a source of pollution Sharp proposed that samples be drawn at one hierarchy and to be analyzed to select those portion of the network which should be sampled at stations of next hierarchy and so on until the pollutant source is found by a process of elimination.

Figure 4.1 shows the location of sampling stations by Sharp's procedure for River Kali in the State of Uttar Pradesh, India.

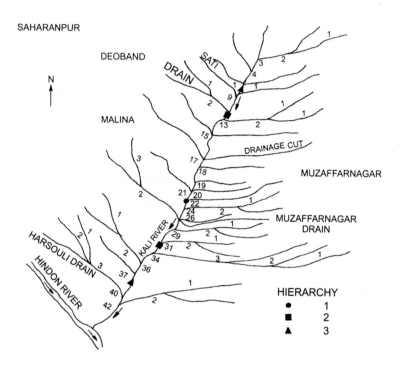

Figure 4.1 Location of Sampling Stations by Sharp's Procedure Based on Number of Tributaries*
(*Courtesy: National Institute of Hydrology, Roorkee, India)

II. Based on Number of Outfalls and BOD Loading

To select the sampling sites based on number of outfalls, all information regarding outfalls discharged into the river and its tributaries need to be surveyed thoroughly. Outfalls for the river to be identified and a cumulative numbering of outfalls has to be indicated. The ordering of the location is based on outfall magnitude. To select sampling sites based on BOD loading, all outfalls and their flow in million gallons/day (MGD) and average 5-day BOD in mg/l need to be ascertained. The sampling locations are then found out following the same procedure as outlined above and are shown in Figure 4.2, for the same river.

The other aspect of considering the selection of sampling sites is the number of variables to be sampled. It has been observed that when several

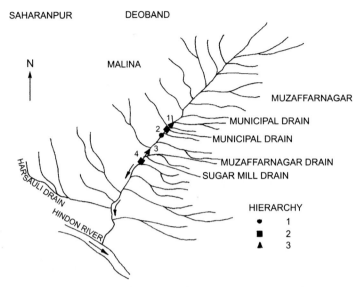

Figure 4.2 Location of Sampling Stations by Sharp's Procedure
Based on Number of Outfalls*

(*Courtesy: National Institute of Hydrology, Roorkee, Garde et al (1983))

are under consideration, the outfall based network or network based on discharge from each outfall seem to be good. The question may arise that the number of outfalls or the quality of discharge from outfalls may change but the number of tributaries contributing to a river system would remain constant over time. The number of tributaries that would remain constant does not mean that there is necessarily a relationship between tributaries and the quantity, quality and location of discharges in the river network. At the same time an outfall based monitoring network designed based on today's information may no longer be correct in a few years. This should not be considered as critical. The aspects should be to address that the system is intended to detect trends in water quality and that trends detected are to be viewed as indicative of the corresponding link.

Once the macro-location of the sampling stations are specified according to criteria as outlined above further exploration of the location is required to ensure representation of water quality variables for a river section. Whether the water quality is truly represented in a sample depends on the uniform distribution of concentration in a river cross-sectional area. A completely mixed region in a river section is a desirable location of sampling and is termed micro-location. The difficulty however lies in the fact that the zone of complete mixing is not well defined. According to Saunders et al (1983) the complete mixing zone can be assumed to be near the point of discharge or to be estimated based on experience. Kittrell (1969) recommends that sampling be done at the quarter and midpoints to give representative values. Another study by Reeder indicates that complete mixing of effluent has taken place about 16 km below the outfall and mixing is incomplete even 5 km below the outfall. Due to the difficulty in the estimation of zones of complete mixing it is suggested that the mixing distance be determined following trial and error procedure. For this multiple samples in the lateral transect be taken at different points downstream from an outfall until the variability of the samples from the river cross section becomes insignificant.

4.2.5   Analysis of Water Quality Data

The various water quality parameters from a river or stream are found out by performing appropriate tests in the laboratory. The parameters are respectively:

pH, conductivity, total dissolved solids, Fe^{+2}, Al^{+3}, NH^+_4, NO^-_3, PO^{-3}_4, oxygen demand in 4-hours, turbidity, Na^+, K^+, Ca^{+2}, Mg^{+2}, CO^{-2}_3, HCO^-_3, Cl^{-2}_2, SO^-_4 and hardness etc., along with those the quantity of river flow are also measured. These water quality data are further analyzed to obtain functional relationships.

I. Quality-Quantity Relationship

To obtain such a relationship a plot of stream discharge against various water quality constituents concentration has to be made in a log-log paper and the best fit curve through the plotted data can be expressed as in power law relationship of the form

$$C = K Q^n \tag{4.1}$$

where Q is the stream discharge, C is the water quality parameter constituent concentration and K and n are coefficient and index respectively, to be determined by regression analysis following method of least squares. The quality quantity relationship as described by the above of equation can be developed for various water quality determinants.

II. Regression-based Modeling

Quite often a dependent variable is dependent on several other quantities. A model for predicting the dependent variable can be developed based on multiple linear regression concepts. A general linear model of the form:

$Y = \beta_1 X_1 + \beta_2 X_2 + \cdots + \beta_p X_p \cdots$ is discussed , where Y is dependent variable, X_1, X_2, X_3 are independent variables and $\beta_1, \beta_2, \cdots, \beta_p$ are unknown coefficients. This model is linear in parameters.

In practice, n observations will be available on Y with the corresponding n observations on each of the p independent variables. Thus set of n equations like Equation 4.2 can be written one for each observation. Essentially we will be solving n equations for p unknown parameters. Thus n must be equal to or greater than p. The n equations in matrix form can be written as

$$
\left.
\begin{aligned}
Y_1 &= \beta_1 X_{1,1} + \beta_2 X_{1,2} + \cdots + \beta_p X_{1,p} \\
Y_2 &= \beta_1 X_{2,1} + \beta_2 X_{2,2} + \cdots + \beta_p X_{2,p} \\
&\ldots \ldots \\
&\ldots \ldots \\
Y_n &= \beta_1 X_{n,1} + \beta_2 X_{n,2} + \cdots + \beta_p X_{n,p}
\end{aligned}
\right\}
\tag{4.2}
$$

The unknown parameters can be estimated following method of least squares and the resultant normal equations can be written as

$$
[X^T][Y] = [X^T][X][\beta] \tag{4.3}
$$

where $[X^T]$ denotes the transpose of matrix $[X]$, $[Y]$ and $[\beta]$ are the column matrices consisting of Y and β terms respectively.

The solution of Equation 4.3 is obtained by pre multiplying by $[X^T X]^{-1}$, which results in the following equation:

$$
[X^T X]^{-1}[X^T][Y] = [X]^{-1}[X^T]^{-1}[X^T][X][\beta] \tag{4.4}
$$

Therefore the parameters can be estimated by

$$
[\beta] = [X^T X]^{-1}[X^T][Y] \tag{4.5}
$$

Once the parameters are obtained the various models of the water quality constituents such as hardness, Cl^-, HCO_3^-, K^+, Na^+, NO_3^-, TDS, conductivity, turbidity etc. can be developed. The development of the multiple regression models is an important step for stream quality forecasting purposes.

4.2.6 Environmental Impact Monitoring

In order to effectively abate pollution one should be able to assess the existing degree of pollution. For this purpose monitoring programmes and methodologies are necessary to sample biotic and abiotic compartments in a standardized way. Besides chemical monitoring, biomonitoring is also required. It requires less sophisticated instrumentation and reflects the integrated monitoring of pollution. Under biomonitoring includes characterization of the quality of the ecosystems, bioaccumulation monitoring and early warning monitoring.

Environmental studies, assessment and the control of impacts are based on the knowledge of the physical processes, physical quantities and parameters, which effect the ecosystems. These are climatic and hydrological data, fluvial and coastal regimes, morphological processes, erosion and transport of soils, groundwater regimes, salinity etc. Monitoring of physical parameters can be done by satellite imagery, atmospheric soundings, optimized network of monitoring stations, data base organization and coping up with insufficient data.

Some of the causes of environmental degradation are due to open cast mining, quarrying, dumping of wastes, tourism, wastewater disposal and pollution of aquifers. The goals and methods to be applied have to be decided upon on the basis of evaluation of various management options and identification of priorities. Restoration goals can range from erosion and pollution control to restoration of amenity values, productivity or ecological values. The measures can include passive restoration and natural regeneration, surface stabilization, nutrient addition, pH adjustment, amelioration of toxicity, species addition and facilitation, selection of tolerant species, re-introduction of species, bio-topic manipulation.

It has been recognized that biological management of eutrophic lakes helps restoration of a degraded ecosystem. The increase in the turbidity by algal blooms, reinforced by bioincubation results in a permanent change to a different and less diverse ecosystem. Reduction of nutrient load is insufficient for recovery of the original system. Intervention at a suitable point in the modified system is required to initiate the return to the original situation. Various options include, the management of the fish population and the macrophyte vegetation. It is thus evident the importance of vegetation as a means for shoreline protection and other applications of habitat protection, restoration or recreation.

4.2.7 Water Quality Management Models

Meaningful water quality management requires an analytical framework that will provide a basis for decision-making. A mathematical model of the water system gives the basic relationship between waste load input and the resulting output. It is therefore necessary that the principles of water quality modeling be understood. Figure 4.3 shows the principal components of a

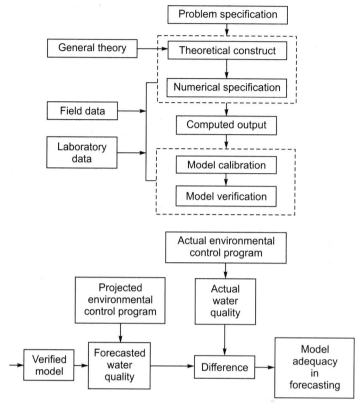

Figure 4.3 Principal Components of a Modeling Framework

mathematical modeling framework. In this context the following definitions may be noted:

Model means a mathematical construct i.e., the differential equation together with assignment of numerical values to model parameters incorporating some prior observations drawn from field and laboratory data. Model calibration means the first stage testing or tuning of a model set of field data preferably a set of field data not used in the original model construction. Model verification means subsequent testing of a calibrated model to additional field data preferably under direct external conditions. The verified model is often used for forecasts of expected water quality under a variety of potential scenarios.

Water Quality Management models incorporate not only the physical cause/effect mechanism but also waste control considerations and economic influences of a given programme. For simple system management model is trivial and for complex multi-waste problems a rational control programme is not always obvious. Effective management requires a dynamic control programme where a sequence of decisions must be made under uncertainty. Steady state management problems are concerned with the strategy and long term planning of water quality improvement programmes.

In a river with one or two waste discharges one can directly estimate the maximum amount of waste that may be discharged so as to attain a specified water quality objective. Figure 4.4 shows the representation of the overall

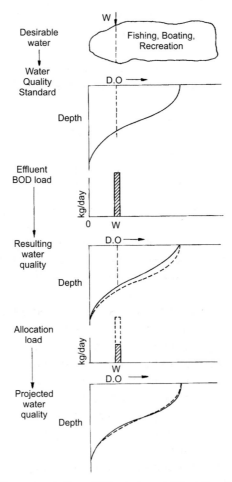

Figure 4.4   Representation of Waste Load Allocation for Dissolved Oxygen

waste load allocation problem when DO is the important parameter for water quality. With a steady state model and establishment of maximum DO deficit from the required DO then the waste load in kg/day of BOD can be calculated by trial and error.

For multiple waste discharges there are numerous combinations of waste reductions at each of the sources that will meet a DO or other water quality goal. It is therefore necessary to specify an additional criterion of choice to uniquely determine the waste removal requirements for multiple sources. In terms of a specific water quality variable such as DO the question to be answered is the degree of waste removal to be made at various waste resources so as to attain a specific level of DO in the water body for the minimum amount of regional expenditure. For this problem only the reductions at the waste source are specifically included.

A major portion of the complexity associated with water quality modeling and prediction is the inherent randomness exhibited through the stream environment. It is not only due to the fact that physical and biological processes are not clearly defined but also due to the imposition of a number of uncertainties associated with the various processes occurring within the stream environment Loucks and Lynn (1966) investigated the effect of inherent uncertainty due to the natural variations in stream flow and waste flow on the probabilty distribution of dissolved oxygen. Padgett and Rao (1979) presented a joint probability distribution of BOD and DO. Kothandaraman and Ewing (1969) and Chadderton et al (1982) have investigated the stochastic nature of the model parameters in assessing the probability distribution of DO deficit. In achieving effective environmental control the procedure of Waste Load Allocation should consider the natural inherent randomness of water quality parameters. The allocation process involves the estimation of stream assimilative capacity, characterization of point and diffuse source inputs, reserve capacity allocation, and a subsequent assignment of available capacity to designated discharges. Determination of the total maximum daily load and a distribution of assimilative capacity in an equitable manner is also required. In view of above the DO deficit must be considered as a random variable. In probabilistic water quality analysis it is typical to deal with the problems of assessing the probability of water quality violation. Uncertainty analysis of stream DO analysis have been concerned with the variability of DO concentrations due to model parameter uncertainty vide Kothandaraman and Ewing (1969), Homberger (1980), Chadderton et al (1982). It was reported in 1967 that, Thayer and Kurtchkoff utilized a stochastic birth and death process to obtain an expression for the probability distribution of DO concentration without considering the uncertainties of model parameters. In 1967, Esen and Rathbun assumed reaeration and deoxygenation rates to be normally distributed and investigated the probability distribution for DO and BOD using a random walk approach. From practical point of view the above-mentioned methods are difficult for most engineers to apply, as they can be derived by using very simple distributions for the model parameters such as uniform and normal. An approximate approach to probabilistic water quality analysis is based on the first order analysis of statistical moments estimate. The statistical moments are then incorporated with an appropriate probability distribution model for the DO deficit. For characterization of the distribution various statistical parameters have to be known and for this the mean and variance of the DO deficit have to be estimated using the first order uncertainty analysis.

To present the general methodology of first order analysis, consider a random variable Y, which is a function of N random variables. So, mathematically Y can be expressed as

$$Y = g(X) \tag{4.6}$$

where,

$X = (X_1, X_2, ..., X_N)$ is a vector containing N random variables X_i. By Taylor's series expansion, the random variable Y can be approximated by:

$$Y^2 = g(\overline{X}) + \sum_{i=1}^{N} \left[\frac{\partial g}{\partial X_i} \right] (X_i - \overline{X}_i) +$$

$$\frac{1}{2} \sum_{i=1}^{N} \sum_{j=1}^{N} \left[\frac{\partial^2 g}{\partial X_i \partial X_j} \right] (X_i - \overline{X}_i)(X_j - \overline{X}_j) \tag{4.7}$$

in which $\overline{X} = (\overline{X}_1, \overline{X}_2, ... \overline{X}_N)$ is a vector containing the mean values of N random variables with second order approximation. The first order approximation of the variable of Y is,

$$\sigma_y^2 = var[Y] = \sum_{i=1}^{N} \sum_{j=1}^{N} \left[\frac{\partial g}{\partial X_i} \right] \left[\frac{\partial g}{\partial X_j} \right] cov[X_i, X_j] \tag{4.8}$$

If the X_i and X_j are uncorrelated, the equation (4.8) reduces to

$$\sigma_y^2 = \sum_{i=1}^{N} \left[\frac{\partial g}{\partial X_i} \right]^2 \sigma_i^2 \tag{4.9}$$

where $\overline{X}$ = means in a first order sense and σ_i^2 is the variance corresponding to random variable X_i.

4.2.8 Basic Water Quality Model

Streeter and Phelps (1925) recognized the capacity of a water resource to receive and assimilate organic waste material depended on the oxygen economy. The first order reactions for deoxygenation and reaeration were combined to give the rate of change of oxygen deficit. The relationship among the parameters affecting the instream dissolved oxygen concentration is given by the following equation

$$D = \frac{K_d L_0}{K_a - K_d} (e^{-k_d t} - e^{-k_a t}) + D_0 e^{-k_a t} \tag{4.10}$$

in which D = dissolved oxygen deficit = (C_s – C) in milligrams per liter; C_s = dissolved oxygen saturation limit, in milligrams per liter; K_a = re-aeration rate co-efficient (per day); K_d = de-oxygenation rate co-efficient (per day), L_0 = initial instream total ultimate biochemical oxygen demand, in mg/l, D_0 = initial instream dissolved oxygen deficit in mg/l, and t is the time of travel from dissolved oxygen deficit of D_0 to D in days.

The first order uncertainty analysis will provide a measure of the uncertainty of the dependent variable, D in terms of the uncertainty in independent variables: K_a, K_d, L_0, D_0 and t; i.e., percentage of the scatter of DO deficit predictions around the true deficit at any point along sag curve can be assigned to each of the independent variable. Taking the partial derivative of D with respect to each of the independent variables.

$$\frac{\partial D}{\partial D_0} = e^{-k_a t} \tag{4.11}$$

$$\frac{\partial D}{\partial L_0} = \frac{K_d}{K_a - K_d} (e^{-k_d t} - e^{-k_a t}) \tag{4.12}$$

$$\frac{\partial D}{\partial K_a} = -\frac{K_d L_0}{(K_a - K_d)^2} (e^{-k_d t} - e^{-k_a t}) + \frac{K_d L_0 t e^{-K_a t}}{(K_a - K_d)^2} - D_0 t e^{-K_a t} \tag{4.13}$$

$$\frac{\partial D}{\partial K_d} = \frac{K_d L_0}{(K_a - K_d)^2} (e^{-k_d t} - e^{-k_a t}) - \frac{K_d L_0 t e^{-K_a t}}{(K_a - K_d)^2} - D_0 t e^{-K_a t} \tag{4.14}$$

$$\frac{\partial D}{\partial t} = \frac{K_d L_0}{K_a - K_d} (K_a e^{-k_d t} - K_d e^{-k_a t}) - K_a D_0 e^{-K_a t} \tag{4.15}$$

The first-order approximation to the total uncertainty in the DO deficit is obtained by applying Equation 4.9, the resulting equation is given below:

$$\sigma_D = \sum_{i=1}^{5} \left(|C_i|^2 \right)^{\frac{1}{2}} \tag{4.16}$$

where

$$C_1 = \frac{\partial D}{\partial D_0} \sigma_{D_0}, \; C_2 = \frac{\partial D}{\partial L_0} \sigma_{L_0}, \; C_3 = \frac{\partial D}{\partial K_a} \sigma_{K_a},$$

$$C_4 = \frac{\partial D}{\partial K_d} \sigma_{K_d} \quad \text{and} \quad C_5 = \frac{\partial D}{\partial t} \sigma_t \tag{4.17}$$

For Equations 4.16 and 4.17, the symbol σ represents the standard deviation of particular variable. Thus Equation 4.16 shows that each of the independent variables contributes to the dispersion of D in a manner proportional to its own variance, σ^2, and proportional to a factor $[(\partial D / \partial)]^2_{mean}$ which is related to the sensitivity of changes in D to changes in the independent variable vide Benjamin and Cornell (1970). Application of the method of first order uncertainty analysis to the BOD-DO system requires estimates of mean parameter values and standard deviations of D_0, L_0, K_d, K_a and t.

The lowest or critical point of DO curve is important as it gives the greatest deficit in dissolved oxygen. The critical time (t_c) can be obtained by differentiating the sag curve equation (Equation 4.10) with respect to time and placing the resulting expression to zero. The duration of minimum oxygen content time (t_c) is thus obtained as:

$$t_c = \frac{1}{K_d(f-1)} \ln \left[f \left(1 + \frac{D_0}{L_0} \right) - f^2 \frac{D_0}{L_0} \right] \tag{4.18}$$

where f = self-purification ratio = K_a / K_d. It is clear from Equation 4.18 that t_c depends on four independent variables namely, K_d, K_a, L_0 and D_0. Taking the partial derivative of t_c with respect to each of the independent variables,

$$\frac{\partial t_c}{\partial K_d} = \frac{1}{K_d^2(f-1)^2} \ln \left[f \left(1 + \frac{D_0}{L_0} \right) - f^2 \frac{D_0}{L_0} \right]$$

$$+ \frac{1}{k_d^2(f-1) \left[1 + \frac{D_0}{L_0}(1-f) \right]} \left(2f \frac{D_0}{L_0} - \frac{D_0}{L_0} - 1 \right) \tag{4.19}$$

$$\frac{\partial t_c}{\partial D_0} = - \frac{1}{K_d L_0 \left[1 - (f-1) \frac{D_0}{L_0} \right]} \tag{4.20}$$

$$\frac{\partial t_c}{\partial K_a} = \frac{1}{K_d^2 (f-1)^2} \ln\left[f\left(1+\frac{D_0}{L_0}\right) - f^2\frac{D_0}{L_0}\right]$$

$$+ \frac{1}{k_d K_a (f-1)\left[1+\frac{D_0}{L_0}(1-f)\right]}\left(1+\frac{D_0}{L_0} - 2f\frac{D_0}{L_0}\right) \qquad (4.21)$$

$$\frac{\partial t_c}{\partial L_0} = -\frac{1}{K_d L_0^2\left[1-(f-1)\frac{D_0}{L_0}\right]} \qquad (4.22)$$

The mean value of t_c is obtained by substituting the mean values of K_d, K_a, L_0 and D_0 in Equation 4.18. The equation for variance in t_c is obtained by substituting various sensitivity coefficients in Equation 4.9, as given below:

$$\sigma_{t_c}^2 = \left(\frac{\partial t_c}{\partial K_d}\right)^2 \sigma_{K_d}^2 + \left(\frac{\partial t_c}{\partial K_a}\right)^2 \sigma_{K_a}^2 + \left(\frac{\partial t_c}{\partial L_0}\right)^2 \sigma_{L_0}^2 + \left(\frac{\partial t_c}{\partial D_0}\right)^2 \sigma_{D_a}^2 \qquad (4.23)$$

The maximum DO deficit (D_c) is obtained by substituting $t = t_c$ in Equation 4.10. From Equation 4.18, the following expression is obtained

$$D_c = \frac{K_d L_0}{K_a} e^{-K_d t_c} \qquad (4.24)$$

It is seen from Equation 4.24 that, D_c is a function of four independent variables namely K_d, K_a, L_0 and t_c. Taking the partial derivative of D_c with respect to each of the independent variables to get the respective sensitive coefficient corresponding to each of the independent variable,

$$\frac{\partial D_c}{\partial K_d} = \frac{L_0(1-K_d t_c)}{K_a} e^{-K_d t_c} \qquad (4.24a)$$

$$\frac{\partial D_c}{\partial K_a} = -\frac{L_d L_0}{K_a^2} e^{-K_d t_c} \qquad (4.24b)$$

$$\frac{\partial D_c}{\partial L_0} = \frac{K_d}{K_a} e^{-K_d t_c} \qquad (4.24c)$$

$$\frac{\partial D_c}{\partial t_c} = -\frac{K_d^2 L_0}{K_a} e^{-K_d t_c} \qquad (4.24d)$$

The variance in D is calculated using the following equation,

$$\sigma_{Dc}^2 = \left(\frac{\partial t_c}{\partial K_d}\right)^2 \sigma_{K_d}^2 + \left(\frac{\partial t_c}{\partial K_a}\right)^2 \sigma_{K_a}^2 + \left(\frac{\partial t_c}{\partial L_0}\right)^2 \sigma_{L_0}^2 + \left(\frac{\partial t_c}{\partial D_0}\right)^2 \sigma_{D_0}^2 \qquad (4.25)$$

Data selection The independent variables of the equation are subject to wide variations from stream to stream or even from reach to reach of the same stream. Global mean values and standard deviations would be difficult to estimate. Fair (1968) defined four classes of streams by the self-purification ratio, $f = K_a/K_d$. This is furnished in Table 4.2.

Table 4.2 Ranges of Water Quality Data by Stream Class

Stream class description	Self purification ratio (f) (dimensionless)	Reaeration rate (K_a) (per day)	Deoxygenation rate (K_d) (per day)	Velocity (V) (m/sec)
Sluggish	1.25 – 1.50	0.05 – 0.10	0.033 – 0.08	0.03 – 0.15
Low-velocity	1.50 – 2.00	0.10 – 1.00	0.050 – 0.67	0.03 – 0.15
Moderate-velocity	2.0 – 3.0	1.00 – 5.00	0.500 – 2.50	0.15 – 0.61
Swift	3.00 – 5.00	1.00 – 10.00	0.20 – 3.33	0.61 – 1.83

An example of uncertainty analysis is furnished herein with the data (Table 4.3) obtained from Burges and Lettenmaier, in 1975.

Table 4.3 Parameters Used in Uncertainty Analysis of Water Quality

Parameters	Mean	Standard deviation	Coefficient of variation
Initial BOD, L_0 (ppm)	12.15	1.00	0.08
Initial DO deficit, D_0 (ppm)	1.00	0.50	0.50
Deoxygen^n. Coeff. K_d (per day)	0.331	0.10	0.32
Reaeration Coeff. K_a (per day)	0.690	0.20	0.29
Travel Time, t (days)	—	—	0.25

The mean DO profile along with their standard deviation obtained by first order analysis is shown in Figure 4.5. The standard deviation increases to a maximum and diminishes in magnitude with distance along the stream. The point of maximum uncertainty is somewhat downstream of the minimum DO level. The mean and standard deviation for minimum DO level and its location on the river determined by first order analysis assuming normal distribution as shown in Table 4.4.

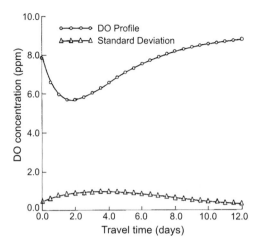

Figure 4.5   DOP Rofile using a First Order Analysis

Table 4.4   Critical DO in the River Reach under Normal Distribution

Dissolved mean	Oxygen (ppm) standard deviation	Travel Time (days) mean	std. deviation
5.770	1.062	1.785	0.138

The probability distribution functions for minimum DO level and its location on the river by first order analysis is shown in Figure 4.6 along with their cumulative distribution functions. It may be mentioned here that the modification of river flow resulting from the construction and operation of a dam or impounding structure has been identified as a significant factor causing water quality and aquatic habitat problems. Accordingly low flow criteria need to be developed for determining the suitability of various flow regimens for fish and wildlife. In order to choose a minimum low flow release which keeps the fishery in good condition and at the same time does not saddle the developer with extra cost the decision maker needs to know the estimated increase in cost of a reservoir to provide minimum flow over that with no such flow for a range of low flows. The extra cost of impoundment may not be considered by the developer as a gift to the fishery and water quality interests, rather it may be considered as a fee that he/she pays for the use of water resources presently enjoyed by the downstream interests and for altering the streamflow regimen to meet his/her particular needs. The Illinois Environmental Protection Agency in USA has provided information on fish suitability or preference as a function of flow velocity and depth for juvenile and adult fish of several species. In a complete cycle of

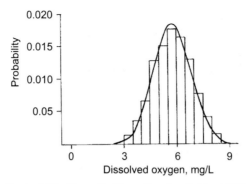

Figure 4.6a Probability Density Function for Critical DO Assuming Normal Distribution with its Parameters Determined by 1st Order Uncertainty Analysis.

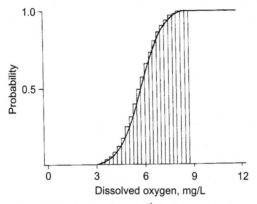

Figure 4.6b Cumulative PDF Using 1st Order Analysis and Normal Distribution

pool riffle sequence the riffle is defined as the portion that has an energy gradient steeper that the average energy gradient of the complete cycle, whereas the pool is the portion that has an energy gradient milder than the cycle average. The riffles act as submerged dam to slow down the release of water from the pools behind them. The varying velocities and depths in riffles and pools provide a range of sub-areas or cells of water more suitable to one fish than another depending on their relative preferences. This variety helps in maintaining different life stages of various fishes and provides a semblance of continuum for their development even with more frequent flow variation. U.S. Fish and Wildlife Service have developed a methodology in this regard.

Incremental Methodology, vide Boven and Milhouse (1978), to describe the effects of incremental changes in stream flow on the instream fishery potential. It allows calculations of weighted usable area (WUA) as an index of habitat suitability. The WUA in a river reach divided into n cells is defined as:

$$WUA = \sum_{i=1}^{n} S(d_i)S(v_i)\ldots A_i \qquad (4.26)$$

in which $S(d_i)$, $S(v_i)$, are suitability indices for depth, velocity, A_i is the surface area of the ith cell which is relatively homogenous with respect to d_i, v_i, the depth, velocity in the cell i. This procedure approximates the total water surface area in a simulated reach to an equivalent area of preferred habitat for fish under consideration. The concept of multiplying the suitability indices or preferences is rather open to question. The preference curves for both velocity and depth are derived with both velocity and depth as independent variable whereas the hydraulic geometry relations indicate a definite relationship between velocity and depth in terms of drainage area and percent flow duration.

4.2.9 Cost Function for Wastewater Treatment

The cost of waste treatment is considered to include capital, operation and maintenance suitably discounted using appropriate interest rates and investment formulae. It is convenient to approximate the non-linear cost function by straight-line segments. Let the marginal costs corresponding to each straight-line segment approximation be V_k (\$/kg/day), where k is an index of the cost curve segment. The quantity f_j represents the total load from source 'j'. Each f_j is now considered to have several parts.

$$f_j = \Sigma f_k \qquad (4.27)$$

which says that f_j equals sum of all f_k parts of the load contained in each segment 'k'. Each segment of cost function can be treated as an individual waste source. The product of V_k and f_k is the cost to remove f_k, kg/day. The total cost (Z) over all k segments is then

$$Z = \Sigma V_k f_k \qquad (4.28)$$

To minimize the cost for wastewater treatment a number of technologies exist or in the process of development. The conventional systems such as trickling filters and activated sludge plants were developed to address the concerns of organic pollution of rivers in temperate climates, rather than the removal of pathogenic micro-organisms which is often a priority. WHO has highlighted the need for research into sewage treatment technology, which can supply effluent of irrigation quality without the limitations of conventional systems. The non-conventional system often utilizes the ecological treatment mechanisms such as aquatic or wetlands system and do not have the mechanical parts or energy requirements of conventional systems. Waste stabilization ponds are one such solution but sometimes constraints occur due to land availabilty and topography. In such situations wetlands could provide an alternative. Gravel bed hydroponic (GBH) reed bed systems consist of channels sealed with geomembrane, concrete or other impermeable material. These channels are filled with gravel and wastewater is percolated horizontally below the surface of the gravel. This subsurface flow reduces the potential for breeding sites of insect vectors and intermediate hosts of excreta related parasitic diseases especially mosquitoes and aquatic snails. Reeds predominantly *Phragmites australis* are planted in the gravel and grown hydrophonically using nutrients in the sewage. The reeds maintain the hydraulic pathways and their rhizospheres support intense microbial activity which ensures sewage treatment. GBH systems are made of a number of these long parallel channels each about 2 m wide, 4 m deep and upto 100 m long. Such beds were built in Egypt and the study shows that the systems removed large quantities of organic matter, ammoniacal nitrogen and both pathogenic microorganisms from domestic sewage. It has also been applied to treat industrial wastes and it has been found that they were able to remove long chain hydrocarbons and fatty acids.

4.2.10 Groundwater Pollution

The major sources of groundwater pollution are due to domestic and industrial wastes. The domestic wastes are sewage and solid wastes. In many cities the sewage is discharged into agricultural lands without adequate treatment thereby contaminating the groundwater. Similarly solid wastes are disposed of in land fills or open dump which if not properly selected from hydrological point of view will result in groundwater pollution due to leaching action as a result of rainfall. Industrial pollution results from effluents of various industries. The chemical and physical characteristics of these chemicals are different. The groundwater quality susceptible to degradation is related to the characteristics of these effluents. Generally, industries discharge their effluents into pits or passed through unlined channels to nearby depressions. The toxic elements therefore get percolated into the groundwater system. The various industries contributing to this adverse effect on ground water quality are respectively, electroplating, tanneries, dyeing and textiles, iron and steel, machine and machine tools, chemicals, fertilizers and insecticides, paints and pigments, paper and pulp, distilleries and breweries, etc. Some of the important parameters from these industries having groundwater pollution potential are: Phenols, cyanide, chromium, nickel, cadmium, copper, iron, mercury, lead, arsenic, titanium, tin, nitrate, phosphorous, fluoride, nitrogen etc.

Landfills are the main groundwater pollution source because of their potential for leaching a variety of hazardous substances. Its potential is more in areas with high rainfall and shallow groundwater.

Apart from above groundwater pollution results from agricultural practices due to application of fertilizers, irrigation salts, and animal wastes and crop waste disposal. High potassium and nitrate in groundwater is due to excessive use of fertilizers, similarly excessive irrigation water as a result of leaching of salts from soils cause increase of several constituents in ground water. Other sources are seawater intrusion due to excessive withdrawls in coastal aquifers and recharge of water contaminated by air pollution also effect groundwater quality. There could be radioactive pollution of groundwater as a result of operations like mining and milling of radioactive ores, chemical reprocessing and radioactive waste disposal. The mass of a contaminant source is the product of solubility or density in the case of infinite solubility and the volume of the contaminant.

4.2.11 Groundwater Quality Monitoring

Water level records are essential to reach conclusion with regard to occurrence and development of groundwater as well as appraising groundwater situation. Groundwater pollution studies generally deal with the problems of identifying pollution sources and contaminant of pollutant spreading from known sources. The other type seeks to protect groundwater from pollution due to overuse. The wide variability of geological materials and their inherent inhomogeneties makes it difficult to choose appropriate network density and representative samples. The quality of groundwater and its temperature may undergo change due to the movement of groundwater induced by operation of well fields. In the vicinity of waste disposal sites including wells, the leachate is the focus of monitoring. The tracking of injected wastes or the tracking of the frontal position of the mixed zone in areas where salt and freshwater meet is of great interest. The water samples from the monitoring grid network well is subjected to complete chemical analysis including trace and minor elements. The data is the source of background information for managerial decisions. It provides an early warning of contaminants through aquifers and assessing the effectiveness of remedial measures as and when adopted. The measures include creation of hydraulic barrier, installation of an impermeable barrier injection of chemical or biochemical reagents to induce chemical or microbial reactions into contaminated aquifer to attenuate the contaminants. The measures are costly and therefore may be tried when alternate sources are not available.

When faced with the groundwater pollution problem a number of questions are to be answered; this is the nature of the pollution, strength of the pollution source, invaded volume of soil, duration of pollution etc. Answer to these questions is possible when chemical quality is studied along with relevant hydrological and hydro-geological parameters. The parameters along with aquifer geometry can help substantially in predicting the movement of dissolved salts in various parts of an aquifer exhibiting different hydraulic parameters and under various pumpage. The geophysical methods also become significant as these methods can delineate sub-surface distribution of resistivity, which reflects on the pollution pattern. Such hydrogeology-based studies for combating pollution caused by liquid radioactive wastes have been carried out in the Snake River Plain aquifer of Idaho in USA. The liquid wastes comprising chlorides, chromium, tritium, strontium and cesium were being introduced in the aquifer of fractured basalt and interbedded sediment. Aquifer geometry and hydraulic permeability helped in determining the dispersion pattern of the contaminating solutes and suitable methods were evolved to arrest the polluting menace.

Agricultural crop fields were recognized as an important source of Nitrogen pollution affecting the Venice Lagoon ecosystem. Till the sixties the agricultural management of this area was characterized by a large diffusion of hedgerow system and vegetated strips. Most of them were pulled out in the last decades to create a modern agricultural cropping system. Venetian Municipality is developing several actions to control lagoon pollution including the planting of windbreaks and hedgerows. These are reintroduced by means of particular landscape planning model accounting for the need of local agriculture. To assess the non-point pollution functionality in the planning system a field test was established. The test was part of a more complex experimentation on non-point pollution in the area of Venice lagoon including control of manure utilization. Grain corn was grown on a calcareous clay-loam soil. One line shelter belt was planted in the field with trees and shrubs. The hedgerow was selected keeping in view of their high capacity to control nutrients leaching. Monitoring of nutrients was done by piezometric wells and traps for run-off. Everyday's sampling was made from wells and after every runoff events from the trap. Fertilization was made with mineral fertilizers compared with liquid animal manure to supply about 300 kg/ha of nitrogen, 120 kg/ha of phosphorous, and 120 kg/ha of potassium. Typical results after two years indicated substantial reduction of pollutants in groundwater with hedgerow.

CASE STUDIES

Matrix Solutions Inc: Soil and groundwater pollution, Calgary, Alberta, Canada has reported solutions of a number of groundwater pollution problems encountered in various parts of Canada. Some of those problems and brief report of the solution suggested are reported herein:

Problem: Dissolved hydrocarbon constituents discovered in household wells in a rural subdivision

Matrix conducted a soil gas survey and oversaw installation of monitoring wells throughout the community and adjacent areas. For monitoring, residents' wells were sampled to delineate the source and extent of the hydrocarbons, which were shown to originate at a nearby gas processing plant. Pilot testing and groundwater modeling were conducted to design a 16-well interception system, which was installed to recover contaminated groundwater. Operation of the system began in 1993. Soil vapor extraction and air spraying systems were installed at the plant in 1995. There has been a decrease of concentrations at the site since remediation began.

Problem: Dissolved hydrocarbons were discovered in piezometers close to the water supply wells for a small town.

As a part of a multiple liability assessment Matrix delineated a hydrocarbon plume near an oil battery that extended to within 20 m of the town well. Groundwater modeling was performed to optimally locate 3 wells, which were situated to provide downgradient interception of the hydrocarbon plume. An air spraying system comprising 6 wells was installed inside the interception zone and has been operating efficiently since 1995.

Problem: Extensive groundwater seepage through soils around a gas well and former drilling sump.

The company had installed a culvert to provide pressure relief, however the culvert flow steadily at 160 cubic meter per day. Matrix performed a resitivity survey and delineated a large buried channel aquifer. Denaturing wells were installed into the channel and pumped to lower the water table, while the Company re-excavated the sump and installed a liner system. The dewatering wells pumped more than 1600 cubic meters per day. After cessation of pumping no additional seepage was noted.

Problem: Identifying alternatives to linear installation to an active brine storage facility near Edmonton, Canada.

Matrix evaluated existing site data and performed solute transport modeling to prevent the import of brine infiltration from the ponds on ground water. Several liner alternatives were proposed on the basis of findings.

4.2.12 Application of Geographic Information System (GIS) Technology in Disaster Management

In the modeling of water resources and environmental engineering projects the data involve geographical components. In addition to the concepts of what and how much a set of data represents the location of the data is also very much important. Information with the geographical components is referenced as spatial data and computer system that has been developed for managing the spatial data are called GIS or geographical information system. It is a computer based system designed to accept large volume of spatial data derived from various sources and efficiently store, retrieve, manipulate, analyze in multivariate fashion and display the data in hard copy or soft copy as maps, statistics or tests. Data of different scales can be linked to each other and that provides a base for generating spatially full covering information for case studies over regions. GIS facilitates analysis of multilayer, multidimensional and multitemporal data and allow that data to be referred to a geodetic grid with applications in the real world of management and its related subjects.

Geographical data consist of spatial data describing real objects with respect to position in co-ordinate system. Spatial data can be reduced to a point, or a line, or as an area with appropriate legends constituting the map. The geographical features are represented internally within the GIS as polygon, lines or points. Illustrations of these basic forms as far as water resources projects are concerned are:

Well locations, stream sampling points may be considered to be represented by points, streams, sewer, water pipes, canals may be considered to be represented by lines and drainage basin, command area by polygons.

The attribute information associated with these features and stored in GIS database could be stream order, stream length, size of command area, pre and post monsoon depth to water table and water qualities etc. Once the geographical features are captured from a map into the data base user can view and print the map in many attractive formats, perform diverse queries to analyze and manipulate the spatial information like which are the areas which may be most affected by flood in the rivers. It makes upgrading easier. The vector and raster (grid) are the two basic geo-coding systems used for storing these information and it is possible to change the data from one to the other. Assessment of water resources potential requires analysis of a large number of parameters of any given catchment. They can be categorized as physiographic characteristics, rainfall, surface water, soil moisture, evapotranspiration, ground water and land use. Physiographic data such as area and shape of catchments, landforms, drainage patterns, stream lengths, stream density, and frequency and channel characteristics are very useful to develop such models to estimate temporal variation of runoff with rapidly changing rainfall events. The drainage density extracted from satellite data from LANDSAT or IRS or SPOT is two to three times greater than that obtained from topographic maps in the scale of 1:250,000. GIS can quantify drainage basin characteristics making it possible to relate many of these factors to surface runoff, leading to improved understanding of the effect of drainage basin characteristics on surface runoff.

Quantification of drainage basin in the past was very cumbersome and time-consuming process. Currently remote sensing and GIS technology can easily quantify morphometric and climatic characteristics of a large drainage basin. GIS can help to establish balance between the various competing demands for water in irrigation, domestic, industrial and recreational uses to achieve better economical use of precious water resources in our country. Using remote sensing data GIS can produce geo-referenced over lays either in map form or in tabular form. Reclassification can also be performed in multiple data layers as part of overlay operation. Thematic mapping consist of overlaying maps of different types but corresponding to the same area. Reclassifying procedure involves operation that reassign thematic values to the categories of an existing map as a function of the initial value, the position, size or shape of the spatial configuration associated with each category for example soil map reclassified into a permeability map.

Measurement of spatial data involves the calculation of point co-ordinates, length of lines, area and perimeter of polygons including volumes. Measurement may be from point to point like distance between different wells in an irrigated area, stream lengths, and area of polygons including areas of catchment or command areas for water resources projects. Total number of points falling within polygon, i.e., the number of wells within irrigated areas required for conjunctive water use for better management of irrigated land is easy, fast and accurate with GIS technology. Using remotely sensed data GIS can produce geo-referenced overlays, which help to carryout impact analysis very quickly and can be used to show how a watershed will be affected by a particular decision. Delineating buffer zone could be overlaid with ideal land capability layers to choose the best possible use of irrigated area or degraded land under a given water resources project. Integrating data on land use with topographic and geologic information can assess the impact of water resources project on the environment.

The basic functions and components of GIS are:

1. Data input/encoding
2. Data storage/management
3. Data manipulation and retrieval
4. Data presentation and output

Map type of data can be encoded and stored either in a vector or roaster grid cell format. Data input is through a process of digitization of the analog map data to digital computer compatible elements. Digitization can be done manually, semi-automatically or through automatic scanning. Applying various data manipulation modules by way of logical queries does extraction of meaningful information from the GIS database. Such examples are basic functions like scale change for overlays, area calculation, proximal search etc. With various layers of the data base elements and the manipulation tools one can optimize suitable models for simulating and analyzing different phenomena. Hence its applicability in multi-theme studies, such as natural disaster impact appraisal is very much obvious. Examples of well-known GIS packages are ARC-INFO, PAMAP, IDRIS, SYNERGIS, etc. Many of these can operate on advanced PC systems and workstations and are menu driven, allowing non-computer professionals to work effectively. One of the most powerful applications of GIS model simulation aimed at answering extremely complex situations and queries with respect to disaster management.

Example (i): If the discharge in a certain river at a location on a given date and time is given then for questions to be answered (a) likely flooded area at a place down stream after say two days (b) which portions of highways are likely to be affected and for how long will they not be serviceable.

Example (ii): If several landslides of specific dimensions have affected highways connecting two important places how many airlifts need to be planned to supply the marooned people.

Example (iii): If an earthquake of certain magnitude occurs in a place what will be the likely extent of damage to buildings and bridges in a city located away from the place.

In many developing regions the intensification of agricultural activities in fragile upland regions has resulted in accelerated rates of soil erosion and a rising trend in the sediment load delivered to rivers. For effective land use planning the effect of soil and water conservation measures on the sediment supply to rivers and reservoirs needs to be predicted. Models based on process basis are available but they require large volume of field data and in many countries such data are not available. The overseas development Unit at the Hydraulic Research Station at Wallingford have developed a user-friendly software CASTLE which is GIS based to predict soil erosion and sediment yields from limited data on soils, crops, sediment yields and rainfall that are available in most countries. It is based on simple erosion predictors such as Universal Soil Loss Equation, which can be applied to land units of similar size plots from which erosion data are usually derived. Predicted source erosion can be adjusted to account for the deficiencies in the soil predictors. The software carries out the sediment routing computations through the river network, taking into account of the sediment sizes being transported. This enables sediment yields, which in large catchments represent only a small portion of the source erosion to be estimated. The sediment routing helps identifying the areas in a catchment making the largest contribution to the sediment supplied to rivers so helping to identify the priority for conservation. The software has been applied in various countries, viz., Thailand, Malaysia, Sri Lanka and South Africa.

For watershed management decision it is necessary to have information regarding

(I) information of physical resources of the watershed,

(II) sediment yield modeling from the watershed,

(III) conservation planning of watershed and its prioritization.

It is known that use of remotely sensed data can be interpreted to derive a number of thematic maps containing information in land use, soil, vegetation, stream network, surface water, snow cover, intensity of erosion etc. The above information can be combined with conventionally measured parameters such as precipitation, evaporation, etc and various topographical features like, height, contour, and slope to provide synthesized information of a watershed. For estimation of average annual sediment yield empirical relations are used. One such empirical models (Garde et al, 1983) can be expressed as follows:

$$V_s = 1.067 \ (10^{-6}) \ p^{1.384} \ A^{1.292} \ D_d^{0.397} \ S^{0.129} \ F_c^{2.51} \qquad (4.29)$$

where, V_s = Sediment yield, mm^3 per year,

A = watershed area, km^2,

S = average slope of watershed,

p = annual precipitation, cm,

D_d = drainage density, km/km^2

F_c = vegetation cover factor, = $(0.21 \ F_1 + 0.2 \ F_2 + 0.6 \ F_3 + 0.8 \ F_4 + F_5)/\Sigma \ F$ (here, F_1, F_2, F_3, F_4 and F_5 are respectively, area under reserved and protected forest, unclassified forest cover, grassland, pasture land and wasteland respectively).

Examination of this regression analysis makes it evident that parameters like, A, D_d, S and F_c are essentially mapping inputs which can be conveniently derived from stream network map, topographic contour map and land use cover information. Land use cover information can be extracted from satellite imagery. The parameters that contribute sediment yield process in the above formula can be stored in a database, which can be updated periodically using satellite imageries. It is therefore possible to have a realistic estimate of erosion rates. It should be kept in mind that not all units of a watershed such as micro, mini, or sub-watersheds contribute sediment yield and at the same rate. It is therefore necessary to identify the erosion prone areas within a watershed or those areas which are likely to contribute the maximum. It is therefore necessary to determine priority indices of a watershed in decreasing order of their sediment yield.

Chakraborti of Indian Institute of Remote Sensing (IIRS), carried out studies in Doon Valley, Uttaranchal, India, for prioritization of various sub-watershed for conservation planning. Resources information of landuse cover and stream network of the watershed are extracted from Landsat terrain map (TM) false colour composite (FCC) image on 1: 50,000 scale following systematic image interpretation techniques. Other information is the topographic contour maps from topographic map of 1:50,000 scale and areal distribution of rainfall based on Thiessen Polygon. The watershed is divided into a number of sub-watersheds. Sub-watershed-wise lumped parameters required for sediment yields are obtained from thematic maps of annual rainfall, stream-network and topographic contours. The parameter vegetation cover factor is determined by regrouping satellite derived land-use cover classes into one of parameters such as area under reserved and protected forest, unclassified forest area, cultivated land, grass and pastureland, wasteland etc.

4.3 FLOOD DISASTER AND ITS MANAGEMENT

Disaster by floods constitute one of the major natural calamities faced by mankind almost every year resulting in substantial loss of life, large scale loss damage to moveable and immovable properties, disruption of communication and community life lines apart from sufferings of various other kinds. Floods and their disastrous impact on the hills and plains of the adjoining rivers and drainage channels have plagued civilizations. Early civilizations flourished along river valleys due to availability of water for drinking, irrigation, navigation etc. Dams on watercourses are normally constructed as a measure to mitigate the effect of flood and to store water for other periods of scanty rainfall and drought. Let us now consider the problem of flood in the Brahmaputra valley in India. The riverine area in the Brahmaputra valley is narrow and confined between the hills in the north and south. The total width of the valley between the foothills is only 80 to 90 km out of which the river occupies average width of 6 to 10 km in most of the places. The foothill area is covered by forest and cultivation of tea is carried out in certain districts, which occupy most of the highland. Cultivated fields with villages cover the remaining width of the valley. There are two national highways in the valley area connecting Assam and other North Eastern States of India. The adverse topography coupled with heavy rainfall and erodible riverbank materials are responsible for disastrous floods in the valley. The river passes through different climatic zones Tibetan plateau region and the hilly ranges and plain areas in India. The mean rainfall over the whole catchment is of the order of 230 cm and in the subcatchments it varies widely from about 174 cm to 635 cm. Most of the rainfall occurs during May to September contributing to serious flood situation.

The problem of flood in the region is further aggravated because of its location in seismic zone and also being subjected to heavy landslides in the hills contributing to heavy sediment load in the river, which completely upsets the river regime and rising of the river bed substantially due to silt deposition. This reduces discharge carrying capacity of the river and consequent flooding.

Some issues related to flood management problems of Ganga-Brahmaputra-Meghna basin are discussed here. The basin is one of the largest river basins in the World and it spreads over countries of India, Bangladesh, Nepal, Tibet (China) and Bhutan. The flood problem of Ganga Brahmaputra Meghna (GBM) deserves special mention as it is difficult to imagine any basin where geopolitical and hydrological conditions have combined to cause such formidable problems in water management. Almost every year major floods affect the areas lying in one or the other part of the basin. It has monsoon climate lasting to an average of four months when the problems of flood becomes very acute. Due to the large basin areas involved the intensity and frequency of flood vary in space and time. This means that the same flood wave can cause heavy damage in one country while another country may not be significantly affected. This is because of the fact that flood discharge depends on the extent of the area, intensity of the rainstorm, and the valley storage in the intervening reach through which the flood water propagates. So a very high flood in the upper reaches of the river may get moderated down to insignificant flood in the lower reaches or the worst flood may be generated due to heavy precipitation in the lower reaches while the upper reaches may be suffering from drought condition.

For flood management purpose first thing required is a study related to hydrology of the river system and the geomorphologic changes consequent to very severe floods. The Ganga (i.e., Ganges) river rises near Tibet-India border collects the snowmelt and flows southwest across India and enters Bangladesh. It continues southeast across the province until it joins the Brahmaputra River in Aricha. Beyond this, the combined river is known as Padma River.

Prior to the 16th century, most of the flow of the Ganga used to pass directly into the Bay of Bengal via Hooghly river near Calcutta (Kolkata). Since that time the channel has migrated progressively to the northeast occupying and abandoning several prominent courses into its present position. River Ganga has got a very long record of hydrological observations. Gage records were initiated in 1910 at the Hardinge Bridge and discharge observations started in the year 1934.

Average annual discharge is of the order of 11,660 cumecs with a maximum of 80,700 cumecs being recorded in 1961 September. Lowest flow recorded is 1,190 cumecs. The flood cycle is quite peaked and the river normally reaches its maximum discharge during the latter half of August or the first half of September. During the rising stage the discharge increases from about 11,330 cumecs to 56,630 cumecs within a period of two months i.e., July and August. This is well illustrated in the water level hydrographs, which show an increase of river stage a total of 6.1 m in two months. The rate of both rise and fall is quite rapid and generally more uniform.

The Brahmaputra River rises in Tibet on the north slope of the Himalayas. It flows eastward for about 1,120 km, turns to the south into the State of Assam and then turns sharply West for about 640 km to the border of Bangladesh. At the border the river curves to the south and on this course to its confluence with the Ganga about 240 km north of the Bay of Bengal. In Bangladesh the width of the river varies from about 1.6 km to more than 12.8 km. Brahmaputra can be classified as a braided river. During low flow the channel shifts back and forth between the main stream banks which are often 6.4 to 12.8 km apart. Gage records and discharge measurements were taken up in 1949 and 1956 respectively. Annual average discharge was about 19,200 cumecs and the maximum discharge recorded was 71,330 cumecs in 1958 and the lowest flow was about 3,285 cumecs in 1960.

Figure 4.7 shows the annual variation of discharges in Brahmaputra and Ganga upstream of their confluence. This discharge hydrograph shows two or three major flood peaks on the Brahmaputra that occur prior to floods in the Ganga. This has got a significant influence on the pattern of river migration downstream of Aricha. Due to the flooding of Brahmaputra scour would occur. However when the latter flood peaks of Brahmaputra coincides with the Ganga flood peak the backwater effect caused by the Ganga current probably causes considerable deposition of sediment in the lower Brahmaputra which changes the river morphology. The Meghna river drains the north east part of Bangladesh and it is predominantly a meandering one, although in several reaches where tributaries joins it shows braided characteristics. The average annual discharge is of the order of 3,510 cumecs and flood flow is of the order of 11,890 cumecs. It however carries heavy sediment load of the order of 200,000 tons during a flood. The above three major rivers have dominated the landscape of Bengal Basin. The sediment brought by them has formed the largest subaerial delta in the world as well as a larger sub-aqueous delta. During a very severe flood around 141,580 cumecs of water flows in the Bay of Bengal via a single outlet and this causes flooding over a large area of land. It is therefore evident that Ganga, Brahmaputra and Meghna (GBM) river flows principally determine the flooding problem of the area. The area is subject to moderate floods at every 4-yr. interval, severe floods at 7-yr. recurring interval and catastrophic floods at a recurrence interval of 50 years.

The rivers in the GBM Basin not only cover interstate boundaries but also encompass international boundaries. Understanding of the national and international issues involved in the proper utilization of water resources

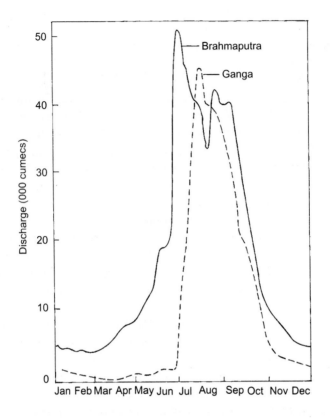

Figure 4.7   Flows in Brahmaputra River at Bahadurabad and Ganga River at Hardinge Bridge in 1981 [Courtesy: Bandyopadhyay, 1995].

potential and flood management under the basin area is very important. Article 262 of the Constitution of India authorizes the Indian Parliament by law for adjudication of any dispute or complaint with regard to use, distribution or control of water in any state river or river valley. Under the existing legal framework the states are responsible for planning, implementing and using of waters of rivers largely according to their own priorities often disregarding uses, existing or future of other states. The international issues involved in India's water resources development schemes can be stated in brief as follows. In the west, it has already been agreed to share the waters of Indus River with Pakistan. India's share is approximately twenty percent of the total annual flow, which is being of the order of 213 km^3. In the eastern side a dispute exists about the minimum flow of Ganga at the India-Bangladesh boundary, i.e., at the Farakka Barrage. Ganga and its numerous tributaries also flow through Nepal. Scientific approach would be development of policy in totality and then after thorough scientific analysis one can arrive at a bargained optimal policy. Three distinguishing features of international or interstate planning are respectively:

(I) Nations and states pursue distinctive, noncomparable and often multiple development objectives

(II) Nations and states are limited in their ability to act autonomously and

(III) Development decisions involve a strategic choice to share a common resource in the presence of both conflicting and mutual interest.

Rogers and King (1985) in a study through a preliminary policy development planning of the lower Ganga basin have shown that India and Bangladesh can adopt six strategies of development. They are respectively, both do nothing, India optimizes, Bangladesh does nothing, Bangladesh optimizes India does nothing, both optimizes independent of each other, regional optimization with distinct sub-regional budget and regional optimization with single regional budget.

Various alternative plans can be analyzed once an agreement regarding overall development of the basin is reached. Many alternative proposals exist for the purpose such as interbasin transfer of water through development of Himalayan and Peninsular rivers through construction of link channels [Rao (1979)] and later on suggested improvements envisaged in National Perspective For Water Resources Development envisaged by the Ministry of Irrigation, Govt. of India. Generally speaking for study of large river basin hierarchical levels for system studies are respectively, at the National level, Regional level, Regional sub-system level and Project level. The first level of study seeks to study the next level study by deciding level of regional allocation and preliminary screening of policies and projects in totality under multiple objectives. It divides the range of interlinkage for analysis in the next level of study.

Most of the land areas covering the Bengal Delta are lowlands and they are also subjected to tidal action along with hurricane surge from time to time. It is important to think of proper planning for water controls similar to such areas. In a study on simulation of the effects of sea level rise in the major rivers of Bangladesh by Haque (1993), it has been reported that the effect of rise in sea level by 0.5 m, 1.0 m or 1.5 m will penetrate about 200, 275 and 325 km upstream and as a result of this almost one third of Bangladesh would be submerged.

4.3.1 Flood Disaster Mitigation

Mitigation of floods includes any activities that prevent an emergency, reduce the chance of an emergency happening or lessen the damaging effects of unavoidable emergencies. Investing now in mitigation steps such as construction of barriers, levees and providing flood insurance will help reduce the amount of structural damage to buildings and crop damage to farms, should a flood or flash flood occur later. People should find out from appropriate agencies whether their property is above or below the flood stage water level and know about the history of flooding in the region. It is also necessary for the people to learn about flood warning signs and also about information on preparation of floods and flash floods. It is also necessary that stockpiling of essential emergency building materials such as plywood, plastic sheeting, lumber nails, hammer and sand pry bars, shovels and sand bags. In the urban areas it is advised to install check valves in building sewer traps to prevent flood waters from baking up in the sewer drains. As a last resort, it is advised to use large corks or stoppers to plug showers, tubs or basins. The people should be encouraged to plan and practice an evacuation route to shelters or high grounds. The people should have disaster supply on hand such as flashlights and extra batteries, first aid kit and emergency food and water, non-electric can operator, essential medicines, cash and sturdy shoes. It is also worthwhile to teach all family members how and when to switch off gas, electricity and water, and also about national flood insurance programmes. During a flood listen to a battery operated radio for the latest storm information, fill bathtubs, sinks and jugs with clean water in case water gets contaminated. It is recommended to move valuable household properties upstairs.

From flood management point of view several measures are available. These are construction of flood embankments, drainage channels, erosion control measures, construction of reservoirs and flood shelters, watershed management, flood plain zoning, flood forecasting etc. In many areas a combination of such measures are required to provide relief from flood. One of the most effective measures is the proper implementation of flood plain zoning. Here it is necessary to demarcate areas liable to floods, preparation of detailed contour plans of such areas to a larger scale with contour intervals of 0.2 to 0.5 m, fixation of reference river gages and then determining the areas likely to be inundated for different water levels and magnitude of floods. Demarcation of areas liable to be affected by floods of different frequencies like say once in 5, 10, 15, 20, 30, 50, 100 years and similar exercise to account for accumulated rain water of different rainfall frequencies say, 5, 10, 25 or 50 years with the help of modern tools like satellite imageries.

The above means that a balance between the values available from the use of flood plains and the potential losses to individuals and society arising from such use There are two schools of flood management group, one school has faith on adjustment and control and the other school has faith on modifying the flood, reducing the susceptibility to damage and reducing the impact of flooding. Flood management has been practiced from ancient times. A flood bank was built on the Hwang Ho river in China more than 2,500 years ago and the diversion of flood waters from the river Nile into lake Moeris was done many years ago.

Modification of the floods can be affected in the atmospheric phase i.e, weather modification, land phase i.e., production of floods, and channel phase i.e., flood control or protective works. Reducing the impact of flooding includes emergency measures like evacuation, flood fighting, disaster relief, tax relief and flood insurance.

4.4 LANDSLIDE HAZARDS AND THEIR MANAGEMENT

Landslide may be defined as rapid draw down and outward movement of slope forming materials composed of rock, soil, artificial fill or combinations of these materials separated from the underlying stationary part by a definite plane of separation. Failure occurs mainly under the action of its own weight in which the displacement has both horizontal and vertical components of considerable magnitude. The rate of movement varies from slow to rapid. Landslides include almost all varieties of mass movement like rock falls, topples, rotational and translational slides, lateral spreads, flows or combinations of these [Varnes (1984)]. Landslides generally have been found to occur on days of intense rainfall subsequent to periods of continuous precipitation. Although rainfall and consequently water is the main triggering factor of landslides, it is necessary to have unstable ground conditions. The catastrophic landslides that can be termed as slump flows occur mostly on escarpments and scap slopes. The other phenomenon responsible is earthquakes which causes ground shaking that induces fissures and help to further widen and extend the existing fracture systems underlying the rocks. The shocks can cause disturbances of slopes which are in temporary equilibrium as in ancient dormant landslides and reactive in their movement. With later rains, more and more water could percolate along the newly developed fractures and an increased incidence of mass movement could be expected in hitherto stable regions. During the process of movement, the landslide extends uphill by gradual head ward caving of the scarp. The release of the overlying debris downhill due to land sliding will help to release the stresses in the underlying rocks that are already characterized by one or two systems of joints. The release triggers wide openings in the joints and promotes rapid downward movement of disjointed blocks. The effect of separation along joint planes could be transmitted on either side of a slide along existing joints and lineaments for considerable distances with widespread subsidence and fracturing of the land surface. This phenomenon which acts like a chain reaction may spread to tens and hundreds of meters from either flank of a slide and trees can be observed to wither away as a result of disturbance to their root system. The cracks so developed will be manifested on floors and walls of dwellings and their orientation significantly reflects the fracture patterns of the underlying rocks. The cracks will also prepare the terrain for the future landslide since during continuous rains, water will disappear in to such fracture systems causing underground erosion within the slopes resulting in eventual downward movement, Sassa (1984). Figures 4.8 and 4.9 show two examples of landslides phenomena.

Landslides and mudflows usually strike without warning. The form of rocks, soil or other debris moving down a slope can devastate anything in its path. Take the following steps to be ready. Get a ground assessment of your property, collect specific information on areas vulnerable to landslides from competent authority. Consult a geotechnical expert for opinions and advice

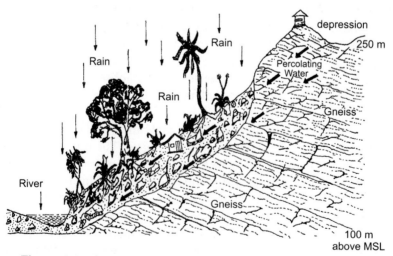

Figure 4.8   Schematic Representation of an Old Landslide*
(*Courtesy, ICODM Guwahati 1998)

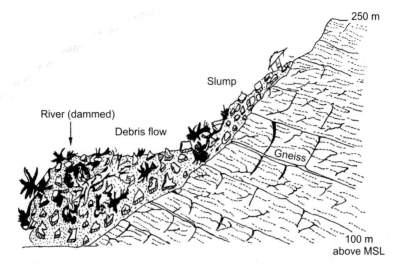

Figure 4.9 Schematic Representation of a New Landslide*
(*Courtesy, ICODM Guwahati 1998)

on landslide problems and on corrective measures. In order to minimize hazards in buildings it is advised to plant ground cover on slopes and build retaining walls. In mudflow areas, build channels or deflector walls to direct the flow around the building. However if you build walls to divert debris flow and the flow lands on a neighbor's property you may be liable for damages. The landslide warning signs are:

(1) doors and windows stick or jam for the first time
(2) new cracks appear in plaster, tile, brick or foundations
(3) outside walls, walls or stairs begin pulling away from the building, slowly developing, widening crack appear on the ground or on paved areas such as streets or driveways, underground utility drains break, bulging ground appears at the base of a slope, water breaks through the ground surface in new locations, fencing retaining walls, utility poles or trees tilt or move, one hears a faint rumbling sound that increases in volume as the landslide nears.

The ground slopes downwards in one specific direction and may begin shifting in that direction under your feet. It is advised that two evacuation routes be at least planned, since roads may become blocked or closed, and also to develop an emergency communication plan and have insurance against mudflow and landslides. During the landslide if one is inside the building he/she is advised to stay indoor and take cover under a desk, table or other piece of sturdy furniture and if outdoor try to set out the path of landslides and mudflow, run to the nearest high ground in a direction away from the path. If it is not possible to escape then it is advised to curl into a tight ball to protect the head. A sinkhole means when groundwater dissolves a vulnerable land surface such as limestone, causing the land surface to collapse from a lack of support. In June 1993, a 30-m wide 8-m deep sinkhole formed under a hotel parking lot in Atlanta, USA which killed two people and caused many injuries. It is advised to stay away from the slide area as there may be danger of additional sliding, also check for the injured and trapped persons near the slide area, and give first aid if required. Help your neighbor, requiring assistance, infants, elderly people and disabled people. One should listen to the battery-operated radio for the latest important information. It should be remembered that flooding might occur after mudflow or landslides and check for any damaged utility, check also the foundations of structures, building and surrounding land for damage. Replant damaged ground as soon as possible. For appropriate designing of correct landslide measures consult a specialist. Mitigation measures include all activities that prevent any emergency, reduce its chance of happening or lessen damaging effects. Preventive steps include planting ground cover, low growing plants on slopes, installing pipe fittings to avoid gas or water leaks.

4.4.1 Landslide Hazard Zone Mapping

Hazard zone map comprises a map demarcating the stretches or areas of varying degree of anticipated slope instability. The map has an in built element of forecasting and hence is probabilistic in nature. It will be able to provide help concerning the following aspects: location, time of occurrence, type of landslide, extent of the slope area likely to be affected and rate of mass movement of the slope mass. The probability of slope instability is dependent on complex interactions among a large number of factors encompassing geotechnical, geological, hydrological and climatic conditions, besides the influence of human activities. Thus methodologies developed in different countries have wide variations arising not only from the nature of slope formation and slope instability problems but also the end uses for which such maps are intended. Preparation of a comprehensive landslide hazard zonation map requires intensive and sustained efforts. The problem is highly interdisciplinary in nature. A large number of data concerning many variables, covering large slope areas has to be evaluated and zonation maps prepared. The use of aerial photographs and adoption of remote sensing techniques help in collection of data. It will of course be necessary to use computer facilities for storage, retrieval and analysis of data. Zonation maps have multifarious uses, some of which are as follows:

Preparation of development plans for townships, dam roads and transport networks, general purpose master plans and landuse plans, discouraging new development in hazardous areas, choice of alignment for new roads and developing programmes for keeping the existing roads free of landslide hazard. For monitoring of surficial movement wooden pegs are generally used as markers over the potential slide zone. Initial position of these markers can be plotted on the contour map with the help of theodolite, steel tape clinometer compass to monitor lateral and vertical movements. For monitoring subsurface movements, inclinometers are generally installed in the active slide zone, which may be identified based on subsoil investigation, geological and geomorphological studies. It is a well known fact that the stability of saturated slopes can be improved significantly by lowering the piezometric level which can be realized by installing horizontal drains or deep trench drains. Horizontal drains could be in the form of PVC pipes with slots and is normally driven at an adverse slope of 5 to 10 degrees into the hillslope. The pipes drain water from deep inside the slide masses, which can be collected and discharged. Deep trench drains consists of deep trenches excavated into the slope and backfilled with clean and well draining rubbles with a geotextile cover. The zone of high permeability thus provided helps to drain water from the soil mass.

To overcome surface due to rain erosion on denuded hill slopes use of natural geogrids made of coir have been found to be very effective since grass cover is normally established within a period of one year and is environment friendly. For promotion of vegetation asphalt mulch technique is very useful as it adds to the stability of slopes. The spray of asphalt mulch on the slope reduces the susceptibility to erosion, it makes the surface impervious, it conserves the nutrient in the soil, it helps the slope surface to absorb more light rays and so encourages growth of vegetation. For controlling rock falls Nelton geogrids is very useful and it also helps promoting vegetation at locations having fragmented rock mass.

4.4.2 Risk Analysis of Water Hazards

Design for control of water involves consideration of risks. A water control structure may fail if the magnitude of the design return period (T) is exceeded within its expected lifetime. This natural hydrologic risk failure can be calculated using the equation

$$\overline{R} = 1 - [1 - P(X \geq X_T)]^n \tag{4.30}$$

where $P(X \geq X_T) = 1/T$, and n is the expected life of the structure (yrs), $\overline{R}$ represents the risk which is the probability that an event $X \geq X_T$ will occur at least once in n years. Plot of the above relationship is shown in Figure 4.10.

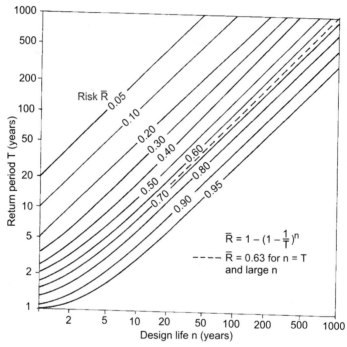

Figure 4.10   Risk of at Least One Exceedance of the Design Event during Design Life.

It can be seen from the figure that for a given risk of failure, the required design return period T increases linearly with the design life n, as T and n become very large. For large values of n, the equation will approximately be equal to $(1 - e^n)$. When T = n, the risk is $(1 - e^{-1}) = 0.632$. To take care of other kinds of uncertainty such as hydraulic, structural, construction, operation, socio-economic, political and environment factor of safety is considered, which means that the actual capacity is larger than the hydrologic design value.

Hydrologic uncertainty may be natural or inherent uncertainty, which arises from the random variability of hydrologic phenomena, model uncertainty due to the approximations made when representing the phenomena by equations with uncertain coefficients. The other type of uncertainties will now be considered. For first order uncertainty analysis, suppose the dependent variable 'z' is expressed as a function of an independent variable 'x'

$$z = f(x) \tag{4.31}$$

there are two sources of error: first in the function $f(x)$ or the model may be incorrect or there could be error in measurement. Kapur and Lamberson (1977) have given analysis when there is model error. Here the effect of discrepancy from the true value of z is determined by expanding $f(x)$ as a Taylor series around the mean value, $x = \bar{x}$. The derivatives in the Taylor series expansion are evaluated at $x = \bar{x}$. Considering only the first order expression, the error in z is

$$z - \bar{z} = \frac{df(x - \bar{x})}{dx} \tag{4.32}$$

The variance of this error is $s_w^2 = E[(z - \bar{z})^2]$, where E is the expectation operator, that is

$$s_w^2 = (df/dx)^2 s_x^2 \tag{4.33}$$

where s_w^2 is the variance of x. Equation (4.33) furnishes the variance of a dependent variable z as a function of the variance of an independent variable x. If z is dependent on several mutually independent variables x_1, $x_2, x_3, \ldots , x_n$, it can be shown that

$$s_w^2 = \left(\frac{\partial f}{\partial x_2}\right) s_{x_1}^2 + \left(\frac{\partial f}{\partial x_2}\right) s_{x_2}^2 + \ldots + \left(\frac{\partial f}{\partial x_n}\right) s_{x_n}^2 \tag{4.34}$$

An overall risk assessment for a particular design, is done by what is known as composite risk analysis. In this analysis the loading or demand placed on a system is the measure of the impact of external events. As for example the magnitude of a flash flood is dependent on the condition of the watershed at the time of the storm producing it and the condition of watershed at the time of storm. The capacity is a measure of the system's ability to withstand the demand.

If demand is designated as L and capacity by C, the risk of capacity failure is then given by the probability that L exceeds C or

$$\bar{R} = P[(C - L) < 0], \tag{4.35}$$

The risk depends on the probability distributions of L and C. Consider that the probability density function of L is f(L). The function could be an extreme value log-Pearson Type III, probability density function. Given f(L) the chance that the loading will exceed a known capacity C^* is

$$P(L > C^*) = \int_{c^n}^{\infty} f(L)\,dL \qquad (4.36)$$

The true capacity is also not known exactly and may be considered to have a probability density function g(C). This could be a normal distribution arising out from the first order analysis of the system. The probability that the capacity be within a small range dC around a value C is g(C)dC. Assuming that L and C are independent random variables, the composite risk is evaluated by finding out the probability that loading will exceed capacity at each value in the range of feasible capacities. By integration one obtains,

$$\overline{R} = \int_{-\infty}^{\infty} \int_{C}^{\infty} [f(L)\,dL]g(C)\,dC \qquad (4.37)$$

The reliability of a system is defined to be the probability that a system will perform its function for a specified period of time under stated conditions. According to Harr (1987), reliability is the probability that the loading will not exceed the capacity,

$$R^* = P(L \leq C) = 1 - \overline{R} \qquad (4.38)$$

So,
$$R^* = \int_{-\infty}^{\infty} \int_{0}^{C} [f(L)\,dL]g(C)\,dC \qquad (4.39)$$

4.4.3   Risk Assessment of Landslides

The landslides are complex and can occur in many ways. These can be expressed as say,

X_1 Due to excessive internal water pressures during heavy rainfall,
X_2 Due to slope angle exceeding a certain value say $30°$,
X_3 Due to erosion at toe plus road cutting at bottom slope,
X_4 Material strength.

According to total probability theorem, total probability of failure due to landslide event E is given by,

$$P[E] = \sum_{i=1}^{i=4} P(E/X_i) \cdot P(X_i) \tag{4.40}$$

Mutually exclusive events X_1, X_2, X_3 and X_4 occur to produce the landslide event E. $P(E/X_i)$ is the probability of E, given that X_i has occurred. $P(X_i)$ is the probability of occurrence of X_i.

Different mathematical methods may be applied to obtain $P(E/X_i)$ and $P(X_i)$. One aspect could be use of historical data within the area of study. If the landslide on the same terrain has occurred already, its probability can be expressed in terms of Bayesian Theorem, which combines conditional probability with total probability as

$$P(X_i/E) = \frac{P(E/X_i)P(X_i)}{P(E)} \tag{4.41}$$

Total Risk (R_i) is the expected number of lives lost, persons injured, damage to property and disruption of economic activity. It is the product of specific risk (R) and elements at risk (E_r) over all landslides and potential landslides in the area

$$R_i = (E_r \cdot R) \tag{4.42}$$

Total risk due to landslide can be calculated by using this procedure.

4.5 DISASTER DUE TO COLLAPSE OF DAMS

Landslides cause massive movement of soil and are of major concern among natural disasters which frequently strike in different parts of the globe and causes extensive damages to life and property. Secondary phenomena like dam break occurs due to massive landslides soil mass falling in the reservoir generating surface waves, which may be responsible for failure of dams. The failure will in turn cause devastation to the people residing in the downstream side. The failure of dams can also be attributed due to torrential rain and consequent flood generated as a result of it especially the old ones, which are under threat. The consequences of dam failure whatever be the cause results in loss of life and damage to property. It may be mentioned here that the failure of Teton dam in USA located at Idaho in the year 1976 caused about four hundred million-dollar loss, eleven people died, about 25,000 people were rendered homeless. The dam was a 93-m high earthen dam with

914.4-m long crest. The breach developed rapidly and subsequent erosion washed away 3 million cubic meters of earth while releasing a peak flow of 70,000 cumecs. The failure of the dam is the result of complex concourse of causes and mechanisms. Generally one can say the failure of dam depends mainly on the type of dam, nature of breach formation and its development, foundation and embankments, type of external disturbances to which the dam is subjected like the forces acting on the dam, incoming flood and the existing operation conditions. Dam failure is classified as sudden or gradual depending on the duration of failure of the dam. If the duration is of the order of 10 to 15 minutes the failure can be termed as sudden otherwise it can be termed as gradual. Arch and gravity dams fail by sudden collapse, overturning or sliding away of the structure due to overstress caused by inadequate design or excessive forces that may result in overtopping of flood flows, earthquakes, deterioration of abutment or foundation materials.

The major causes of earth dam failure are sliding, overtopping, seepage, human intervention and earthquake. Sliding can occur to the reservoir banks, embankment or foundation when the shear stress due to external loads along a plane in the soil mass exceeds the shear strength that can be sustained in that plane. In such a situation the failure along the plane is imminent. Periodic cycle of the saturation of the porous material and the increased pore water pressure reduce the shear strength which leads to loss of stability and produce major landslides. Due to this sudden arrival of soil mass in the reservoir, generally produces a high solitary wave propagating against the earth dam. An initial breach can form immediately in the dam and progressive erosion caused by flowing water will lead to a partial or total failure of the dam. Improvements in the analysis of probable maximum storm have caused significant increase in the predicted probable maximum flood (PMF). Many earlier dams have not been designed on PMF values. As a result many dams once considered safe are now considered unsafe due to inadequate spillway capacity. In the year 1982, United States Corps of Engineers inspected 8,639 high hazard dams under Federal Dam Safety Programme. Of these 2,884 dams were found to be potentially unsafe due to inadequate spillway capacity. Seepage of water can be through the embankment, foundation or abutment of the dam. Uncontrolled or controlled seepage through the body of the dam or foundation may lead to piping or sloughing and the subsequent failure of the dam. Due to continuous seepage breach will be formed in the embankment resulting in the outflow through the breach. The breach size will continuously grow as material is removed by outflow from the storage and storm water runoff. The size, shape and time required for the development of breach are dependent on the embankment material and the characteristics of the flow forming the breach. Breaches of this type can occur fairly rapidly or can take several hours to develop. Earthquake at the dam site can produce waves or landslides and consequently leads to a partial or total failure of dam. Depending on the severity of the earthquake the entire dam may be washed off or only a part of it may be removed due to breach failure. Such failure is sudden in nature. Other causes are due to differential settlement of the foundation of the dam. Londe has reported the findings of Middlebrooks who provided a statistical analysis of earth dam failure using 200 case histories. The record covers a period of hundred years. Figure 4.11 provides the result in terms of percentages of each category of failures. From the diagram it is obvious that only 15% of the total failure numbers is by sliding which could be evaluated by usual concept of factor of safety. The remaining 85% do not come under conventional stability analysis and are relevant to design, construction and operation procedures, which can hardly be computed in the conventional manner. A case study related to dam break analysis as a result of the likely failure of Maithon Dam located in the Damodar Valley area in Eastern India is discussed now.

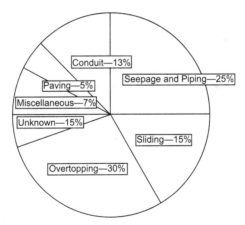

Figure 4.11 Statistics of Earth Dam Failures

A CASE STUDY

The Damodar Valley area is approximately located within the latitude 22° 20′ to 24° 30′ and longitude 84° 45′ to 88° 30′ in the states of West Bengal and Jharkhand in India. The Damodar river rises in Chhota Nagpur (Jharkhand State) watershed and after a south easterly course of about 563.27 km falls into the river Hooghly. The Maithon and Panchet Hill dams are earth dams under Damodar Valley Corporation (DVC). The main purpose of these reservoirs is flood control. Other two dams are Tilaiya and Konar. Maithon and Panchet Hill dams have got a capacity of 1.23 km^3 each. The Maithon is located in Barakar River and Panchet in Damodar River both in the Dhanbad district of the State of Jharkhand. The location plan of DVC reservoirs is shown in Figure 4.12.

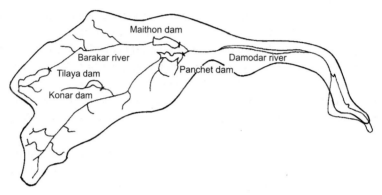

Figure 4.12 Location Plan of Damodar Valley Corporation (DVC) Reservoirs

The outflow hydrograph resulting from the possible failure of Maithon Dam has been analyzed assuming inflow to the reservoir to be zero and considering the breach outflow. The elevation storage relationship for the dam is shown in Figure 4.13 and the other physical features of the dam are as follows:

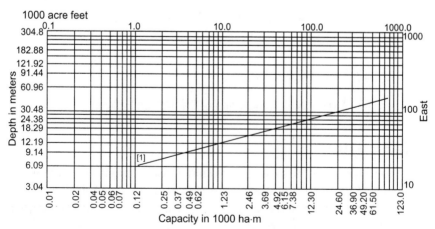

Figure 4.13 Depth-Capacity Relation for Maithon Reservoir

Location: Latitude: 23° 47', Longitude: 86° 49' E, State: Jharkhand, District: Dhanbad. River: Barakar, Drainage area: 6,293.67 km^2, Year of construction: 1957, length of earth dam: 643.43 m, height above river bed: 49.37m, crest width: 9.0 m, elevation of the top of dam above MSL: 156.05 m, spillway crest level above MSL: 140.21 m.

The details regarding the side slopes of the embankment are assumed as 1 V: 2.5 H for upstream side since the same are not available for the analysis. The mean particle diameter and specific gravity of the materials of the dams are assumed as 12 mm and 2.65 respectively. The result after analysis is shown in Figure 4.14. It can be seen that the predicted outflow hydrograph has steep rising limb in the form of S-shape and gradual falling limb of exponential nature. The computed peak discharge is 25,519 cumecs and the time to reach the peak discharge is found to be 35 min. Further it can be seen that the outflow after 3 hrs of breach formation is negligibly small.

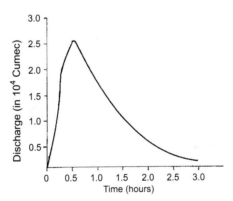

Figure 4.14   Predicted Outflow from Failure of Maithon Dam, Jharkhand, India

4.6   HAZARDS DUE TO DROUGHTS

Drought is a general term implying a deficiency of precipitation of sufficient magnitude so as to hamper the economic activity. Agricultural drought occurs when plant growth gets seriously affected on account of prolonged shortage of moisture in soil. Hydrological drought occurs with prolonged deficiency of rainfall, resulting in marked depletion of surface water and consequent drying up of reservoirs, lakes, streams/rivers, cessation of stream flows and fall in groundwater levels. Studies on the incidence of droughts as well as on their prediction is vital especially for countries like India having a predominant agricultural economy.

Human influence in drought incidence is generally felt either through anthropogenic effects on the climate or the way the operation and land use practice is carried out thereby resulting changes in the hydrological regime. By this change in regime it is meant interfering with infiltration process, modification of stream flows and altering its quality. It is therefore necessary to adopt appropriate measures to mitigate adverse effects of drought on human, livestock and agricultural crop. Another important problem in the drought-prone areas is the provision of drinking water stock.

Another aspect to be considered is the green house effect causing global warming, which will result in severe water crisis in semi-arid regions. Since the year 1900 the global annual average temperature has increased by 0.3° to 0.6° Celsius and average rise in sea level is of the order of 10 to 20 cm. There has also been significant increase in the incidence of storms, heat waves, and floods during the last several decades.

Even though the total amount of water on earth is considered to be intact the rapid growth in population and irrigated agriculture, industrial development is causing severe stress on the quantity and quality of natural system. It is therefore necessary to adopt consistent policy for rational management of water resources especially to tackle serious shortages during drought. Such management practice can be evolved by artificial modification of surface water flows and by better management and utilization of groundwater resource. The latter is more important in arid zone, which is always drought prone.

With the above background in view we will now discuss some engineering measures to be adapted to store and conserve water for optimal use in drought-prone areas.

4.6.1   Rainwater Harvesting for Drought Mitigation

Water harvesting was practiced as early as 4500 B.C. in many parts of the world. It can be source of water for variety of uses in arid and semi-arid zones when common sources like streams, springs or wells will fail. In addition to supplying drinking water for people, livestock and wildlife it can also provide supplementary water for growing food and fiber crops. In rainfed areas a small additional increment of harvesting water can increase the crop output significantly and lower the chance of crop failure. Sometimes it makes a difference between crop and no crop in drought years. It may also be useful for fish production and for growing forage crops to relieve grazing pressure on rangeland.

4.6.2 Rainwater Harvesting for Domestic Use

It is common practice to harvest rainwater from house roofs, especially surfaced areas, rock faces etc and to store in cisterns or tanks for domestic uses. Cisterns are excavated in soil and water proofed with a cement plaster. Traditional water jars are used for rainwater collection in Thailand. These innovations have their origin in traditional grain storage bins. Rooftop rainwater harvesting is of common use in India, particularly in Rajasthan, Lakshadweep and Andaman and Nicobar Islands. Two typical rainwater harvesting structures viz., percolation pit and dug-cum-recharge borewell are shown in Figures 4.15 and 4.16.

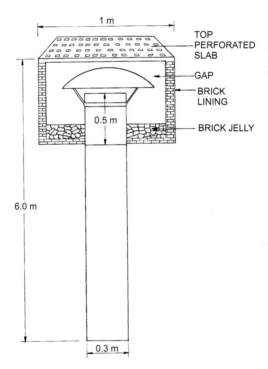

Figure 4.15 Rainwater Harvesting by Percolation Pit (Courtesy: K.R.G. Rainwater Harvesting Company, Chennai, India).

4.6.3 Rainwater Harvesting for Agriculture

Water for agriculture has been already discussed in Section 2.3. In that section various techniques such as raised bed irrigation, sprinkler

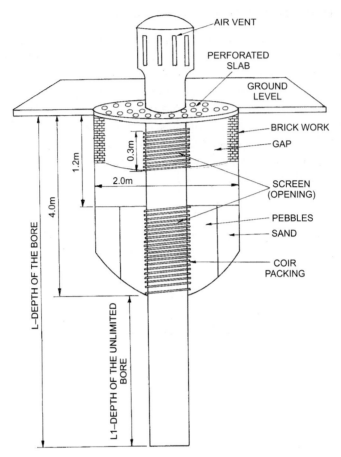

Figure 4.16 Rainwater Harvesting by Dug-cum-Recharge Bore Well (Courtesy: K.R.G. Water Harvesting Company, Chennai, India).

irrigation, drip irrigation etc. have been dealt with. In this section, agriculture in arid, low rainfall areas is discussed. Modification of ground surface to induce runoff is the main aspect of rainwater harvesting in arid lands where rain falls infrequently and is insufficient for good crop growth. So, treating of non-cropped catchment and diverting it to cropped areas or storage reservoirs is practiced. Such types of practices were adopted for precipitation harvesting about 500 years ago in USA and other parts of the world. It is possible to harvest precipitation in the areas having as low as 50-mm average annual rainfall by treating catchment. In Israel, usable runoff had been induced on a catchment even in a very dry year having 24 mm of rain. Reducing the infiltration rate into the soil surface and increasing hydraulic efficiency of the catchment are the main approaches to increasing runoff. Soil compaction, adding additives, sprayed asphalt-coating, and use of membranes, concreting and using iron sheets and adding hydrophobic chemicals can make infiltration reduction. Hydraulic efficiency can be improved by decreasing depression storage, reducing wetted area during recession hydrograph. When rain stops the duration of recession and the speed of reduction of flow area govern the water losses. By changing slope, runoff length, cross section areas, shape and gradient of channel, the time of flow recession and wetted area can be altered. Other methods are by roaded catchment, flat batter tanks, and contour catchments.

4.6.4 Natural Groundwater Recharge— Methods for Estimation

Rainfall is the most important source of natural groundwater recharge. Various methodologies have been developed for computation of natural groundwater recharge. To name a few they are empirical methods, hydrological budgeting method and groundwater fluctuation method. Hydrological budgeting methods have been adopted for computation of groundwater resources in the states of Rajasthan and Haryana in India. Empirical formulae are derived based on the fluctuations of water table and rainfall amounts. One such formula proposed by Chaturvedi (1992) can be expressed as

$$R_{GW} = 3.429 \ (P/2.54 - 14)^{0.5} \tag{4.43}$$

where R_{GW} = natural groundwater recharge (cm), and P = annual rainfall (cm), the relationship being valid for annual rainfall exceeding 38 cm.

4.6.4.1 *Hydrological Budgeting Method*

The method is based on the assumption that rainfall in excess of evapotranspiration losses is utilized in bringing the soil moisture upto its field capacity and the rest is available for groundwater recharge and runoff. Mathematically the relationship can be expressed as

$$P + I = Q + ET + \Delta SW + \Delta SM + \Delta GW \tag{4.44}$$

where, P = rainfall, I = applied irrigation water, Q = runoff, ET = evapotranspiration, ΔSW = change in surface water storage, ΔSM = change in soil moisture, and ΔGW = change in groundwater storage.

4.6.4.2 Groundwater Level Fluctuation Method

The rise and fall of groundwater level over an area is a measure of change in ground water storage, which is computed as a product of specific yield, the average rise in level and the area over which the change occurs. The change in groundwater storage can be expressed by following relationship

$$\Delta GW = \Delta H \cdot S_y \tag{4.45}$$

where, ΔH = change in groundwater stage, and S_y = specific yield.

Estimation of groundwater recharge can also be done by other methods namely, analysis of base flow hydrograph, hydrochemical methods, tracer techniques, etc.

4.6.5 Drought Management

General Water management practices to be undertaken for management of drought:

Irrigation priorities: Irrigate highly visible and intensively marginal areas first. Drought sensitive plants should have been the priorities, but turf should have lower priority. Although turf is drought sensitive it is cheaper to replace turf than replace trees and shrubs.

Time of the day: Water early in the morning when less water loss occurs from evaporation and wind drift in the morning, because of water temperature and less wind.

Irrigation frequency: Irrigate deeply at long intervals rather than frequent shallow watering. Deep watering improves drought resistance by promoting deeper and extensive root systems. Depth of watering should be 15 to 30 cm for turf and bedding plants, 30 cm for perennial shrubs and trees. 2.5 cm of irrigation wets a sandy soil of about 30 cm depth.

Maintenance: Examine the irrigation system and repair leaks promptly. Keep weeds under control as weeds steal water from plants. Do not apply fertilizer, if you do, do so with nitrogen fertilizer. Fertilization stimulates growth and increases water needs. Avoid unnecessary application of pesticides that require watering in.

The management practices that need to be followed for protection of turf are as follows. Irrigate turf only after about 30% of the lawn starts to wilt. Signs of wilting include footprints that remain in the grass long after being made, a bluish grey appearance to the lawn. Raise the cutting height of turf. Although taller grasses use slightly more water than shorter grass a higher cutting height promotes deeper rooting and maintains turf quality longer.

The management practices for bedding plants, shrubs and trees are as follows:

Add mulch to beds to reduce evaporation from soil and to moderate soil temperature, reducing stress on roots. Final depth of mulch should be 7.5 to 10 cm after settling. If possible do not use overhead sprinkler for shrub and flowerbeds, handwater, flood irrigate, or trickle irrigation. Greater water loss can occur with overhead irrigation because of evaporation and wind drift. Irrigate trees and shrubs after they start to wilt drooping leaves and a change in leaf color are signs of wilting. Many trees and shrubs can survive drought without irrigation, provided they are well established and were irrigated prior to the drought. It is suggested that container plants be moved to shaded areas so that their water needs will be reduced. The recommendations made herein should be followed when drought is so severe and water use is restricted that landscape plant survival is in question.

   (i) Only irrigate plants when they start to wilt;
  (ii) apply chemical wilting agents to soil so it will absorb water uniformly and prevent dry spots;
 (iii) for bahai grass lawns stop irrigating and allow the grass to go dormant. Bahai grass will turn brown but it recovers well when irrigation resumes;
  (iv) prune plants severely to reduce leaf area;
   (v) remove weak plants;
  (vi) make the dense beds of plants thinner, to reduce competition among plants.

4.6.5.1 Application of Advanced Technology for Drought Management

The National Remote Sensing Agency (NRSA), Govt. of India under the auspicious of the National Technology Mission For Drinking Water and with the active collaboration of State Departments has prepared hydrogeomorphlogical maps (Scale 1:250,000) for the whole of India utilizing landsat terrain maps (TMs)/Indian Remote Sensing (IRS) satellite imageries. The identification of lineaments has immense importance in hard rock hydrogeology as they can identify rock fractures that localize groundwater. It has been reported by several investigators that yields of wells from lineaments are about fourteen times more than that of wells away from lineaments in the case of Warangal district of Andhra Pradesh, in Southern India. Image rectification and preparation of a geographic information system (GIS) file through visual interpretation of standard false color composite (FCC) data is performed to extract surfacial expression of subsurface water accumulation.

238

4.6.5.2 *Human Issues in Drought Management*

Drought has to be viewed in totality affecting the entire ecosystem and not from meteorological and agricultural angles only. While delineating the drought-affected areas, location conditions of available natural water resources like rivers, streams or ponds should be taken into account. The distances of these water sources from the habitat centers and the dependence of these sources on rainfall should be taken care of. For larger losses from these water resources due to higher evaporation and larger withdrawal rates may occur during durations of drought. Human population, cattle population and plant life is affected most from drought conditions. As for example a 25% less rainfall than the normal in an area with higher population density will do more damage than the same in an area having a small population density. The severity of the effects of drought on farming community would depend on their economic conditions also. The other aspect to be considered is that after the drought has been established in an area there seems to be a tendency for it to persist and expand in adjacent areas. Little is known about the physical mechanism involved in this expansion and persistence. The same feature is noticed in the crop production also which is affected during a number of crop seasons following drought period. In view of the same the population density becomes an important factor. Even if drought incidence is of less frequency in a densely populated area its effect will be serious because of the larger number of people affected and in view of the above mentioned persistence factor. Further during many of the drought periods, shortages of drinking water also occur. While it may be possible to transport food grains to drought affected areas, transport of drinking water is not only a costly proposition but also would involve difficult management problems. It is therefore necessary to give due weightage to the population density while judging the severity of drought in specific regions especially because of the heterogenity in population density of India.

4.7 INFORMATION AND SYSTEM ORGANIZATION FOR DISASTER MITIGATION

Identification of the diverse types of information, terrain attributes required for mitigation and prevention of natural hazards is of great significance. The broad information requirement for natural disaster management is shown in Table 4.5 for few of the most frequently occurring disasters like cyclones, storms, floods, landslides and earthquakes. Most of the information needs have a bias to spatial location. It is thus evident that the speed at which

Table 4.5   Impact of Natural Disasters and Information Requirement for Damage Mitigation[†]

Disaster type	Impact on land and infrastructure	Relief measures	Information requirement for damage containment
STORMS and CYCLONES	Loss of life and livestock; Building collapse; Crop Damage; Land degradation; Flood damage to road/rail network and communication lines.	Shifting of men, movable property and livestock (before disaster); Land reclamation; Provision of food/shelter; Repair to dwellings/transport/ communication; Compensation for damages/losses.	Storm and rainfall monitoring; *Rainfall and tidal wave modeling, *Location and likely extent of damages to dwellings, transport and communication links; *Inundation area modeling; *Land-use pattern; *Population and livestock distribution; *Transportation capacity and bottlenecks; *Damage assessment; *Cost of relief measures.
FLOODS	As above	As above	All the above + Monitoring river flow, reservoir discharge and status of flood control structures*
LANDSLIDES	Drainage dislocations; Damage to roads, buildings power and water lines; Damage to crops; Land degradation; Loss of life and livestock.	As above	Landslide Hazard Zonation; Monitoring drainage and slope conditions in critical areas; Land-use pattern; Transport Capacity and bottlenecks; Population and livestock distribution; Damage assessment;* Unit cost of different relief measures.
EARTH-QUAKES	All the above + Damage to manmade constructions (likely to be very large and widespread). Drying up/ emergence of wells/springs.	All the above + Refugee rehabilitation (No pre-disaster measure possible).	All the above + Seismological and other data on precursors and aftershocks

Note: * Indicates spatial information.
[†Courtesy: ICODM, Guwahati (1998)]

information on nature and magnitude of the disasters and their locations become available has an important bearing on mitigation planning. The risk mitigation planning system is basically an information system with a major part of its elements having a geographical connotation. Depending on the nature and extent of the likely disaster information required for mitigation planning could vary considerably. The information system efficiency depends on the nature and types of information elements, their reliability status and the system of organizing data along with user's perception of the

240

inter-relationships between the diverse data elements and extent of data manipulation capability for scenario-building. Table 4.6 shows identification of data types, sources for collection and realistic upgradation frequency. Most of the areas of information need are common. The storm and rainfall related parameters like cloud, wind speed, rainfall are generally monitored 2 to 3 times daily on a regular basis through satellite surveillance. Terrain parameters, habitations, other constructions, transport and communication elements etc do not change rapidly so an update frequency of 5 to 10 years

Table 4.6   Different Types of Data, Their Sources and Updation Frequency*

Disaster type	Information need	Data requirement	Types of data	Sources	Updation frequency
	Storm and rainfall monitoring	Cloud cover/Wind speed/Rainfall data	M.A.	Ind. Met-Dept.	H/D/W/M/Y
	Rainfall and tidal wave modeling	Cloud cover/Wind Speed/Rainfall/ Ground elevation/ Bathymetry	M.A.	Ind. Met-Dept Surv. India: Naval Hydro office	H/D for Met. Data: Every 5/10 years for Elevn. data
	Location and likely damage to Dwellings	Building census: Cadastral maps	M.A.	State Land Records	Every 5 years
STORMS, CYCLONES, FLOODS	Transport and Commn. links	Road/Rail maps (Classified): Power and telephone line route, maps; Cost of Construction; Repair cost/km	M.A.	State P.W.D; Elect. Boards; Railways; Telecom Dept.	Every 5 years
	Inundable area modeling	River Gage records: Rainfall data; Reservoir storage/ release data; Details of surface drainage channels (location, flow, capacity); Likely obstructions; Topographic details	M.A.	Irrgn. Deptt; CWC; Elect. Boards; River Projects; Surv. of India	H/D for gaging and water release; others annual to 5 yearly
	Land use	Topographic maps; Air Photos Sat. imagery; Crop Survey data (type, area yield estimate) Damages; Costs	M.A.	Survey of India NRSA, State Land user Boards	Maps and Photos 10 yearly; Others seasonal

Contd.

Population and livestock distribution	Census data, Birth and Death register; Livestock inventory; Cadastral maps	M.A.	Census Dept. Vet. Dept. Mun- icipal/Village records; State Land Records	Annual population and livestock; 10 years for others
Transportation capacity and bottlenecks	Road/Rail maps with Classification; Loco/ wagon trucks availability/carrying capacity; Likely obstruction in route	M.A.	State PWD. Road Transport Autho- rity; Railways	Annual for wagon/loco/ trucks availability; 5 years for others
Damage assessment	Loss of life/livestock/ land; Damage to land crop/buildings/roads/ drainage structure/ bridge and culverts. Estimates for repairs of each	M.A.	District, Block & village level officials; NGOs; Agriculture Deptt/NRSA	With each disaster
Unit cost of relief measures	Construction cost for different types of buildings/roads/rail liens/culverts/bridges/ drains/slope protection/ desilting/dredging/ embankments etc.	M.A	PWD; Irrigation Deptt. Agriculture Deptt; Railways; Telecom Deptt.	5 years
LANDSLIDES, EARTH- QUAKES Landslide Hazard Zonation	Landslide incidence; Slope characters; Slope-forming material; Lithology; Bedding and structures; Drainage and possible obstructions; Rainfall data (Including frequency); Topo maps; Air photos; Satellite image	M.A.	Survey of India; NRSA, Geological survey; Met. Deptt.	Annual for landslide Incidence, Drainage obstacles and satellite images; 10 years for others
Monitoring drainage and slope conditions	Topographic maps; Air photos;Satellite images; Rainfall/Stream gage/ Reservoir discharge data; Cuts along critical slopes (roads/ buildings etc.)	M.A.	Survey of India; NRSA; Geological Survey; Met. Deptt.; Irrigation Deptt.; PWD Municipal and village records	Seasonal/ Annual
Seismological	Seismograph records;	M.A.	Met. Deptt;	Round the year

Contd.

data on precursors and after-shocks	Changes in resistivity, animal behaviour, groundwater levels etc.	Village Headman
Land use		Same as for Storms/Cyclones/Floods
Transportation capacity and bottlenecks		Same as for Storms/Cyclones/Floods
Population and livestock distribution		Same as for Storms/Cyclones/Floods
Damage assessment		Same as for Storms/Cyclones/Floods
Unit cost of different relief measures		Same as for Storms/Cyclones/Floods
Location and likely damage to buildings		Same as for Storms/Cyclones/Floods
Transport and commn. links		Same as for Storms/Cyclones/Floods

Note: Data types (Col.-4) depicted as M—Map type and A—Attribute type Updation frequency (Col-6), H—Hourly; D—Daily; W—Weekly; M—Monthly; Y—Yearly.
*Courtesy: ICODM, Guwahati (1998)

would be adequate. Delineation of blocks and assessment of potential damages in each due to the disaster can be made by experts along with probability of occurrence of different level of disasters. Such assessments of the affected areas modeling will be the most used component of operational information system.

The system is to be so designed so that flow of information is effective. A two-level distributed system with the first level comprising state capitals, major towns and the second at the district level. The second level nodes should have data storage capacity to cover up not only all the terrain specific parameters of its command area but also abstract of the parameters pertaining to all the adjacent units including disaster relief material storage details. Update of the dynamic parameters should also be attended to from these nodes.

A high speed data network having high volume data transfer capacity with the use of satellite communication is required for the operation of such a system.

4.8 REFERENCES

Bandyopadhyay, J., (1995), 'Water Management in the Ganges-Brahmaputra Basin: Emerging Challenges for the 21st Century', *Water Resources Development*, 11(4), 411-442.

Benjamin, J.R., and Cornell, C.A., (1970), *Probability, Statistics, and Decision for Civil Engineers*, McGraw-Hill Book Co, Inc, New York, USA.

Boven, K.D. and Milhouse, R.T., (1978), 'Hydraulic Simulation in Instream Flow Studies: Theory and Techniques', *Co-operative Instream Flow Service Group*, Paper No. 5, 121 pp., Fort Collins, Colorado, USA.

Chadderton, R.A., Miller, A.C. and McDonnell, A.J., (1982), 'Uncertainty analysis of dissolved oxygen model', *J. Environmental Engineering Division*, ASCE, 108(5), 1003-1012, New York, USA.

Chaturvedi, M.C. (1992), *Water Resources System Planning and Management*, Tata McGraw Hill Publishing Co. Ltd, New Delhi, India.

Chaturvedi, M.C. and Rogers. P., (1975), 'Large scale Water Resources System Planning with Reference to Ganga Basin', *Proc. Second World Congress*, International Water Resources Association, New Delhi, India.

Coleman, J.M., (1969) 'Brahmaputra river channel processes and sedimentation', *Sedimentation Geology*, Elsevier Publishing Co., Amsterdam, The Netherlands.

Fair, G.M., Geyer, J.C., and Okun, D.A., (1968), *Water and Wastewater Engineering*, Vol. 2, John Wiley and Sons, Inc., New York, N.Y., USA.

Franco, D. Perelli. M. and Scottolin, M. (1996), 'Buffer strips to protect the Venice lagoon from non-point source of pollution' *Proc. Intl. Conference on Buffer Zones: Their Processes and Potential in Water Protection*, Aug-Sept., Heythrop, U.K.

Garde, R.J., Ranga Raju, K.G., Swamee, P.K., Miraki, G.D. and Molanezhad, M. (1983) Mathematical Modelling of Sedimentation Process in Reservoirs and upstream Reaches. Hyd. Engg. Report, Civil Engg. Deptt. University of Roorkee, Roorkee, India

Gumbel, E.J., (1958), *Statistics of Extremes*, Columbia University Press, New York, USA.

Gutenberg, B. and Richter, C.F. (1954) Seismicity of the earth and associated phenomena, Princeton University Press, New Jersey, USA.

Haque, M., (1993), 'Simulation of the effects of sea level rise in the major rivers of Bangladesh' , *Jourl. of the Institution of Engineers (I)*, Vol. 74, Calcutta, India.

Harr, M.E., (1987), Reliability based Design in Civil Engineering, McGraw-Hill, New York, USA.

Hormberger, G.M., (1980), 'Uncertainty in dissolved oxygen prediction due to variabilty in algal photosynthesis', *Water Research*, 14(4): 355-361.

Kameswara Rao (1998), *Vibration Analysis and Foundation Dynamics*, Wheeler Publications, USA.

Kapur, K.C. and Lamberson, L.R., (1977), *Reliability in Engineering Design*, Wiley, New York, USA.

Kittrell, F.W., (1969), 'A practical guide to water quality studies of streams', Federal Pollution Control Administration, Publication, No: CWR-5, U.S. Department of Interior, Washington, DC, USA.

Kothandaraman, V., and Ewing, B.B., (1969), 'A probabilistic analysis of dissolved-biochemical oxygen demand relationship in stream', *Jourl. of Water Pollution. Cont. Fed.*, 41(2): R73-R(90)

Londe, P., (1980), 'Lessons from earth dam failure', *Symposium on Problems and Practices of Dam Engineering*, pp. 17-28, Bangkok, Thailand.

Loucks, D.P., and Lynn, W.R., (1966), 'Probabilistic models for predicting stream quality', *Water Resources Research*, 2(3): 593-605.

Padgett, W.J. and Rao, A.N.V., (1979), 'Estimation of BOD and DO probability distribution', *Jourl. Envir. Engg. Div.*, ASCE, 105(3): 525-533, New York, USA.

Proceedings of International Conference on Disaster Management (1998), Guwahati, Assam, Organized by Tezpur University, Tezpur, Assam, India.

Rao, K.L., (1979), *India's Water Wealth*, Orient Longman, Hyderabad, India.

Robinatte, G.O. (1984), Water conservation in landscape design and maintenance; Von Nostrand Reinhold Comp. Inc, New York, USA.

Rogers, P., and King, S.L., (1985), 'Decentralized planning for the Ganga Basin, decomposition by river basins', *Water Resources System Planning: Some Case Studies for India*, Indian Academy of Science, Bangalore, India.

Sachs, R.M., Kretchum, T. and Mock, T., (1975), 'Minimum Irrigation Requirements for Landscape Plants', *Jourl. American Society of Horticulture Societies*, 100(5): 499-502.

Sasikumar, K. (1993), *A new approach for the analysis and simulation of gradual earth dam failure*, M.Tech Thesis, Civil Engineering Department, IIT, Kharagpur, W. Bengal, India.

Sinha, B.P.C. Sharma, S.K. and Pal, O.P (1996), *Ground water pollution studies in India*, INCOH/SAR-12/96, National Institute of Hydrology, Roorkee, India.

Sassa, K., (1984), Monitoring of a crystalline schist landslide-compressive creep affected by underground erosion: Proceedings IV International Symposium on Landslides, Toronto, Canada.

Saunders, T.G., Ward, W.C., Lofts, J.C., Steels, T.D., Adrain, D.D. and Yevjevich, V. (1983), 'Design of Networks for Monitoring of Water Quality', *Journal of Water Resources Research*, USA.

Sharp, W.E., (1970), 'Stream order as a measure of sample source of uncertainty', *Journal of Water Resources Research*, Vol. 6, No. 3, USA.

Singh, Krishan P. (1982), Desirable Low Flow Releases below Dams: Fish Habitats and Reservoir Costs Water International, Vol. 7, No. 3, Autumn, IWRA.

Streeter, H.W., and Phelps, E.B., (1925), 'A study of the pollution and natural purification of the Ohio River', *Public Health Bulletin*, U.S. Public Health Service, Washington, DC, USA.

Varnes J.D., (1984), 'Landslide hazard zonation; A review of principles and practice', *Natural Hazards 3*, UNESCO Publication, Paris, France.

Verma, H.N. and Tiwari, K.N. (1995), *Current Status and Prospects of Rainwater Harvesting*, INCOH/SAR-3/95, National Institute of Hydrology, Roorkee, India.

Zaruba, Q. and Menel, V., (1969) *Landslides and Their Control*, Elsevier, Amsterdam, The Netherlands.

Chapter 5

Eco-technological Practices for Sustainable Development

5.1 GENERAL

The previous chapters in this book have highlighted various interactions between natural hydrological processes and many human-induced processes. The effect of unsustainable human activities on water availability and use has been briefly discussed in Section 1.2 of Chapter 1. Similarly the strategies required for achieving sustainable municipal water supply or the techniques which can be adopted for sustainable agriculture/pisciculture have been described in Chapter 2. In continuation with this, Section 3.7 of Chapter 3 enlists the basic considerations for eco-friendly design of water systems. While in Chapter 4, the mitigation and management of water hazards like floods, landslides, dam breaks and droughts as well as strategies like rainwater harvesting have been discussed. The present chapter summarizes all such positive activities under a single umbrella of *eco-technology* leading to *sustainable development*.

Eco-technology involves blending frontier technologies such as information technology, space technology and biotechnology with—ecological prudence and practices of local communities. Such technology blending is done jointly with rural families in a participatory research mode in whole villages, termed 'Biovillages', which are laboratories for 'eco-technology in action'. The objective of eco-technological practices is to achieve *sustainable development* through conservation of natural resources such as air, water, land, etc.

The most widely quoted definition of the term 'Sustainable Development' [Brundtland Report, 1988] links the two concepts of 'environment' and 'development'. It refers to the 'development seeking to meet the needs of the present generation without compromising the ability of future generations to meet their own needs'. It aims at assuring the on-going productivity of exploitable natural resources and conserving all species of fauna and flora'.

This concept has been further developed within the UN system, which tends to use the term: *sustainable human development*. The introduction of the 'human' dimension places human development at the forefront of regional integration and sustainable development of a country or continent. 'In recognition of the centrality of the human dimension to development, the UN general Assembly in 1986 adopted a 'Declaration on the right to development' stating that the human person is the central subject of development', and called upon member states 'to ensure access to the basic resources, education, health services, food, housing, employment and a fair distribution of income'.

Therefore, water conservation is essential to achieve sustainable agriculture and food production, which is a major component of sustainable development. Since ancient times, human beings have depended on rainwater followed by river water and groundwater for domestic and agricultural needs.

In areas where rain-fed agriculture is not feasible, the water used for farming and domestic purposes stems usually from running waters and dams or wells. Any method that saves water, especially the application of sources other than river water, reservoir water or pumped groundwater, can be termed a water conservation method. Traditional water conservation techniques have been practiced in many dry areas since millennia. Some examples mentioned here are water harvesting, quanats, artesian wells and horizontal wells. The revival and improvement of these methods would provide excellent opportunities to ease the water scarcity in the arid and semi-arid areas of the world.

5.2 TRADITIONAL WATER CONSERVATION PRACTICES

When it comes to supplying water for irrigation and domestic purposes, most people think of rivers and large reservoirs or tube wells, which tap groundwater resources. Indeed, a large proportion of the water for irrigation is taken either from streams and rivers or from reservoirs and aquifers, but this is not the full picture. In India, for example, about 40 million hectares are irrigated using these sources, whereas some 6 million hectares are irrigated from 'other sources'. In Indian statistics, these 'other sources' are summarized as 'sources other than government canals, wells and tube wells' (Sengupta, 1993). Behind this term are various water conservation techniques and traditional irrigation methods, mainly various forms of water harvesting. These traditional techniques played a much greater role in the past and were the backbone of ancient civilizations in arid and semi-arid areas around the world. Besides water harvesting, other traditional techniques, which tap into groundwater resources, like the quanat systems, are still used widely. These two techniques make it possible for plants to utilize the water immediately or for it to be stored in reservoirs for later use in the dry periods of the year. A third group of traditional techniques which ease water scarcity are related to in-situ water conservation. Water scarcity will be one of the major threats to humankind in the next century. As the available water resources taken from streams, rivers and groundwater will not be sufficient in dry areas to cover the needs of agricultural and urban areas, we have to reassess the value of traditional water conservation techniques and find out their potential to ease water scarcity. During the last two decades, traditional techniques have gained importance, and renewed interest in them can be observed among researchers as well as practitioners.

5.2.1 Historical Overview

Since ancient times, farmers and herders in dry areas of the tropics and subtropics have, under a wide range of ecological conditions, attempted to conserve water and to increase agricultural production. A number of indigenous water conservation techniques can be found in areas which have between 100 and 1500 mm annual precipitation and with population densities varying from 10-500 persons/km^2. To ease water scarcity in dry areas without using irrigation water from permanent rivers, reservoirs or lifted groundwater, several methods can be applied (including the ancient techniques) as described below:

1. Making better use of rainfall by
   - Minimizing runoff losses (in combination with increased infiltration)
   - Collection and concentration of rainfall (including storage)
   - Minimizing evaporation losses
   - Minimizing transpiration losses
   - Improving rainwater usage by plants
2. Making use of groundwater without water lifting by
   - quanat systems
   - artesian wells
   - horizontal wells

5.2.1.1 *Making Better Use of Rainfall*

i). Minimizing runoff losses (in combination with increased infiltration):

Examples of such techniques are:
- ridges, constructed along the contour
- stone lines made of single or multiple rows of stones
- trash lines made of organic residues such as straw, maize stokes, weeds, etc.
- furrows, with crops being grown at the bottom of the furrows or at the edges;
- tied ridges
- pitting systems: either manually prepared (the crops are grown in small depressions of 30 cm diameter with soil improved by organic material) or by tractor-pulled implements
- terraces: induced terraces, constructed terraces, conservation terraces
- mulching: covering the soil with organic material
- contour farming
- strip cropping
- conservation tillage or no tillage
- agro-forestry techniques, e.g., planting of *Faidherbia albida*.

Minimization of runoff losses (in combination with increased infiltration) and the collection and concentration of rainfall (including storage) are the most important approaches in making better use of the scanty rainfall in arid and semi-arid areas. In tropical and subtropical countries, many indigenous agricultural practices are directed towards minimizing runoff losses in combination with protecting the soil resources, i.e., they contain certain elements of water and soil conservation. In other cases, the moisture conservation effect is a secondary effect of soil conservation, as shown in traditional techniques such as constructing dry stone bunds or terraces. There are also a number of traditional techniques, which are primarily directed to in-situ moisture conservation. For example we have the application of the 'zay' technique, which is common in Mali and Burkina Faso. The 'zay' system consists of small pits, 5-15 cm in depth and 10 - 30 cm in diameter. The pits are spaced out, leaving 50-100 cm between them. Manure and grasses are mixed with some of the soil and put into the zay to increase nutrient status and water holding capacity (Figure 5.1). The pits fill up when it rains, and the water infiltrates into the soil where it is stored. With modification of the area between the pits, such that it serves as a 'catchment', the 'zay' could be regarded as a water harvesting technique (see below). Zays are often applied in combination with bunds to conserve runoff.

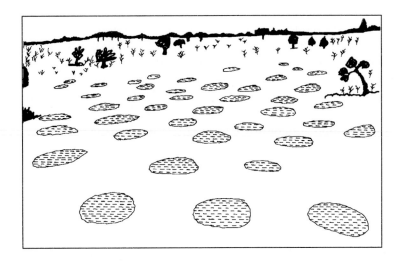

Figure 5.1 'Zay' Pitting Holes in Burkina Faso, Serving Improved Soil Moisture and Soil Fertility, After Rainfall

[Source: www.ubka.uni-karlsruhe.de/indexer-vvv/1998/bau-verm/6]

Projects in a number of countries have shown that moisture conservation techniques are accepted by farmers as they can decrease the production risk considerably and increase the average yield significantly (Critchley et al., 1992b). Experiments in Kanguessanou, Kayes Province, Mali (550 mm/yr precipitation), have shown, that bunding, i.e., the construction of an earth dike around the field, could increase the yield of Sorghum from 0.8 t/ha to 1.8 t/ha, even in dry years.

ii). *Collection and concentration of rainfall (including storage):* The collection and concentration of rainfall and its use for the irrigation of crops, pastures, trees, livestock consumption and household purposes is often called *water harvesting.* In the past, water harvesting played an important role for agricultural societies in arid and semi-arid areas worldwide. After a decline, it picked up new interest in recent decades. Each water harvesting system requires a:

- 'runoff area' (catchment) with a sufficiently high run off coefficient and
- 'run-on' area for utilization and/or storage of the accumulated water, in case of agricultural use the cropping area

Water Harvesting has been practiced in many dry regions of the world since millennia:

Asia: The importance of the Middle East in the development of ancient water harvesting techniques is unquestioned. Archeological evidence of water harvesting structures appears in Jordan, Syria, Iraq, the Negev and the Arabian Peninsula, especially Yemen. In Jordan, there is indication of early water harvesting structures believed to have been constructed over 9,000 years ago. The internationally best known runoff-irrigation systems have been found in the semi-arid to arid Negev desert region of Israel (Evenari et al., 1971). Runoff agriculture in this region can be traced back as far as the 10th century BC when it was introduced by the Israelites. It continued throughout Roman rule and reached its peak during the Byzantine era. In northern Yemen, a system dating back to at least 1,000 BC diverted enough floodwater to irrigate 20,000 hectares producing agricultural products that may have fed as many as 300,000 people (Eger, 1988). Farmers in the same area are still irrigating with floodwater, making the region perhaps one of the few places on earth where runoff agriculture has been continuously used since the earliest time of settlement. In the South Tihama of Saudi Arabia, flood irrigation is traditionally used for sorghum production. Today, approximately 35,000 ha land, supporting 8,500 to 10,000 farm holdings, are still being irrigated with flood waters (Wildenhahn, 1985). In Baluchistan, Pakistan, two water harvesting techniques were already applied in ancient times: the 'Khuskaba' system and the 'Sailaba' system. The first employs bunds being built across the slope of the land to increase infiltration. The latter utilizes floods in natural water courses which are captured by earthen bunds (Oosterbaan, 1983).

India in particular is ecologically very diverse—from the dry cold deserts of Ladakh to the dry, hot desert of Rajasthan; from the sub-temperate mountains of the Nilgiris to the alluvial monotonous plains of the Gangetic valley and the plateaus of the Deccan and Chhotanagpur; and from the dry slopes of the Aravallis to the humid slopes of Meghalaya, Nagaland and Mizoram in North Eastern India. Over the centuries, Indians have developed a range of techniques to harvest every possible form of water—from rainwater to groundwater, river water and floodwater [Agarwal and Narain, (1997 and 2001)]. Some of the major traditional water harvesting structures from different parts of India are described in the following paragraphs.

In India's arid and semi-arid areas, the 'tank' system is traditionally the backbone of agricultural production. Tanks are constructed either by bunding or by excavating the ground and collecting rainwater. In western Rajastan, with desert-like conditions having only 167 mm annual precipitation, large bunds were constructed as early as the 15th century to accumulate runoff. These 'Khadins' create a reservoir which can be emptied at the end of the monsoon season to cultivate wheat and chickpeas (Kolarkar et al., 1983). A *khadin*, also called a *dhora* (Figure 5.2), is an ingenious construction designed to harvest surface runoff water for agriculture. Its main feature is a very long (100-300 m) earthen embankment built across the lower hill slopes lying below gravelly uplands. Sluices and spillways allow excess water to drain off. The *khadin* system is based on the principle of harvesting rainwater on farmland and subsequent use of this water-saturated land for crop production. First designed by the Paliwal Brahmins of Jaisalmer, western Rajasthan in the 15th century, this system has great similarity with the irrigation methods of the people of Ur (present Iraq) around 4500 BC and later of the Nabateans in the Middle East. A similar system is also reported to have been practiced 4,000 years ago in the Negev desert, and in southwestern Colorado 500 years ago.

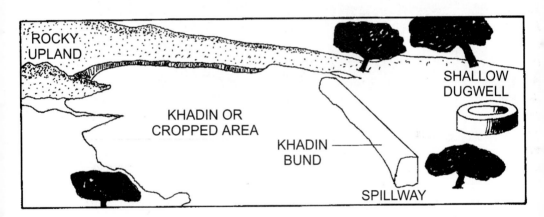

Figure 5.2 A Typical *Khadin* in Rajasthan, Western India

[*Source:* http://www.rainwaterharvesting.org/]

A similar system called 'Ahar' (Figure 5.3) was developed in the state of Bihar [UNEP (1983), Pacey and Cullis (1986)]. An *ahar* is a catchment basin embanked on three sides, the 'fourth' side being the natural gradient of the land itself. Ahar beds were also used to grow a *rabi* (winter) crop after draining out the excess water that remained after *kharif* (summer) cultivation. Ahars are often built in a series. It was observed that brackish groundwater in the neighborhood of Ahars became potable after the Ahar was built. *Pynes* are articifial channels constructed to utilize river water in agricultural fields. Starting out from the river, *pynes* meander through fields to end up in an ahar. Most *pynes* flow within 10 km of a river and their length is not more than 20 km.

Figure 5.3   An *Ahar* and *Pyne* Type Water Harvesting System in Bihar State, Eastern India

[Source: : http://www.rainwaterharvesting.org/; Agarwal and Narain, 2001]

This traditional floodwater harvesting system is indigenous to south Bihar. In south Bihar, the terrain has a marked slope—1 m per km—from south to north. The soil here is sandy and does not retain water.

Groundwater levels are low. Rivers in this region swell only during the monsoon, but the water is swiftly carried away or percolates down into the sand. All these factors make floodwater harvesting the best option here, to which this system is admirably suited. The ahar-pyne system received a death-blow under the nineteenth-century British colonial regime. The post-independent state was hardly better. In 1949, a Flood Advisory Committee investigating continuous floods in Bihar's Gaya district came to the conclusion that 'the fundamental reason for recurrence of floods was the destruction of the old irrigational system in the district'.

Of late, though, some villages in Bihar have taken up the initiative to re-build and re-use the system. One such village is Dihra. It is a small village 28 km southwest of Patna city. In 1995, some village youths realized that they could impound the waters of the Pachuhuan (a seasonal stream passing through the village that falls into the nearby river Punpun) and use its bed as a reservoir to meet the village's irrigation needs. Essentially, this meant creating an *ahar-pyne* system. After many doubts, the village powers-that-be gave the go-ahead. Money was collected and work began in May 1995. After a month of voluntary labor the villagers completed their work mid-June. Their efforts have borne fruit. By 2000 AD, the *ahar* was irrigating 80 ha of land. The people grow two cereal crops and one crop of vegetables every year. The returns from the sale of what they produce are good. The village is no longer a poor one.

Zings (Figure 5.4) are water harvesting structures found in Ladakh, Jammu & Kashmir, northern India. They are small tanks, which collect melted glacier water. Essential to the system is the network of guiding channels that brings the water from the glacier to the tank. As glaciers melt during the day, the channels fill up with a trickle that in the afternoon turns into flowing water. The water collects towards the evening, and is used the next day. A water official called the *churpun* ensures that water is equitably distributed.

Kuls (Figure 5.5) are water channels found in precipitous mountain areas. These channels carry water from glaciers to villages in the Spiti valley of Himachal Pradesh in northern India. Where the terrain is muddy, the *kul* is lined with rocks to keep it from becoming clogged. In the Jammu region too, similar irrigation systems called *kuhls* are found.

Johads (Figure 5.6) are small earthen check dams that capture and conserve rainwater, improving percolation and groundwater recharge. Starting 1984, the last sixteen years have seen the revival of some 3000 *johads* spread across more than 650 villages in Alwar district, Rajasthan. This has resulted in a general rise of the groundwater level by almost 6 meters and a

Figure 5.4   A *Zing* in Jammu & Kashmir State, North India
[Source: http://www.rainwaterharvesting.org/]

Figure 5.5   A Typical *Kul* in Himachal Padesh, North India
[Source: http://www.rainwaterharvesting.org/]

33 percent increase in the forest cover in the area. Five rivers that used to go dry immediately following the monsoon have now become perennial, such as the River Arvari, has come alive.

West Bengal in Eastern India and Bangladesh once had an extraordinary system of *inundation canals*. Sir William Willcocks, a British irrigation expert

Figure 5.6   A *Johad* Revived in Rajasthan, India
[Source: http://www.rainwaterharvesting.org/]

who had also worked in Egypt and Iraq, claimed that inundation canals were in vogue in the region till about two centuries ago. Floodwater entered the fields through the inundation canals, carrying not only rich silt but also fish, which swam through these canals into the lakes and tanks to feed on the larva of mosquitoes. This helped to check malaria in this region. According to Willcocks, the ancient system of overflow irrigation had lasted for thousands of years. Unfortunately, during the Afghan-Maratha war in the 18th century and the subsequent British conquest of India, this irrigation system was neglected, and was never revived. According to Willcocks, the distinguishing features of the irrigation system were:

1) the canals were broad and shallow, carrying the crest waters of the river floods, rich in fine clay and free from coarse sand;
2) the canals were long and continuous and fairly parallel to each other, and at the right distance from each other for purposes of irrigation;
3) irrigation was performed by cuts in the banks of the canals, which were closed when the flood was over.

Bandharas (Figure 5.7) are check dams or diversion weirs built across rivers. A traditional system found in Maharashtra, Western India, their presence raises the water level of the rivers so that it begins to flow into channels. They were also used to impound water and form a large reservoir. Where a *bandhara* was built across a small stream, the water supply would usually last for a few months after the rains. They were built either by villagers or by private persons who received rent-free land in return for their public act. Most Bandharas are defunct today. Very few are still in use and about eight were revived by the Irrigation Department.

Figure 5.7 A *Bandhara* Revived in Maharashtra State, Western India

[Source: http://www.rainwaterharvesting.org/]

The community-managed *phad irrigation system*, prevalent in north western Maharashtra, probably came into existence some 300-400 years ago. The system operated on three rivers in the Tapi basin—Panjhra, Mosam and Aram—in Dhule and Nasik districts (still in use in some places here). The system starts with a *bandhara* (check dam or diversion-weir) built across a rivers. From the *bandharas* branch out *kalvas* (canals) to carry water into the fields. The length of these canals varies from 2-12 km. Each canal has a uniform discharge capacity of about 450 liters/second. *Charis* (distributaries) are built for feeding water from the kalva to different areas of the *phad*. *Sarangs* (field channels) carry water to individual fields. *Sandams* (escapes), along with *kalvas* and *charis*, drain away excess water. In this way water reaches the *kayam baghayat* (agricultural command area), usually divided into four *phads* (blocks). The size of a *phad* can vary from 10-200 ha, the average being 100-125 ha. Every year, the village decides which phads to use and which to leave fallow. Only one type of crop is allowed in one phad. Generally, sugarcane is grown in one or two phads; seasonal crops are grown in the others. This ensures a healthy crop rotation system that maintains soil fertility, and reduces the danger of waterlogging and salinity. The *phad* system has given rise to a unique social system to manage water use.

Bamboo Drip Irrigation in the Meghalaya state in north eastern India (Figure 5.8) has an ingenious system of tapping of stream and springwater by using bamboo pipes to irrigate plantations. About 18-20 liters of water entering the bamboo pipe system per minute gets transported over several hundred meters and finally gets reduced to 20-80 drops per minute at the site of the plant. This 200-year-old system is used by the tribal farmers of Khasi and Jaintia hills to drip-irrigate their black pepper cultivation. The wettest

Figure 5.8   Bamboo Drip Irrigation System in Meghalaya State,
North-eastern India
[Source: http://www.rainwaterharvesting.org/]

place of the world Cherrapunji is located here. Bamboo pipes are used to divert perennial springs on the hilltops to the lower reaches by gravity. The channel sections, made of bamboo, divert and convey water to the plot site where it is distributed without leakage into branches, again made and laid out with different forms of bamboo pipes. Manipulating the intake pipe positions also controls the flow of water into the lateral pipes. Reduced channel sections and diversion units are used at the last stage of water application. The last channel section enables the water to be dropped near the roots of the plant. Bamboos of varying diameters are used for laying the channels. About a third of the outer casing in length and internodes of bamboo pieces have to be removed while fabricating the system. Later, the bamboo channel is smoothened by using a dao, a type of local axe, a round chisel fitted with a long handle. Other components are small pipes and channels of varying sizes used for diversion and distribution of water from the main channel. About four to five stages of distribution are involved from the point of the water diversion to the application point.

Africa: Since at least Roman times, water harvesting techniques were applied extensively in northern Africa. Archeological research revealed that the wealth of the 'granary of the Roman empire' was largely based on runoff irrigation (Gilbertson, 1986). In Morocco's Anti Atlas region, Kutsch (1982) investigated the traditional water harvesting techniques, some of which are still practiced today. He found a wealth of experience and a great variety of locally well adapted systems. In Algeria, the 'lacs collinaires', the rainwater storage ponds, are a traditional means of water harvesting for agriculture. The open ponds are mainly used for watering animals. In Tunisia, the 'Meskat', the 'Jessour' and the 'Mgoud' systems have a long tradition, and are also still practiced. The 'Meskat' is a microcatchment system which provides fruit tree plantations with about 2,000 m^3 extra water during the rainy season (Figure 5.9); the 'Jessour' system is a terraced wadi system with earth dikes ('tabia'). Lastly, the 'Mgouds' are channel systems used to divert flood water from the wadi to the fields (Tobbi, 1993).

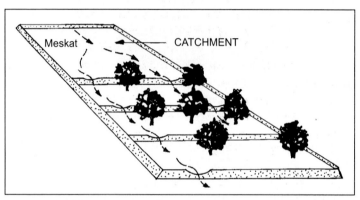

Figure 5.9 The Tunisian 'Meskat' Water Harvesting System (Adapted from www.ubka.uni-karlsruhe.de/indexer-vvv/ 1998/bau-verm/6)

In Egypt, the North West coast and the northern Sinai areas have a long tradition in water harvesting. Some wadi terracing structures have been used there for over centuries. Traditional techniques of water harvesting have been reported from many regions of Sub Saharan Africa (Critchley et al., 1992a): The 'Caag' and the 'Gawan' systems are practiced in Somalia. The former is a technique used to impound runoff from small water courses, gullies or even roadside drains; the latter is made up of small bunds which divide plots into 'grids' of basins of 500 m^2 or above in size. In Sudan, various types of 'Hafirs' have been in use since ancient times. Their water is used for domestic and animal consumption as well as for pasture improvement and paddy cultivation (UNEP, 1983). The 'Haussa' in Niger and the 'Mossi' in Burkina Faso traditionally divert water to their fields with rock bunds, stalks and earth or construct rock bunds and stone terraces.

America: The Papago Indians and other groups practiced traditional water harvesting in the Sonoran desert. Brush weirs were used to spread the floodwaters. Elsewhere, fields were irrigated by gravity-fed channels (arroyos) leading water from earth and stick or rock diversion weirs. A highly sophisticated distribution system was demonstrated by the flood water diversion system of Chaco Canyon, New Mexico, USA.

According to the size of the catchment and the relation between the size of the catchment and that of the cropping area, one can distinguish different types of water harvesting: microcatchment water harvesting, macrocatchment water harvesting and large catchment water harvesting. The higher the aridity of an area, the more catchment area is required in relation to the cropping area in order to yield the same amount of water. Microcatchment water harvesting (Figure 5.10) is a method of collecting surface runoff (sheet or rill flow) from a small catchment area and storing it in the root zone of an adjacent infiltration basin. The basin is planted with a single tree or bush or with annual crops. Water harvesting from macrocatchments is also called *water harvesting from long slopes* or *harvesting from external catchment systems* (Pacey and Cullis, 1988). In this case, the catchment located outside the cropping area collects mainly turbulent runoff and channel flow. This is conveyed to the cropping area located below the foot of the hill on flat terrain. As the name suggests, the catchments of large catchment water harvesting systems can be many square kilometers in size and give rise to runoff water flowing through major wadis. Large catchment water harvesting is also called *floodwater harvesting* by many authors, and comprises two forms: In case of *floodwater harvesting within the stream bed* the water flow is dammed and, as a result, inundates the valley bottom of the entire flood plain. The water is forced to infiltrate and the wetted area can be used for agriculture or pasture improvement. In case of *floodwater diversion*, the wadi water is forced to leave its natural course and conveyed to nearby cropping areas. Large catchment water harvesting requires more complex structures of dams and distribution networks and a higher technical input than the other two water harvesting methods.

iii) Minimizing evaporation losses: Only part of the rainfall or irrigation water can be used by the plants, the rest percolates into the deep groundwater, or is lost by evaporation from surfaces and evapotranspiration by plants. There are numerous methods to reduce such losses and to improve soil moisture, of which the most important ones are mulching, i.e., the application of organic or inorganic material, such as peat, plant debris, straw, etc., which slows down the surface runoff, improves the soil moisture and reduces evaporation losses. Soil cover

Figure 5.10 Microcatchment Water Harvesting for Growing Trees
(Adapted from www.ubka.uni-karlsruhe.de/indexer-vvv/
1998/bau-verm/6)

by crops slows down runoff, and minimizes evaporation losses. An additional advantage is the possibility of nitrogen fixation by specific plant species or the use of the cover for grazing. Ploughing can hinder the capillary movement of water in the soil and, as a consequence, reduce the evaporation. Shelter belts consisting in most cases of bushes or trees slow down the wind speed near the crops, thereby reducing evaporation, erosion and physical damage to crops. At the same time, they may increase precipitation and serve as fodder, fuelwood etc. Some chemicals such as emulsions of water and oil or wax (hexadecanol, bitumen, asphalt, latex), reduce evaporation when sprayed on soil or water surfaces.

iv) Minimizing transpiration losses: The aforementioned shelterbelts, which reduce evaporation losses due to a reduction of wind speed, also reduce water loss through transpiration in plants. If crops are sprayed with Kaolinite and water, they are heated up less by the sun and therefore loose less water by transpiration. Also wax emulsions, silicone or latex applied directly on the plant, may reduce transpiration. An optimal solution would be to select crops, which have a low evapotranspiration by nature, as many of those endemic to dry areas.

v) Improving rainwater usage by plants: The improvement of rainwater usage by plants means the improvement of soil moisture retention, together with decreased percolation, and an increase in the availability of stored water for plants. The addition of organic matter such as compost, peat, seaweed, paper and crushed coal to the soil leads to higher moisture retention. A positive side effect of the application of organic matter is improved soil fertility. Some artificially produced compounds such as starch copolymers or granular polymers (PAM = Polyacrylamide) have been applied successfully as well. Special care should be given to the selection of crop species. One should seek varieties suitable to the site, for example, those that develop a fine and deep rooting-system.

5.2.1.2 *Making Use of Groundwater without Water Lifting*

i) Quanat systems A Quanat (Figure 5.11) is a horizontal tunnel that taps underground water in an alluvial fan without pumps or equipment, and brings it to surface so that the water can be used (National Academy of Sciences, 1974). A quanat system is composed of three parts:

- one or more vertical head wells, dug into the water-bearing layers of the alluvial fan, to collect the water
- a gently downward-sloping underground horizontal tunnel leading the water from the head wells to lower point at the surface
- a series of vertical shafts between the ground surface and the tunnel, for ventilation and removal of excavated debris.

Quanats, bringing groundwater from mountainous areas to arid plains, have an inclination of 1–2‰ and a length of up to 30 km, yielding water in the range of 5-60 lit/s, in extreme cases up to 270 lit/s (National Academy of Sciences, 1974).

The Quanat systems (as shown in Figure 5.12) and the use of water from Artesian wells have a very long history: The origin of the "quanat" technique is Persia (i.e., ancient Iran), where it was developed about 3,000 years ago. The Persians of that time learned to dig tunnels to bring mountain groundwater to arid plains. This knowledge spread to the neighboring countries and was distributed by the Muslim Arab invaders throughout North Africa all the way up to Spain. Though new quanats are seldom built today, many old ones are still maintained and deliver water steadily to fields and villages. In Iran there are still some 40,000 quanats comprising more than 270,000 km of underground channels that supplied about 35 percent of the country's water two decades ago (National Academy of Sciences, 1974). Rehabilitation programes were started in a number of countries such as Oman and Morocco.

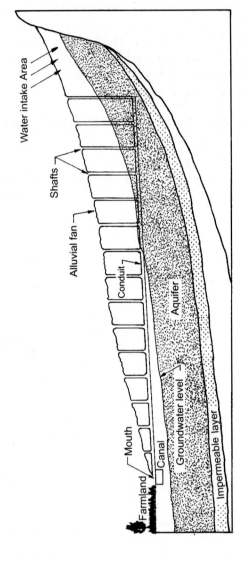

Figure 5.11 *Quanat* System Conveying Water by Gravity to the Ground Surface (Adapted from National Academy of Sciences, 1974)

When digging bore-holes in (semi-)arid areas, ancient farmers detected the special nature of artesian wells, which deliver groundwater without any need to lift it. Some of them have been in use for more than 2,000 years.

Figure 5.12   Craters, Each Marking the Mouth of a Quanat Shaft in Southern Morocco. Groundwater from the Higher Atlas Mountains is Directed to an Oasis in the Plains (Adapted from www.ubka.uni-karlsruhe.de/indexer-vvv/1998/bau-verm/6).

ii) **Artesian wells** *Artesian wells* (Figure 5.13) tap groundwater of aquifers where the water is confined by an overlying, relatively impermeable layer and the pressure is sufficient to raise the water in a well without pumping. This type of spring can have a high water yield.

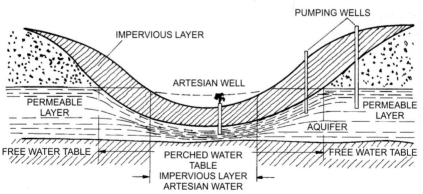

Figure 5.13   Artesian Well (Adapted from www.ubka.uni-karlsruhe.de/indexer-vvv/1998/bau-verm/6)

iii) Horizontal wells: Horizontal wells (Figure 5.14) tap underground water, which is trapped by an impervious geological barrier. They are installed by boring a hole with a horizontal boring rig and inserting a steel pipecasing.

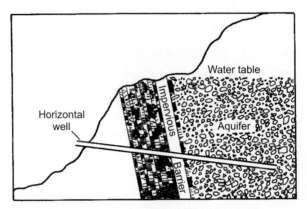

Figure 5.14 Horizontal Well [Adapted from National Academy of Sciences, 1974]

5.2.2 Added Values due to Traditional Water Conservation Systems

All the above-mentioned techniques have the advantage of increasing the amount of water available for agricultural and other purposes, and easing water scarcity in arid, semi-arid areas as well as hilly areas. They require relatively low input and, if planned and managed properly, can contribute to the sustainable use of the precious water. Water Harvesting has the potential to increase the productivity of arable and grazing land by increasing the yields and by reducing the risk of crop failure. It also facilitates re- or afforestation, fruit tree planting or agroforestry. With regard to tree establishment, water harvesting can contribute to the fight against desertification. Water harvesting is relatively cheap and can therefore be a viable alternative where irrigation water from other sources is not readily available or too costly. Unlike pumping water for animal consumption, water harvesting saves energy and maintenance costs. Using harvested rainwater helps to decrease the use of other valuable water sources, such as groundwater. Properly applied methods of evaporation loss reduction, such as the application of mulch and the careful ploughing or planting of shelter belts have proved to be beneficial for plant establishment, plant growth and yield, but also for erosion control and prevention of further land degradation. Since quanats are a traditional method, the local people are aware of this technique and more ready to implement or revive it. Artesian wells provide a low-cost possibility for steady access to sufficient water resources. Horizontal wells are relatively low cost; no pumps are needed and the yield can be controlled. There is no danger of contamination by dust or animals.

5.2.3 Limitations of Traditional Water Conservation Systems

It is well acknowledged that the above-mentioned techniques have more advantages than disadvantages. Nevertheless, some drawbacks may occur. Although water harvesting can increase the water available for plants, climatic risks still exist. Further, in years with extremely low rainfall, it cannot compensate for the shortage. Water harvesting provides no guarantee for high yields, since yields do not depend on water only, but also on other factors such as soil fertility, pest control, etc. The labor input required for water harvesting is often higher than that of rainfed farming. Successful water harvesting projects are often based on field experience and trial and error rather than on scientifically well-established techniques, and can therefore not be reproduced easily. Agricultural extension services have often limited experience with it. Further disadvantages are the possible conflicts between users of upstream and downstream, and a possible harm to fauna and flora adapted to running waters and other wetlands. The application of chemicals for the purpose of lowering evaporation and transpiration is relatively expensive and might not be suitable for developing countries. In addition, the environmental impacts have not been fully researched. Quanats require relatively high labor input and maintenance. They bear a maximum flow during the rainy season and a minimum flow during the dry season, as opposed to the demand for irrigation water. Under certain circumstances, they may dry up completely for a certain time. The areas that can be supplied with water from quanats lie at lower elevations of alluvial fans and provide less fertile soil conditions than those areas, which are higher up. As for all situations, where groundwater is extracted, there is a risk of exploiting more water than can be recharged by natural processes in the same year. Artesian wells depend on the presence of perched water tables and might not be available. They have the drawback that they provide water all the time, the volume per second depending on the natural pressure. Unused water might therefore be wasted. Horizontal wells are a good and simple method, but they can only be used where the geological conditions exist. The technical installation requires some experience.

5.2.4 Outlook on Traditional Water Conservation Systems

Although the results of traditional water conservation methods are encouraging and should be promoted, the ancient and current methods to ease water scarcity might not be efficient enough to compensate for extreme climatic conditions, global climate change and the increasing demand for agricultural production due to an excessive population growth. In some countries, the water is so scarce, that every available measure is used to preserve literally every drop of rain. Costs increase with the effectivity of the application of water conservation methods. In water harvesting, for example, it might be easy and cheap in a specific setting, to collect 50% of the runoff, but with increasing harvest, the costs per unit collected water increase considerably. Nevertheless, in countries like Morocco, high investment is made in order to meet the water demand. During recent years some methodological and technological developments took place with regards to rainfed farming in combination with a water conservation technique, or the combination of two or more different water conservation techniques. Water harvesting, for example, can be used as supplemental water source for rainfed agriculture, when the runoff is stored for application after the rainy season. The stored water allows for a lengthening of the cropping season or for a second crop. Water harvesting has been used in combination with underground dams that recharge groundwater. In Morocco, for example, water harvesting and quanats are used for water supply for the same area, thereby reducing the risks involved and saving some groundwater. Much research and much practical experience is still required before scientists will be able to recommend a particular method off-hand. As soon as more information on the hydrological, soil and crop parameters is available, models could be applied for comprehensive water management plans in certain areas. The potential is there, and hopefully, sustainable traditional water conservation methods will also receive the required backing of planners, who are aware of the scarcity of the resource water and the need to preserve this resource, which is so essential for life on earth.

5.3 RECENT ECO-TECHNOLOGICAL PRACTICES

In the previous Section 5.2, major types of traditional water conservation systems in different parts of the world have been discussed. In this section, recent eco-technological practices will be dealt with. These practices view sustainable development in a holistic manner and bring about water conservation through multiple dimensions such as soil conservation, water demand management, water supply augmentation etc. Some of the important practices among these are afforestation, rainwater/fog harvesting, wastewater treatment/reuse, other water conservation techniques and groundwater conservation/recharge. The following sections describe these practices.

5.3.1 Afforestation

Afforestation is the practice of growing different kinds of trees to create a plantation forest. It results in absorption of atmospheric heat, conservation of bio-diversity, conservation of soil, retention of soil moisture, increase in infiltration and reduction of runoff. Most of the highly developed nations have recorded an increase in their forest percentage of area due to a strong afforestation program. Norway is one of the best examples, wherein school children are trained in such activities.

Plantation forests now comprise around 135 million ha globally, with annual plantation afforestation and reforestation rates nearing 10% of the total area. Some 90% of plantation forests have been established primarily to provide industrial wood, and their relative global importance in this role is increasing rapidly. Most of the remaining 10% of plantation forests were established primarily to supply fuel or wood for non-industrial use. About 75% of the existing plantation forest estate is established in temperate regions, but it is in the tropics that the rate of expansion is the greatest. The expanding tropical plantation forest estate includes trees grown primarily as agricultural plantation crops and which now also supply wood to forest industries. Almost all existing plantation forests were established and are managed as even-aged monocultures; species and inter-specific hybrids of a few genera dominate plantation forestry worldwide.

Effective research and development, based on appropriate genetic resources and good silviculture (i.e., cultivation of forest trees), are the foundations of successful plantation forestry production. Resolving relatively fundamental issues remain the priority in many young plantation programs; in more advanced programmes, the application of more sophisticated technologies—particularly in biotechnology and processing—is necessary to maintain improvements in production. Many plantation forests, particularly in the tropics, are not yet achieving their productive potential.

The sustainability of plantation forestry is an issue of wide interest and concern. The evidence from industrial plantation forestry suggests that biological sustainability, in terms of wood yield, is likely to be realized provided good practice is maintained. The relative benefits and costs of plantation forestry in broader environmental terms, and in terms of its social impacts, are the subject of greater controversy, and pose the greatest challenge to plantation foresters as we enter the new millennium. The experience with plantation forestry as it has developed in the 20th century offers us an excellent platform for rising to these challenges.

5.3.1.1 Plantation Forests

It is difficult to define either 'Afforestation' or 'plantation forests' precisely. In particular, it is often not easy to distinguish between afforestation and either rehabilitation of degraded forest ecosystems or enrichment planting, or between plantation forests and various forms of trees on farms. The definition proposed by the Food and Agricultural Organization (FAO) to the 1967 World Symposium on 'Manmade Forests and their Industrial Importance', which uses as its criterion land use changes associated with afforestation or reforestation, has been the basis of subsequent official estimates (e.g., Pandey, 1995), and is adopted here for the sake of consistency. However, any consideration of plantation forests should acknowledge that the distinction between them and some other forms of forestry is not always clear; thus, definitions, discussion and estimates vary.

The global extent of plantation forests in 1990 was estimated at around 135 million ha [FAO (1993), Pandey (1995)]. About 75% of these plantation forests are in temperate regions and about 25% in the tropics and subtropics; some 5% of are found in Africa, a little more than 10% in each of the American continents, some 20% in the CIS (Commonwealth of Independent States, i.e., former USSR), and around 25% in each of Asia-Pacific and Europe [Kanowski and Savill (1992)]. Species and interspecific hybrids of only a few genera—*Acacia, Eucalyptus, Picea* and *Pinus*—dominate plantation forests, with those of a few others, e.g., *Araucaria, Gmelina, Larix, Paraserianthes, Populus, Pseudotsuga* or *Tectona*—of regional importance [Evans (1992), Pandey (1995), Savill and Evans (1986)]. The ownership of plantation forests extends from governments and large industrial corporations to individual farmers, and their management varies considerably, from relatively simple and low-input to highly sophisticated and intensive. Most plantation forests have been established as even-aged monoculture crops of trees with the primary purpose of wood production (Evans, 1997). Around 90% of existing plantations have been established for the production of wood for industrial use, and most of the remainder to produce wood for use as fuel or roundwood. Some plantation forests are grown and managed, either primarily or jointly, for non-wood products such as essential oils, tannins, or fodder. The provision of a diverse range of other forest benefits and services, including environmental protection or rehabilitation, recreational opportunities, and CO_2 sequestration are also primary or secondary objectives for many plantation forests.

Trees grown as agricultural plantation crops—e.g., rubber or coconut—have not traditionally been considered as forest plantations. However, the distinction between the two forms of plantation culture is diminishing from two perspectives: from that of the forest manager, as rotation ages reduce and the intensity of forest plantation management increases; and from that of the agricultural tree estate manager, as these crops begin to be used for wood products. The recent example of forest industry development based on wood supply from Asian rubber plantations exemplifies the latter, and provides a striking example of how shifting supply factors and improved processing technologies can offer opportunities to non-traditional supply sources, and thus expand the plantation base. Rubberwood recovered from rubber estate re-establishment programs now substitutes for many traditional industrial uses of natural forest woods from South East Asia, and provides the raw material for newer products such as medium-density fibreboard. Similar processing developments are in train, though as yet less advanced, for the other major tropical estate tree crops, oil palm and coconut. Given the substantial areas of these plantation crops worldwide—estimated at around 7 M ha of rubber estate, 4 M ha of coconut, and 3 M ha of oil palm—they have considerable potential to both supplement and compete with production from more conventional plantation forests.

The harvest rotations of forest plantations vary enormously, from annual or sub-annual for some non-wood products, to around 200 years for traditionally-managed high-value temperate hardwoods. With few exceptions so far, shorter rotation plantations—typically of 5 to 15 years—have been grown for fuel, fiber, and longer rotation plantations—typically upwards of 25 years—principally for sawn or veneer wood products.

Notwithstanding successful antecedents in both temperate (e.g., oak in Europe) and tropical (e.g., teak in India, rest of Asia) [Keh, 1997] environments, plantation forests on large scale are a twentieth-century phenomenon. The majority of the world's plantation forests have been established in the past half-century, and the rate of plantation afforestation has been increasing progressively during this period. Global rates of forest plantation establishment and reestablishment are poorly known, but are estimated at around 2.6 million ha annually in the tropics [FAO (1993), Pandey (1995)], and perhaps 10 million ha in the temperate zones [Mather (1990 and 1993)].

Recent plantation expansion has been greatest in the southern hemisphere: in South America (principally Argentina, Chile and Brazil), Asia (principally Indonesia) and New Zealand, where particular coincidences of public policies, opportunities and market forces have been most conducive to afforestation. In some countries, e.g., Indonesia or Chile, plantation establishment remains concentrated on sites converted directly from natural ecosystems; in others, e.g., New Zealand or Portugal, plantation establishment has shifted entirely to sites formerly used for agriculture. The quality of plantation afforestation varies widely, and has been especially problematic in some tropical environments [Pandey (1995 and 1997)].

Plantation forests currently provide around 10% of the world's wood harvest; this proportion is rising and will continue to rise rapidly, as the area of natural forest available for harvesting diminishes, as economic pressures and technological change favor plantation crops, and as the plantation forest estate matures and expands. The contribution of plantations to wood production within domestic economies varies enormously, reflecting different forest endowments and policies—from, for example, nearly 100% in New Zealand or South Africa, to around 50% in Argentina or Zimbabwe, to negligible levels in Canada or Papua New Guinea.

Given the wood production objectives of most plantation forests, and the commodity nature of most wood markets, plantation growth rates are of fundamental importance because of their implications for the cost of wood at harvest. Only around 10% of existing plantations can be classified as "fast-growing" yielding more than 14 m^3/yr; most of these plantations are in the southern hemisphere, with around 40% in each of South America and Asia-Pacific. The majority of "fast-growing" plantations are of species such as *Acacia* or *Eucalyptus* grown on short rotations for the relatively low-value uses of fuel, fiber or roundwood; perhaps a third are longer-rotation crops, of either softwood or hardwood species, grown principally for sawn- or veneer- wood.

Global supply and trade forecasts, for both plantation production and its share of total wood harvest, are imprecise and complicated by the uncertainties of demand growth within developing economies as Apsey and Reed (1996) comment, " ... the challenge is to sort out the hype from the reality with respect to fast growing plantations. Until this is done, a good share of strategic planning rests on a whirlpool of speculation". Imprecision notwithstanding, it is apparent that fast-growing plantation forests are already the most cost-competitive source of pulpwood globally, and that the expansion of the plantation resource is likely to constrain pulpwood price increases over the next decade. As the availability and relative importance in trade of higher-value wood products from plantation forests increase, so too will the influence of the plantation harvest on both supply and demand options for these products.

Plantation forestry at a global or semi-global scale has been the subject of a number of recent reviews. These reviews highlight some important common elements and trends:

- Use of well-adapted genetic resources, and good silviculture at all stages from nursery to harvest, are the two technical foundations of successful plantation forestry; each can make the difference between resounding success and abject failure. Many tropical plantations are not achieving their production potential because of inadequate attention to these fundamental elements (Pandey, 1997). Successful plantation forestry is also based on sound and substantial research and development, its implementation in operational management, and the maintenance of close links between research and practice as each evolves. There is ample evidence of the adverse consequences of failing to link adequately research and practice [Evans (1992), Kanowski and Savill (1992)].

- Many plantation forestry programmes have been founded on and developed through international and regional cooperation; the century-long history of cooperative research under International Union of Forest Research Organizations (IUFRO) auspices (Burley and Adlard, 1992), and the more recent role of FAO, demonstrate the many benefits of collaboration to plantation researchers and managers.

- The appropriate level of research varies with the stage of development of the plantation programme. For example, there remain many fundamental questions which must be resolved to support new plantation programmes. The continuing expansion of plantation forests onto sites for which there is as yet little plantation forestry experience will continue to demand such fundamental research. In contrast, for programmes that are already well established, increasingly sophisticated research and development will be necessary to deliver or maintain gains.

- As in other primary production enterprises, advanced technologies are playing an increasingly important role in plantation forestry:

- Applications of biotechnology in forestry have recently been reviewed by Haines (1994); those currently of most relevance are genomic mapping, molecular markers, and transformation and micro propagation. Their application in the production and propagation of inter-specific hybrids is of particular interest to many plantation programmes. The implementation of many biotechnologies are interdependent, and most are dependent for delivery on successful clonal propagation techniques, which are now in operational use in many programmes. The optimal integration of biotechnologies with plantation forestry is programme-specific.

- Advances in processing technologies are allowing the use of smaller and younger trees, and of species not previously considered suitable for value-added processing.

- Adequate planning and decision support systems are central to successful plantation enterprises. Appropriate systems range from the relatively simple (Ahlbäck, 1997) to the sophisticated (e.g. Pritchard, 1989); the lack of effective systems has been a major constraint to, in particular, many tropical plantation enterprises (Pandey, 1997).

5.3.1.2 *Environmental Sustainability of Plantation Forestry*

There is long history of concern for the environmental sustainability of plantation forestry (Evans, 1997). As plantation forests expand, so too have concerns for their sustainability in the broader sense. The sustainability of plantation forestry is now an issue in terms of each of its biological, economic and social dimensions, as well as in the more holistic sense of their conjunction (Barbier, 1987); sustainability concerns in plantation forestry have a number of manifestations, as outlined below. Discussion of the biological or environmental sustainability of plantation forests has three principal strands:

- The first strand is the broad debate concerning the environmental costs and benefits associated with afforestation, particularly where it is preceded by conversion of natural ecosystems. Argument around this topic ranges across the spectrum of issues and views, from the imperative of meeting the wood products needs of growing populations in the face of declining natural forest resources (Pandey, 1995), to the environmental impacts of forest conversion and plantation afforestation.

- The second strand has a narrower focus, on concerns for the biological sustainability of plantation forests *per se*, particularly their composition as monocultures. This topic has been reviewed, and the evidence is encouraging. As Evans (1997), concludes, "plantation forestry is likely to be sustainable in terms of wood yield in most situations provided good practice is maintained".

- A third strand is manifested by increasing research into alternative plantation regimes, principally the feasibility, advantages and disadvantages of mixed-species plantations as a means of enhancing sustainability. Although experience remains limited, there are clearly circumstances, which favor mixed-species stands. Some of these are social and economic, as discussed below.

5.3.1.3 *Economic Dimensions of Plantation Forestry*

The economic dimensions of plantation forestry have two principal manifestations:

- Firstly, the commodity nature of most forest plantation products—either fibre for pulp production, or utility grade timber—and the increasing globalization of markets for these products maintains strong price pressure in favor of the lowest cost producers. Production costs are determined by the inescapable trio of land, labor and capital costs, and by forest productivity. The inevitable consequence of these pressures is the trend towards shorter crop rotations, which have been facilitated by advances in processing technologies, and the search for enhanced productivity. However, particularly for solid wood products, there remain trade-offs between harvest age, growth rates and product quality.
- Secondly, as a consequence of imperatives which are at least as ideological as economic, the ownership of forest plantations is shifting from the public to the private sector as governments divest themselves of public assets. Many researchers have explored issues associated with this transition in the ownership of plantation forests. The appropriate role of government in plantation forestry remains an issue of debate, regardless of the level of public ownership of plantation forests, reflecting the various responsibilities of government, for example, in fostering an environment conducive to investment in tree growing, in the regulation of industry and of land use, and as steward of the environment and other community values.

5.3.1.4 *Social Dimensions of Plantation Forestry*

The social dimensions of plantation forestry also have a number of manifestations as under:

- The fuel and wood needs of the rural poor were the primary motivating force for the establishment of non-industrial plantations. Afforestation with this intent began on a large scale in the late 1970s (Pandey, 1995), as the international forestry community began to focus on how trees could better meet the needs of the world's poorest people. Non-industrial plantation establishment has been greatest in Africa and Asia (Pandey, 1995); whilst the concern (e.g., as exemplified in the 1978 World Forestry Congress, *Trees for People*) and intent were genuine, the social consequences of non-industrial forestry have been mixed. However, the sometimes bitter experiences gained in non-industrial plantation forestry have helped foresters develop the means to better assess and address the needs of the poor and rural people.

- An emerging discussion of the social implications of industrial plantation forestry, which acknowledges that these are not necessarily positive, and may indeed be quite adverse. This discussion has occurred and continues at a range of scales, for example, focused on particular projects [e.g., Cavalcanti (1996), for the case of Aracruz, Brazil), in terms of national policy [e.g., Wahli and Ylbhi (1992), for the case of Indonesia; Roche (1992), for the case of New Zealand], or more global terms [Kanowski (1997), Shiva (1993)]. It is likely that the progress of this discussion will mirror, in many senses, that which has preceded it for non-industrial plantations.

5.3.1.5 *The Future of Plantation Forestry*

There is evidence of an emerging dichotomy in plantation forestry concept and practice, between what has been characterized as relatively simpler and relatively more complex production systems. Plantation forests, as we know them, are relatively simple production systems, typically even-aged monocultures, with the capacity to produce wood yields many times, often at least tenfold, greater than most natural forests. The importance of simple plantation forests in meeting the wood needs of societies will continue to increase; providing they are well managed, these plantation forests should satisfy sustainability criteria (Evans, 1997).

This plantation forestry for commodity production benefits considerably from economies of scale and integration with industrial processing; it is also under strong cost and profit pressure, thus both demanding and permitting relatively high levels of resource inputs. Consequently, it will be increasingly concentrated on those sites, which are inherently more productive than on those which are marginal, and from which the costs of transport to processors are least. The implication is of plantation programmes which are more intensive silviculturally and less extensive geographically, located where the forest land base is stable, secure and productive, and where the economics of wood production—in terms both of cost structures within forestry and of relativities with other land uses—are most favorable. Prevailing political ideologies suggest these plantations will increasingly be under private, or quasi-private, ownership and management. Whilst successful—sometimes outstandingly—in producing wood, simple plantation systems do not necessarily address well the other needs of societies in which they are embedded. Where much of the less economically-developed world—land is scarce, time horizons short, or demand strong for the non-industrial products and services of forests, the outputs of simple production systems are unlikely to meet the more complex needs of societies. In these circumstances, a broader conception of plantation forestry and range of plantation objectives, and a more intimate integration with other land uses, is essential if plantation forestry is to prosper and be sustained. More complex plantation forestry explicitly recognizes that wood is not the only product that people demand of forests, and seeks to maximize social benefits rather than just wood production.

The particular expression of plantation forestry—where it lies along the continuum from simple to complex—will depend on the particular context; in developing a more complex plantation forestry, we have much to gain from our experiences of a wide spectrum of forestry activities, including agro-forestry, community forestry, and simpler plantation forestry.

More complex plantation forestry will be characterized variously by:

- A more intimate association between forests and other land uses. Simple plantation forestry is typified by a sharp distinction between plantation forest and other land use. The boundary between plantation forest and non-forest use will become less distinct as plantation forestry becomes more complex. The various taungya systems, widely practiced as means of afforestation in the tropics (Evans, 1992), are an example of this complexity at the early stages of plantation forestry; much farm forestry demonstrates such integration at the level of the farm enterprise, irrespective of the particular configuration of tree growing; more direct involvement of local people in the conception and implementation of plantation forestry, and in the sharing of its benefits and products. The variety of joint venture or share farmer schemes, which recognize landowners' interests and priorities as well as those of the forest industry's, exemplify this for the case of farm forestry. There is increasing understanding of how participatory planning, management and use might be developed and practiced in a forestry context and this approach now characterizes some programmes involving plantation forestry. As the presence or absence of trees is important in determining land tenure in many societies, locally-appropriate tenure arrangements are essential to facilitate more complex plantation forestry (Sargent, 1990); more diverse species composition and plantation structure, yielding an earlier and more continuing flow of a wider range of products and services than result from simple plantation forests. This does not necessarily imply that tree species will be grown as polycultures, though this may offer advantages in particular circumstances. In others, a mosaic of relatively small blocks of different tree species may be more easily managed, but still yield the desired range of outputs.

Although its rationale is broader, more complex plantation forestry may also represent an effective strategy for risk minimization, as Sargent (1990) demonstrated for the case of proposed eucalyptus plantations in Thailand. Their conclusions—that the cost of *not* implementing complex plantation systems will exceed the cost of doing so—are likely to apply increasingly elsewhere, and have instructive parallels in other land use contexts. There are many examples of how foresters have responded to social and environmental imperatives by developing more complex plantation forestry systems that still meet wood production objectives. These include:

- The silvicultural manipulation of *Pinus* plantations in Nepal, principally to promote the development of native broadleaved species, to increase species diversity and the range of forest products of more direct benefit to local people (Gilmour et al, 1990).
- Britain's National Forest and Community Forests (Countryside and Forestry Commissions, 1991), in which plantation forests for wood production are designed and managed to emphasize amenity, conservation and landscape values. These new forests are paralleled by the "restructuring" of the British Forest Enterprise's simple plantation forests, to enhance non-wood production functions, at an opportunity cost to simple wood production of around 10%.
- The integrated farm production systems, such as those associated with Spanish (Wilson et al, 1995) or Australian (Inions, 1995) forest industries, in which a variety of out-grower arrangements are employed to generate an enhanced income stream for farmers and assured wood supply for industrial use.
- Recognition of the capacity of integrated tree growing and farming systems to deliver substantial non-market benefits to both the landowner and the wider community, in addition to the direct returns to the owner. For example, in Australia's lower rainfall zones,

plantation forests integrated with the farm enterprise are playing the leading role in limiting the salinization of agricultural land as well as helping to diversify farm incomes (Robins et al, 1996); in many environments where catchment protection, stabilization or restoration are priorities, appropriate integration of tree growing and agricultural practices is an important element of watershed management and rehabilitation strategies.

The adoption of complex plantation practice does not preclude use of new technologies or innovative management regimes, as Wilson et al's (1995) description of integrated production systems based on genetically engineered fiber demonstrates. On the contrary, as numerous examples demonstrate, innovation in forestry practice is more likely to follow from involving a greater number of growers and allowing a diversity of management regimes.

Successful plantation forestry will continue to depend on effective research, development and management, and on innovation and technological advances. It will also depend increasingly on recognition of and respect for the principle of sustainability, in its full sense. As Evans (1997) comments, plantation forestry is merely a technology for delivering the benefits of trees to society; the appropriate form of that technology will vary with social, environmental and economic circumstance. What is clear is that the sustainability of plantation forestry will be enhanced, and the benefits of investments most fully realized, where plantation purpose and practice are embedded within the broader social and economic contexts.

In realizing the considerable potential of plantation forestry to benefit society, one of the principal challenges to plantation forest owners, managers and scientists is to progress from a narrow focus, which Shiva (1993) has characterized as "monocultures of the mind", to a broader appreciation of plantation purpose and practice. We are well-placed to do so, by building on the considerable body of experience and information we have gained relevant to plantation and other forms of forestry in many environments. It is in doing so that we shall sustain plantation forestry in the next century, and maximize its benefits.

5.3.2 Rainwater Harvesting

Rainwater harvesting has been already mentioned briefly in Section 4.6 in the previous Chapter. Major water harvesting techniques have been explained in Section 5.2 of this Chapter. Now let us understand rainwater harvesting from a very basic point of view.

Rainwater harvesting, in its broadest sense, can be defined as the collection of runoff for human use. The collection processes involve various techniques such as the collection of water from rooftops and the land surface, as well as within water courses. These techniques are widely used globally both for meeting drinking water supply needs and for irrigation purposes.

5.3.2.1 Technical Description of a Rainwater Harvesting System

Rainwater harvesting is a technology used for collecting and storing rainwater from rooftops, the land surface or rock catchments using simple techniques such as jars and pots as well as more complex techniques such as underground check dams. The techniques usually found in Asia and Africa arise from practices employed by ancient civilizations within these regions and still serve as a major source of drinking water supply in rural areas. Commonly used systems are constructed of three principal components, namely, the catchment area, the collection device, and the conveyance system.

- *Catchment Areas*
  - i) *Rooftop catchments:* In the most basic form of this technology, rainwater is collected in simple vessels at the edge of the roof. Variations on this basic approach include collection of rainwater in gutters which drain to the collection vessel through down-pipes constructed for this purpose, and/or the diversion of rainwater from the gutters to containers for settling particulates before being conveyed to the storage container for the domestic use. As the rooftop is the main catchment area, the amount and quality of rainwater collected depends on the area and type of roofing material. Reasonably pure rainwater can be collected from roofs constructed with galvanized corrugated iron, aluminium or asbestos cement sheets, tiles and slates, although thatched roofs tied with bamboo gutters and laid in proper slopes can produce almost the same amount of runoff less expensively (Gould, 1992). However, the bamboo roofs are least suitable because of possible health hazards. Similarly, roofs with metallic paint or other coatings are not recommended as they may impart tastes or color to the collected water. Roof catchments should also be cleaned regularly to remove dust, leaves and bird droppings so as to maintain the quality of the product water. Figure 5.16 shows a schematic representation of a rooftop collection system.

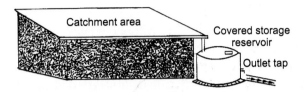

Figure 5.15   Rooftop Rainwater Catchment System

[Source: http://www.unep.or.jp/ietc/Publications/TechPublications/TechPub-8e/rainwater1.asp]

ii) *Ground surface catchments:* Rainwater harvesting using ground or land surface catchment areas is a less complex way of collecting rainwater. It involves improving runoff capacity of the land surface through various techniques including collection of runoff with drain pipes and storage of collected water (Figure 5.16). Compared to rooftop catchment techniques, ground catchment techniques provide more opportunity for collecting water from a larger surface area. By retaining the flows (including flood flows) of small creeks and streams in small storage reservoirs (on surface or underground) created by low cost (e.g., earthen) dams, this technology can meet water demands during dry periods. There is a possibility of high rates of water loss due to infiltration into the ground, and, because of the often marginal quality of the water collected, this technique is mainly suitable for storing water for agricultural purposes.

Techniques available for increasing the runoff within ground catchment areas involve: i) clearing or altering vegetation cover, ii)

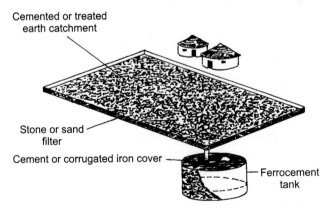

Figure 5.16   Ground Rainwater Catchment System

[Source: http://www.unep.or.jp/ietc/Publications/TechPublications/TechPub-8e/rainwater1.asp]

increasing the land slope with artificial ground cover, and iii) reducing soil permeability by the soil compaction and application of chemicals.

Clearing or altering vegetation cover: Clearing vegetation from the ground can increase surface runoff but also can induce more soil erosion. Use of dense vegetation cover such as grass is usually suggested as it helps to both maintain an high rate of runoff and minimize soil erosion.

Increasing slope: Steeper slopes can allow rapid runoff of rainfall to the collector. However, the rate of runoff has to be controlled to minimize soil erosion from the catchment field. Use of plastic sheets, asphalt or tiles along with slope can further increase efficiency by reducing both evaporative losses and soil erosion. The use of flat sheets of galvanized iron with timber frames to prevent corrosion was recommended and constructed in the State of Victoria, Australia, about 65 years ago (Kenyon, 1929; cited in UNEP, 1982).

Soil compaction by physical means: This involves smoothing and compacting of soil surface using equipment such as graders and rollers. To increase the surface runoff and minimize soil erosion rates, conservation bench terraces are constructed along a slope perpendicular to runoff flow. The bench terraces are separated by the sloping collectors and provision is made for distributing the runoff evenly across the field strips as sheet flow. Excess flows are routed to a lower collector and stored (UNEP, 1982).

Soil compaction by chemical treatments: In addition to clearing, shaping and compacting a catchment area, chemical applications with such soil treatments as sodium can significantly reduce the soil permeability. Use of aqueous solutions of a silicone-water repellent is another technique for enhancing soil compaction technologies. Though soil permeability can be reduced through chemical treatments, soil compaction can induce greater rates of soil erosion and may be expensive. Use of sodium-based chemicals may increase the salt content in the collected water, which may not be suitable both for drinking and irrigation purposes.

iii) *Rock catchments systems*: The presence of massive rock outcrops provides suitable catchment surfaces for freshwater augmentation (Figure 5.17). In these systems, runoff is channelled along stone and cement gutters, constructed on the rock surface, to reservoirs contained by concrete dams. The collected water then can be transported through a gravity fed pipe network to household standpipes.

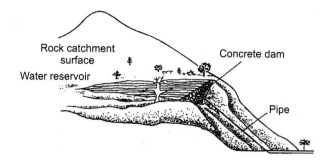

Figure 5.17   Rock Catchment System for Rainwater

[Source: http://www.unep.or.jp/ietc/Publications/TechPublications/TechPub-8e/rainwater1.asp]

• **Collection Devices**

  i) *Storage tanks:* Storage tanks for collecting rainwater harvested using guttering may be either above or below the ground. Precautions required in the use of storage tanks include provision of an adequate enclosure to minimize contamination from human, animal or other environmental contaminants, and a tight cover to prevent algal growth and the breeding of mosquitos. Open containers are not recommended for collecting water for drinking purposes. Various types of rainwater storage facilities can be found in practice. Among them are cylindrical ferrocement tanks and mortar jars. The ferrocement tank consists of a lightly reinforced concrete base on which is erected a circular vertical cylinder with a 10 mm steel base. This cylinder is further wrapped in two layers of light wire mesh to form the frame of the tank. Mortar jars are large jar shaped vessels constructed from wire reinforced mortar. The storage capacity needed should be calculated to take into consideration the length of any dry spells, the amount of rainfall, and the per capita water consumption rate. In most of the Asian countries, the winter months are dry, sometimes for weeks on end, and the annual average rainfall can occur within just a few days. In such circumstances, the storage capacity should be large enough to cover the demands of two to three weeks. For example, a three person household should have a minimum capacity of 3 (Persons) × 90 (lit) × 20 (days) = 5 400 lit.

  ii) *Rainfall water containers:* As an alternative to storage tanks, battery tanks (i.e., interconnected tanks) made of pottery, ferrocement, or polyethylene may be suitable. The polyethylene tanks are compact but have a large storage capacity (of 1,000 to 2,000 lit.), are easy to clean and have many openings which can be fitted with fittings for connecting pipes. In Asia, jars made of earthen materials or ferrocement tanks are commonly used. During the 1980s, the use of rainwater catchment technologies, especially roof catchment systems, expanded rapidly in a number of regions, including Thailand where more than ten million 2 m^3 ferrocement rainwater jars were built and many tens of thousands of larger ferrocement tanks were constructed between 1991 and 1993. Early problems with the jar design were quickly addressed by including a metal cover using readily available, standard brass fixtures.

The immense success of the jar programme springs from the fact that the technology met a real need, was affordable, and invited community participation. The programme also captured the imagination and support of not only the citizens, but also of government at both local and national levels as well as community based organizations, small-scale enterprises and donor agencies. The introduction and rapid promotion of bamboo reinforced tanks, however, was less successful because the bamboo was attacked by termites, bacteria and fungus. More than 50,000 tanks were built between 1986 and 1993 (mainly in Thailand and Indonesia) before a number started to fail, and, by the late 1980s, the bamboo reinforced tank design, which had promised to provide an excellent low-cost alternative to ferrocement tanks, had to be abandoned.

The *design considerations* vary according to the type of tank and **various other factors** have to be considered while designing the rainwater tanks (Gould, 1992) which are:

- *A solid, secure cover* to keep out insects, dirt and sunlight which will act to prevent the growth of algae inside the tank.
- *A coarse inlet filter* for excluding coarse debris, dirt, leaves, and other solid materials.
- *An overflow pipe.*
- *A manhole, sump and drain* for cleaning.
- *An extraction system* that doesn't contaminate the water (e.g., a tap or pump).
- *A lock on the tap.*
- *A soakaway* to prevent spilled water from forming puddles near the tank.
- *A maximum height of 2 m* to limit the water pressure acting on the container to minimize burst tanks.
- *A device to indicate the level of water* in the tank.
- *A sediment trap,* tipping bucket or other fouled flush mechanism.
- *A second, clear water storage tank* if the rainwater has to be subjected to some form of water treatment, such as desalination using a density stratification process in the first tank.

• **Conveyance Systems**

Conveyance systems are required to transfer the rainwater collected on the rooftops to the storage tanks. This is usually accomplished by making connections to one or more down-pipes connected to the rooftop gutters. When selecting a conveyance system, consideration should be given to the fact that, when it first starts to rain, dirt and debris from the rooftop and gutters will be washed into the down-pipe. Thus, the relatively clean water will only be available some time later in the storm. There are several possible choices to selectively collect clean water for the storage tanks. The most common is the down-pipe flap. With this flap it is possible to direct the first flush of water flow through the down-pipe, while later rainfall is diverted into a storage tank. When it starts to rain, the flap is left in the closed position, directing water to the down-pipe, and, later, opened when relatively clean water can be collected. A great disadvantage of using this type of conveyance control system is the necessity to observe the runoff quality and manually operate the flap. An alternative approach would be to automate the opening of the flap as described below.

A simple and effective method of diverting rainwater without the need for supervision is depicted in Figure 5.18. A funnel-shaped insert is integrated into the down-pipe system. Because the upper edge of the funnel is not in direct contact with the sides of the down-pipe, and a small gap exists between the down-pipe walls and the funnel, water is free to flow both around the funnel and through the funnel. When it first starts to rain, the volume of water passing down the pipe is small, and the "dirty" water runs down the walls of the pipe, around the funnel and is discharged to the ground as is normally the case with rainwater guttering. However, as the

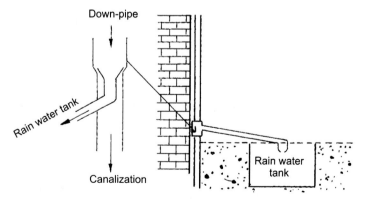

Figure 5.18   A Typical Rainwater Conveyance System

[Source: http://www.unep.or.jp/ietc/Publications/TechPublications/TechPub-8e/rainwater1.asp]

rainfall continues, the volume of water increases and "clean" water fills the down-pipe. At this higher volume, the funnel collects the clean water and redirects it to a storage tank. The pipes used for the collection of rainwater, wherever possible, should be made of plastic, PVC or other inert substance, as the pH of rainwater can be low (acidic) and could cause corrosion, and mobilization of metals, in metal pipes.

In order to safely fill a rainwater storage tank, it is necessary to make sure that excess water can overflow, and that blockages in the pipes or dirt in the water do not cause damage or contamination of the water supply. The design of the funnel system, with the drain-pipe being larger than the rainwater tank feed-pipe, helps to ensure that the water supply is protected by allowing excess water to bypass the storage tank. A modification of this design is shown in Figure 5.19, which illustrates a simple overflow/bypass system. In this system, it also is possible to fill the tank from a municipal drinking water source, so that even during a prolonged drought the tank can be kept full. Care should be taken, however, to ensure that rainwater does not enter the drinking water distribution system.

Calculating the amount available: When using rainwater for water supply purposes, it is important to recognize the fact that the supply is not constant

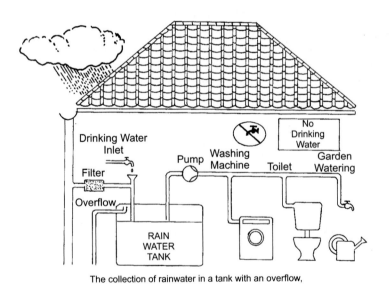

The collection of rainwater in a tank with an overflow,
transmission to the several users

Figure 5.19   A Typical Rainwater Distribution System

[Source: http://www.unep.or.jp/ietc/Publications/TechPublications/TechPub-8e/
rainwater1.asp]

throughout the year and plan an adequately-sized storage system to provide water during dry periods. A knowledge of the rainfall quantity and seasonality, the area of the collection area and volume of the storage container, and quantity and period of use during which water is required for water supply purposes is critical. For example, in Tokyo, the average annual rainfall is 1,800 mm, and, assuming that the effective collection area of a house is equal to its roof area, the typical collection area is about 100 m². Thus, the average annual volume of rainwater falling on the roof may be calculated as the product of the collection area, 100 m², and rainfall amount, 1,800 mm, or 180 m³. However, in practice, this volume can never be achieved since a portion of the rainwater evaporates from the rooftop and a portion, including the first flush, may be lost to the drainage system. Additional rainwater volume may be lost as overflow from the storage container if the storage tank is of insufficient volume to contain the entire volume of runoff. Thus, the net usable or available amount of rainwater from a tiled roof would be approximately 70% to 80% of the gross volume of rainfall, or about 130 m³ to 140 m³ if the water container is big enough to hold that quantity of rainwater available. Such a volume would be sufficient to save a significant amount of freshwater and money.

Estimation of the required volume of water: The individual daily rate of water consumption per person tends to be variable and may be difficult to calculate. Statistics vary from 130 to 175 liters per capita per day (lpcd) in developing countries. Of this volume, at least half is used for purposes for which water of a lesser quality would suffice. Indicative volumes are shown in Table 5.1, which summarizes the volumes of water used for household purposes, and indicates possibilities for the use of rainwater to supplement a municipal supply. Table 5.1 clearly shows that approximately 80 to 95 liters of the average daily volume of water consumed per person could be provided by the use of rainwater.

Table 5.1 Typical Per Capita Volume of Daily Water Consumption

Municipal water utilization	Possible rainwater utilization	
Highest quality	High quality	Low quality
Drinking /Cooking: 3–6 l	Washing dishes: 8–10 l	Toilet/sanitation: 40–50 l
Body care: 8 l	Washing clothes: 16 l	Other uses: 12 l
Shower/Bath: 40–50 l	Watering garden: 7 l	
Total = 50–65 l	Total = 30–33 l	Total = 52–62 l

[Source: http://www.unep.or.jp/ietc/Publications/TechPublications/TechPub-8e/rainwater1.asp]

5.3.2.2 Extent of Use of a Rainwater Harvesting System

The history of rainwater harvesting in Asia can be traced back to about the 9th or 10th Century and the small-scale collection of rainwater from roofs and simple brush dam constructions in the rural areas of South and Southeast Asia. Rainwater collection from the eaves of roofs or via simple gutters into traditional jars and pots has been traced back almost 2,000 years in Thailand. Rainwater harvesting has long been used in the Loess Plateau regions of China. More recently, however, about 40,000 well storage tanks, in a variety of different forms, were constructed between 1970 and 1974 using a technology which stores rainwater and stormwater runoff in ponds of various sizes. A thin layer of red clay is generally laid on the bottom of the ponds to minimize seepage losses. Trees, planted at the edges of the ponds, help to minimize evaporative losses from the ponds (UNEP, 1982).

5.3.2.3 Operation and Maintenance of a Rainwater Harvesting System

Maintenance is generally limited to the annual cleaning of the tank and regular inspection of the gutters and down-pipes. Maintenance typically consists of the removal of dirt, leaves and other accumulated materials. Such cleaning should take place annually before the start of the major rainfall season. However, cracks in the storage tanks can create major problems and should be repaired immediately. In the case of ground and rock catchments, additional care is required to avoid damage and contamination by people/animals, and proper fencing is required.

5.3.2.4 Level of Involvement

Various levels of governmental and community involvement in the development of rainwater harvesting technologies in different parts of Asia were noted. In Thailand and the Philippines, both governmental and household-based initiatives played key roles in expanding the use of this technology, especially in water scarce areas such as northeast Thailand.

5.3.2.5 Costs

The capital cost of rainwater harvesting systems is highly dependent on the type of catchment, conveyance and storage tank materials used. However, the cost of harvested rainwater in Asia, which varies from $0.17 to $0.37 per m^3 of water storage (Table 5.2), is relatively low compared to many countries in Africa.

Table 5.2 Costs of Rainwater Catchment Tanks in Two Typical Asian Countries

System	Vol m^3	Cost $	Annual equivalent cost $/m^3$	Country
Reinforced Cement Jar	2	25	0.17	Thailand
Concrete Ring	11.3	250	0.29	Thailand
Wire Framed Ferrocement	2	67	0.37	Philippines
Wire Framed Ferrocement	4	125	0.35	Philippines

[Source: http://www.unep.or.jp/ietc/Publications/TechPublications/TechPub-8e/rainwater1.asp]

Compared to deep and shallow tubewells, rainwater collection systems are more cost effective, especially if the initial investment does not include the cost of roofing materials. The initial per unit cost of rainwater storage tanks (jars) in northeast Thailand is estimated to be about $1 per liter, and each tank can last for more than ten years. The reported operation and maintenance costs are negligible.

5.3.2.6 Effectiveness of the Technology

The feasibility of rainwater harvesting in a particular locality is highly dependent upon the amount and intensity of rainfall. Other variables, such as catchment area and type of catchment surface, usually can be adjusted according to household needs. As rainfall is usually unevenly distributed throughout the year, rainwater collection methods can serve as only supplementary sources of household water. The viability of rainwater harvesting systems is also a function of: the quantity and quality of water available from other sources; household size and per capita water requirements; and budget available. The decision maker has to balance the total cost of the project against the available budget, including the economic benefit of conserving water supplied from other sources. Likewise, the cost of physical and environmental degradation associated with the development of available alternative sources should also be calculated and added to the economic analysis.

Assuming that rainwater harvesting has been determined to be feasible, two kinds of techniques—statistical and graphical methods—have been developed to aid in determining the size of the storage tanks. These methods are applicable for rooftop catchment systems only, and detail guidelines for design of these storage tanks can be found in Gould (1992) and Pacey and Cullis (1989).

Accounts of serious illness linked to rainwater supplies are few, suggesting that rainwater harvesting technologies are effective sources of water supply for many household purposes. It would appear that the potential for slight contamination of roof runoff from occasional bird droppings does not represent a major health risk; nevertheless, placing taps at least 10 cm above the base of the rainwater storage tanks allows any debris entering the tank to settle on the bottom, where it will not affect the quality of the stored water, provided it remains undisturbed. Ideally, storage tanks should be cleaned annually, and sieves should be fitted to the gutters and down-pipes to further minimize particulate contamination. A coarse sieve should be fitted in the gutter where the down-pipe is located. Such sieves are available made of plastic coated steel-wire or plastic, and may be wedged on top and/or inside gutter and near the down-pipe. It is also possible to fit a fine sieve within the down-pipe itself, but this must be removable for cleaning. A fine filter should also be fitted over the outlet of the down-pipe as the coarser sieves situated higher in the system may pass small particulates such as leaf fragments, etc. A simple and very inexpensive method is to use a small, fabric sack, which may be secured over the feed-pipe where it enters the storage tank.

If rainwater is used to supply household appliances such as the washing machine, even the tiniest particles of dirt may cause damage to the machine and the washing. To minimize the occurrence of such damage, it is advisable to install a fine filter, which is used in drinking water systems in the supply line upstream of the appliances. For use in wash basins or bath tubs, it is advisable to sterilize the water using a chlorine dosage pump.

5.3.2.7 Suitability of a Rainwater Harvesting System

The augmentation of municipal water supplies with harvested rainwater is suited to both urban and rural areas. The construction of cement jars or provision of gutters does not require very highly skilled manpower.

5.3.2.8 Advantages

Rainwater harvesting technologies are simple to install and operate. Local people can be easily trained to implement such technologies, and construction materials are also readily available. Rainwater harvesting is convenient in the sense that it provides water at the point of consumption, and family members have full control of their own systems, which greatly reduces operation and maintenance problems. Running costs, also, are almost negligible. Water collected from roof catchments usually is of acceptable quality for domestic purposes. As it is collected using existing structures not specially constructed for the purpose, rainwater harvesting has few negative environmental impacts compared to other water supply project technologies. Although regional or other local factors can modify the local climatic conditions, rainwater can be a continuous source of water supply for both the rural and poor. Depending upon household capacity and needs, both the water collection and storage capacity may be increased as needed within the available catchment area.

5.3.2.9 Disadvantages

Disadvantages of rainwater harvesting technologies are mainly due to the limited supply and uncertainty of rainfall. Adoption of this technology requires a "bottom up" approach rather than the more usual "top down" approach employed in other water resources development projects. This may make rainwater harvesting less attractive to some governmental agencies tasked with providing water supplies in developing countries, but the mobilization of local government and NGO resources can serve the same basic role in the development of rainwater-based schemes as water resources development agencies in the larger, more traditional public water supply schemes.

5.3.2.10 Cultural Acceptability

Rainwater harvesting is an accepted freshwater augmentation technology in Asia. While the bacteriological quality of rainwater collected from ground catchments is poor, that from properly maintained rooftop catchment systems, equipped with storage tanks having good covers and taps, is generally suitable for drinking, and frequently meets World Health Organization (WHO) drinking water standards. Notwithstanding, such water generally is of higher quality than most traditional, and many of improved, water sources found in the developing world. Contrary to popular beliefs, rather than becoming stale with extended storage, rainwater quality often improves as bacteria and pathogens gradually die off. Rooftop catchment, rainwater storage tanks can provide good quality water, clean enough for drinking, as long as the rooftop is clean, impervious, and made from non-toxic materials (lead paints and asbestos roofing materials should be avoided), and located away from over-hanging trees since birds and animals in the trees may defecate on the roof.

5.3.2.11 Further Development of the Technology

Rainwater harvesting appears to be one of the most promising alternatives for supplying freshwater in the face of increasing water scarcity and escalating demand. The pressures on rural water supplies, greater environmental impacts associated with new projects, and increased opposition from non-governmental organizations (NGOs) to the development of new surface water sources, as well as deteriorating water quality in surface reservoirs already constructed, constrain the ability of communities to meet the demand for freshwater from traditional sources, and present an opportunity for augmentation of water supplies using this technology.

5.3.2.12 Water Quality Considerations and Local People's Preferences

Rainwater harvesting systems, especially those sourced from rooftop catchments, can provide clean water for drinking purposes. The quality of the water, however, is largely dependent on the type of roofing materials used and the frequency of cleaning of the surface. A study carried out on 189 rainwater tanks and jars in Thailand showed that only 2 of the 89 tanks sampled, and none of the 97 rainwater jars sampled, contained pathogens. Based on the results of bacterial analyses, 40% of the 189 tanks and jars sampled met the WHO drinking water standards. All of the tanks and jars sampled met the WHO standards for heavy metals, including the standards for cadmium, chromium, lead, copper and iron.

In northeast Thailand, the groundwater—the only readily available source of water, is highly saline. The local people are aware of the water quality benefits to be had by using rainwater. Before the Thai government launched the rainwater harvesting programme in 1986, the local people made use of rainwater harvested from thatched roofs as well as groundwater obtained from shallow tubewells. During a recent field visit to the area, the local people stated that they were afraid of drinking water from deep tubewells, even though the groundwater abstracted from the deep tubewells was reported to be less saline and suitable for drinking water purposes. When asked, the local people mentioned that they preferred shallow tubewell water because it had a nicer taste than the water from deep tubewells; however, they preferred water from the thatched roofs because of sweet taste. After the Thai government launched the rainwater harvesting programme in 1986, many villagers in this region of Thailand replaced the thatched roofs with zinc sheets to increase the volume of rainwater harvested. Every house now has 6,000 liter capacity jars for rainwater collection, and the jar manufacturing industry has been commercialized in the area. The demand for jars remains greater than the ability of the manufacturing firm's capacity to supply.

5.3.3 Fog Harvesting

This innovative technology is based on the fact that water can be collected from fogs under favorable climatic conditions. Fogs are defined as a mass of water vapor condensed into small water droplets at, or just above, the Earth's surface. The small water droplets present in the fog precipitate when they come in contact with objects. The frequent fogs that occur in the arid coastal areas of Peru and Chile are traditionally known as *camanchacas*. These fogs have the potential to provide an alternative source of freshwater in this otherwise dry region if harvested through the use of simple and low-cost collection systems known as fog collectors. Present research suggests that fog collectors work best in coastal areas where the water can be harvested as the fog moves inland driven by the wind. However, the technology could also potentially supply water for multiple uses in mountainous areas should the water present in stratocumulus clouds, at altitudes of approximately 400 m to 1,200 m, be harvested.

5.3.3.1 Technical Description

Full-scale fog collectors are simple, flat, rectangular nets of nylon supported by a post at either end and arranged perpendicular to the direction of the prevailing wind. The one used in a pilot-scale project in the El Tofo region of Chile consisted of a single 2 m × 24 m panel with a surface area of 48 m². Alternatively, the collectors may be more complex structures, made up of a series of such collection panels joined together. The number and size of the modules chosen will depend on local topography and the quality of the materials used in the panels. Multiple-unit systems have the advantage of a lower cost per unit of water produced, and the number of panels in use can be changed as climatic conditions and demand for water vary.

The surface of fog collectors is usually made of fine-mesh nylon or polypropylene netting, e.g., 'shade cloth', locally available in Chile under the brand name Raschel. Raschel netting (made of flat, black polypropylene filaments, 1.0 mm wide and 0.1 mm thick, in a triangular weave) can be produced in varying mesh densities. After testing the efficiency of various mesh densities, the fog collectors used at El Tofo were equipped with Raschel netting providing 35% coverage, mounted in double layers. This proportion of polypropylene-surface-to-opening extracts about 30% of the water from the fog passing through the nets.

As water collects on the net, the droplets join to form larger drops that fall under the influence of gravity into a trough or gutter at the bottom of the panel, from which it is conveyed to a storage tank or cistern. The collector itself is completely passive, and the water is conveyed to the storage system by gravity. If site topography permits, the stored water can also be conveyed by gravity to the point of use. The storage and distribution system usually consists of a plastic channel or PVC pipe approximately 110 mm in diameter which can be connected to a water hose for conveyance to the storage site/ point of use. Storage is usually in a closed concrete cistern. A 30 m^3 underground cistern is used in the zone of Antofagasta in northern Chile. In the most common type of fog collector (see Figure 5.20), the water is collected in a 200 liter drum.

Storage facilities should be provided for at least 50% of the expected maximum daily volume of water consumed. However, because the fog phenomenon is not perfectly regular from day to day, it may be necessary to store additional water to meet demands on days when no fog water is collected. Chlorination of storage tanks may be necessary if the water is used for drinking or cooking purposes.

5.3.3.2 Extent of Use

Fog harvesting has been investigated for more than thirty years and has been implemented successfully in the mountainous coastal areas of Chile, Ecuador, Mexico, and Peru. Because of a similar climate and mountainous conditions, this technology also can be implemented in other regions as shown in Figure 5.21.

In Chile, the National Forestry Corporation (CONAF), the Catholic University of the North, and the Catholic University of Chile are implementing the technology in several regions, including El Toro, Los

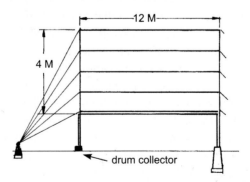

Figure 5.20 Section of a Typical Flat, Rectangular Nylon Mesh Fog Collector

[Source: http://www.oas.org/usde/publications/Unit/oea59e/ch12].

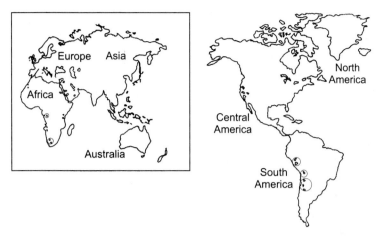

Figure 5.21   Existing and Potential Fog Harvesting Locations in the
World (shown encircled)

[Source: W. Canto Vera, et al. 1993. *Fog Water Collection System*. IDRC, Ottawa,
Canada; also available at http://www.oas.org/usde/publications/Unit/oea59e/ch12].

Nidos, Cerro Moreno, Travesía, San Jorge, and Pan de Azúcar. The results of
the several experiments conducted in the northern coastal mountain region
indicate the feasibility and applicability of this technology for supplying
good quality water for a variety of purposes, including potable water and
water for commercial, industrial, agricultural, and environmental uses.
These experiments were conducted between 1967 and 1988 at altitudes
ranging from 530 m to 948 m using different types of fog water collectors. The
different types of *neblinometers* (devices to measure liquid water content of
fog) and fog collectors resulted in different water yields under the same
climatic conditions and geographic location. A neblinometer or fog collector
with a screen containing a double Raschel (30%) mesh was the most
successful and the one that is currently recommended.

In Peru, the National Meteorological and Hydrological Service
(SENAMHI) has been cooperating with the Estratus Company since the
1960s in implementing the technology in the following areas: Lachay,
Pasamayo, Cerro Campana, Atiquipa, Cerro Orara (Ventinilla-Ancón),
Cerro Colorado (Villa María de Triunfo), and Cahuide Recreational Park
(Ate-Vitarte), and in southern Ecuador the Center for Alternative Social
Research (CISA) is beginning to work in the National Park of Machalilla on
Cerro La Gotera using the Chilean installations as models.

5.3.3.3 *Operation and Maintenance*

Operating this technology is very simple after once the fog collection system and associated facilities are properly installed. Training of personnel to operate the system might not be necessary if the users participate in the development and installation of the required equipment. A very important factor in the successful use of this technology is the establishment of a routine quality control programme. This programme should address both the fog collection system and the possible contamination of the harvested water, and include the following tasks:

- *Inspection of cable tensions.* Loss of proper cable tension can result in water loss by failing to capture the harvested water in the receiving system. It can also cause structural damage to the collector panels.
- *Inspection of cable fasteners.* Loose fasteners in the collection structure can cause the system to collapse and/or be destroyed.
- *Inspection of horizontal mesh net tensions.* Loose nets will lead to a loss of harvesting efficiency and can also break easily.
- *Maintenance of mesh nets.* After prolonged use, the nets may tear. Tears should be repaired immediately to avoid having to replace the entire panel. Algae can also grow on the surface of the mesh net after one or two years of use, accumulating dust, which will cloud the collected water and cause offensive taste and odor problems. The mesh net should be cleaned with a soft plastic brush as soon as algal growth is detected.
- *Maintenance of collector drains.* A screen should be installed at the end of the receiving trough to trap undesirable materials (insects, plants, and other debris) and prevent contamination of water in the storage tank. This screen should be inspected and cleaned periodically.
- *Maintenance of pipelines and pressure outlets.* Pipelines should be kept as clean as possible to prevent accumulation of sediments and decomposition of organic matter. Openings along the pipes should be built to facilitate flushing or partial cleaning of the system. Likewise, pressure outlets should be inspected and cleaned frequently to avoid accumulation of sediments. Openings in the system must be protected against possible entry of insects and other contaminants.
- *Maintenance of cisterns and storage tanks.* Tanks must be cleaned periodically with a solution of concentrated calcium chloride to prevent the accumulation of fungi and bacteria on the walls.
- *Monitoring of dissolved chlorine.* A decrease in the concentration of chlorine in potable water is a good indicator of possible growth of

microorganisms. Monitoring of the dissolved chlorine will help to prevent the development of bacterial problems.

5.3.3.4 Level of Involvement

It is strongly recommended that the end users participate in the construction of the project. Community participation will help to reduce the labor cost, provide the community with operation/maintenance experience, and develop a sense of community ownership/responsibility for the success of the project. Government subsidies, particularly in the initial stages, might be necessary to reduce the cost of constructing/installing the facilities. A cost-sharing approach could be adopted so that end users will pay for the pipeline and operating costs, with the government/external agency assuming the cost of providing storage/distribution to homes.

5.3.3.5 Costs

Actual costs of fog harvesting systems vary from location to location. In a project in the region of Antofagasta, Chile, the installation cost of a fog collector was estimated to be $90/m^2$ of mesh, while, in another project in northern Chile, the cost of a 48 m^2 fog collector was approximately $378 ($225 in materials, $63 in labor, and $39 in incidentals). This latter system produced a yield of 3.0 l/m^2 of mesh/day. The cost of a fog harvesting project constructed in the village of Chungungo, Chile, is shown in Table 5.3. The most expensive item in this system is the pipeline that carries the water from the fog collection panel to the storage tank located in the village.

Table 5.3 Capital Investment Cost and Life Span cf Fog Water Collection System Components

Component	Cost ($)	% of total cost	Life span (years)
Collection	27680	22.7	12
Main pipeline	43787	35.9	20
Storage (100 m^3 tank)	15632	12.8	20
Treatment	2037	1.7	10
Distribution	32806	26.9	20
Total	121,942	100.0	

[Source: http://www.oas.org/usde/publications/Unit/oea59e/ch12].

Maintenance and operating costs are relatively low compared to other technologies. In the project in Antofagasta, the operation and maintenance cost was estimated at $600/year. This cost is significantly less than that of the Chungungo project: operating costs in that project were estimated at

$4,740, and maintenance costs at $7,590 (resulting in a total cost of $12,330/year).

Both the capital costs and the operating and maintenance costs are affected by the efficiency of the collection system, the length of the pipeline that carries the water from the collection panels to the storage areas, and the size of the storage tank. For example, the unit cost for a system with an efficiency of 2.0 lit/m^2/day was estimated to be $4.80/1,000 liters. If the efficiency was improved to 5.0 lit/m^2/day, then the unit cost would be reduced to $1.90/1,000 litres. In the Antofagasta project, the unit cost of production was estimated at $1.41/1,000 litres with a production of 2.5 lit/m^2/day.

5.3.3.6 Effectiveness of the Technology

Experimental projects conducted in Chile indicate that it is possible to harvest between 5.3 lit/m^2/day and 13.4 lit/m^2/day depending on the location, season, and type of collection system used. At El Tofo, Chile, during the period between 1987 and 1990, an average fog harvest of 3.0 lit/m^2/day was obtained using 50 fog collectors made with Raschel mesh netting. Fog harvesting efficiencies were found to be highest during the spring and summer months, and lowest during the winter months. The average water collection rates during the fog seasons in Chile and Peru were 3.0 and 9.0 lit/m^2/day, respectively; the lengths of the fog seasons were 365 and 210 days, respectively. While this seems to indicate that higher rates are obtained during shorter fog seasons, the practical implications are that a shorter fog season will require large storage facilities in order to ensure a supply of water during non-fog periods. Thus, a minimum fog season duration of half a year might serve as a guideline when considering the feasibility of using this technology for water supply purposes; however, a detailed economic analysis to determine the minimum duration of the fog season that would make this technology cost-effective should be made. In general, fog harvesting has been found more efficient and more cost-effective in arid regions than other conventional systems.

5.3.3.7 Suitability

In order to implement a fog harvesting programme, the potential for extracting water from fogs first must be investigated. The following factors affect the volume of water that can be extracted from fogs and the frequency with which the water can be harvested:

- **Frequency of fog occurrence,** which is a function of atmospheric pressure and circulation, oceanic water temperature, and the presence of thermal inversions.

- **Fog water content,** which is a function of altitude, seasons and terrain features.
- **Design of fog water collection system,** which is a function of wind velocity and direction, topographic conditions, and the materials used in the construction of the fog collector.

The occurrence of fogs can be assessed from reports compiled by government meteorological agencies. To be successful, this technology should be located in regions where favorable climatic conditions exist. Since fogs/clouds are carried to the harvesting site by the wind, the interaction of the topography and the wind will be influential in determining the success of the site chosen. The following factors should be considered in selecting an appropriate site for fog harvesting:

Global Wind Patterns: Persistent winds from one direction are ideal for fog collection. The high-pressure area in the eastern part of the South Pacific Ocean produces onshore, southwest winds in northern Chile for most of the year and southerly winds along the coast of Peru.

Topography: It is necessary to have sufficient topographic relief to intercept the fogs/clouds; examples, on a continental scale, include the coastal mountains of Chile, Peru, and Ecuador, and, on a local scale, isolated hills or coastal dunes.

Relief in the surrounding areas: It is important that there be no major obstacle to the wind within a few kilometers upwind of the site. In arid coastal regions, the presence of an inland depression or basin that heats up during the day can be advantageous, as the localized low pressure area thus created can enhance the sea breeze and increase the wind speed at which marine cloud decks flow over the collection devices.

Altitude: The thickness of the stratocumulus clouds and the height of their bases will vary with location. A desirable working altitude is at two-thirds of the cloud thickness above the base. This portion of the cloud will normally have the highest liquid water content. In Chile and Peru, the working altitudes range from 400 m to 1 000 m above sea level.

Orientation of the topographic features: It is important that the longitudinal axis of the mountain range, hills, or dune system be approximately perpendicular to the direction of the wind bringing the clouds from the ocean. The clouds will flow over the ridge lines and through passes, with the fog often dissipating on the downwind side.

Distance from the coastline: There are many high-elevation continental locations with frequent fog cover resulting from either the transport of upwind clouds or the formation of orographic clouds. In these cases, the distance to the coastline is irrelevant. However, areas of high relief near the coastline are generally preferred sites for fog harvesting.

Space for collectors: Ridge lines and the upwind edges of flat-topped mountains are good fog harvesting sites. When long fog water collectors are used, they should be placed at intervals of about 4.0 m to allow the wind to blow around the collectors.

Crestline and upwind locations: Slightly lower-altitude upwind locations are acceptable, as are constant-altitude locations on a flat terrain. But locations behind a ridge or hill, especially where the wind is flowing downslope, should be avoided.

Prior to implementing a fog water harvesting programme, a pilot-scale assessment of the collection system proposed for use and the water content of the fog at the proposed harvesting site should be undertaken. Low cost and low maintenance measurement devices to measure the liquid water content of fog, called *neblinometers*, have been developed at the Catholic University of Chile. Figure 5.22 illustrates four different types of neblinometers:

(A) a pluviograph with a perforated cylinder;
(B) a cylinder with a nylon mesh screen;
(C) multiple mesh screens made of nylon or polypropylene mesh; and
(D) a single mesh screen made of nylon or polypropylene mesh.

The devices capture water droplets present in the fog on nylon filaments that are mounted in an iron frame. The original neblinometer had an area of 0.25 m^2 made up of a panel with a length and width of 0.5 m, and fitted with a screen having a warp of 180 nylon threads 0.4 mm in diameter. The iron frame was 1.0 cm in diameter and was supported on a 2.0 m iron pole. These simple devices can be left in the field for more than a year without maintenance and can be easily modified to collect fog water samples for chemical analysis.

In pilot projects, use of a neblinometer with single or multiple panels having a width and length of one meter, fitted with fine-mesh nylon or polypropylene netting is recommended. It should be equipped with an anemometer to measure wind velocity and a vane to measure wind direction. The neblinometer can be connected to a data logger so that data can be made available in computer-compatible formats.

5.3.3.8 Advantages

- A fog collection system can be easily built or assembled on-site. Installation and connection of the collection panels is quick and simple. Assembly is not labor-intensive and requires little skill.

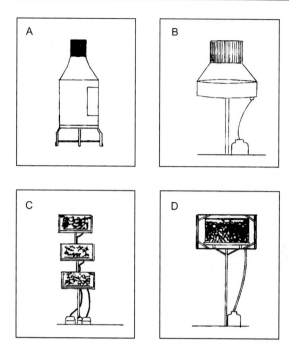

Figure 5.22 Types of Neblinometers

[Source: G. Soto Alvarez, National Forestry Corporation, Antofagasta, Chile; also available at http://www.oas.org/usde/publications/Unit/oea59e/ch12].

- No energy is needed to operate the system or transport the water.
- Maintenance and repair requirements are generally minimal.
- Capital investment and other costs are low in comparison with those of conventional sources of potable water supply used, especially in mountainous regions.
- The technology can provide environmental benefits when used in national parks in mountainous areas, or as an inexpensive source of water supply for reforestation projects.
- It has the potential to create viable communities in inhospitable environments and to improve the quality of life for people in mountainous rural communities.
- The water quality is better than from existing water sources used for agriculture and domestic purposes.

5.3.3.9 Disadvantages

- This technology might represent a significant investment risk unless a pilot project is first carried out to quantify the potential rate and yield that can be anticipated from the fog harvesting rate and the seasonably of the fog of the area under consideration.

- Community participation in the process of developing and operating the technology in order to reduce installation and operating and maintenance costs is necessary.
- If the harvesting area is not close to the point of use, the installation of the pipeline needed to deliver the water can be very costly in areas of high topographic relief.
- The technology is very sensitive to changes in climatic conditions which could affect the water content and frequency of occurrence of fogs; a backup water supply to be used during periods of unfavorable climatic conditions is recommended.
- In some coastal regions (e.g., in Paposo, Chile), fog water has failed to meet drinking water quality standards because of concentrations of chlorine, nitrate, and some minerals.
- Caution is required to minimize impacts on the landscape and the flora and fauna of the region during the construction of the fog harvesting equipment and the storage and distribution facilities.

5.3.3.10 Cultural Acceptability

This technology has been accepted by communities in the mountainous areas of Chile and Peru. However, some skepticism has been expressed regarding its applicability to other regions. It remains a localized water supply option, dependent on local climatic conditions.

5.3.3.11 Future Development of the Technology

To improve fog harvesting technology, design improvements are necessary to increase the efficiency of the fog collectors. New, more durable materials should be developed. The storage and distribution systems needs to be made more cost-effective. An information and community education programme should be established prior to the implementation of this technology.

5.3.4 Wastewater Treatment Technologies

Relatively simple wastewater treatment technologies can be designed to provide low cost sanitation and environmental protection while providing additional benefits from the reuse of water. These technologies use natural aquatic and terrestrial systems. They are in use in a number of locations throughout Latin America and the Caribbean.

These systems may be classified into three principal types, as shown in Figure 5.23. Mechanical treatment systems, which use natural processes within a constructed environment, tend to be used when suitable lands are unavailable for the implementation of natural system technologies. Aquatic systems are represented by lagoons; facultative, aerated, and hydrograph controlled release (HCR) lagoons are variations of this technology. Further, the lagoon-based treatment systems can be supplemented by additional pre- or post-treatments using constructed wetlands, aquacultural production systems, and/or sand filtration. They are used to treat a variety of wastewaters and function under a wide range of weather conditions. Terrestrial systems make use of the nutrients contained in wastewaters; plant growth and soil adsorption convert biologically available nutrients into less-available forms of biomass, which is then harvested for a variety of uses, including methane gas production, alcohol production, or cattle feed supplements.

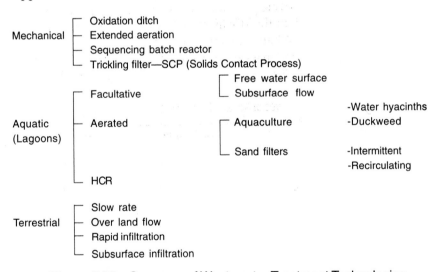

Figure 5.23 Summary of Wastewater Treatment Technologies

[Source: http://www.oas.org/usde/publications/Unit/oea59e/ch25.htm].

5.3.4.1 Technical Description

• Mechanical Treatment Technologies

Mechanical systems utilize a combination of physical, biological, and chemical processes to achieve the treatment objectives. Using essentially natural processes within an artificial environment, mechanical treatment technologies use a series of tanks, along with pumps, blowers, screens, grinders, and other mechanical components, to treat wastewaters. Various types of instrumentation control flow of wastewater in the system. Sequencing batch reactors (SBR), oxidation ditches, and extended aeration systems are all variations of the activated-sludge process, which is a suspended-growth system. The trickling filter solids contact process (TF-SCP), in contrast, is an attached-growth system. These treatment systems are effective where land is at a premium.

• **Aquatic Treatment Technologies**
Facultative lagoons are the most common form of aquatic treatment-lagoon technology currently in use. The water layer near the surface is aerobic while the bottom layer, which includes sludge deposits, is anaerobic. The intermediate layer is aerobic near the top and anaerobic near the bottom, and constitutes the facultative zone. Aerated lagoons are smaller and deeper than facultative lagoons. These systems evolved from stabilization ponds when aeration devices were added to counteract odors arising from septic conditions. The aeration devices can be mechanical or diffused air systems. The chief disadvantage of lagoons is high effluent solids content, which can exceed 100 mg/l. To counteract this, hydrograph controlled release (HCR) lagoons are a recent innovation. In this system, wastewater is discharged only during periods when the stream flow is adequate to prevent water quality degradation. When stream conditions prohibit discharge, wastewater is accumulated in a storage lagoon. Typical design parameters are summarized in Table 5.4.

Table 5.4   Typical Design Features of Aquatic Treatment Units

Technology	Treatment goal	Detention time (days)	Depth (m)	Organic loading (kg/ha/day)
Oxidation pond	Secondary	10-40	0.9-1.35	40.9-125
Facultative pond	Secondary	25-180	1.35-2.3	22.7-68.2
Aerated pond	Secondary, polishing	7-20	1.8-5.4	51.1-204.5
Storage pond, HCR pond	Secondary, storage, polishing	100-200	2.7-4.5	22.7-68.2
Root zone Treatment, Hyacinth pond	Secondary	30-50	<1.35	<51.1

Source: http://www.oas.org/usde/publications/Unit/oea59e/ch25.htm].

Constructed wetlands, aquacultural operations, and sand filters are generally the most successful methods of polishing the treated wastewater effluent from the lagoons. These systems have also been used with more traditional, engineered primary treatment technologies such as Imhoff tanks, septic tanks, and primary clarifiers. Their main advantage is to provide additional treatment beyond secondary treatment where required. In recent years, constructed wetlands have been utilized in two designs: systems using surface water flows and systems using subsurface flows. Both systems utilize the roots of plants to provide substrate for the growth of attached bacteria, which utilize the nutrients present in the effluents and for the transfer of oxygen. Bacteria do the bulk of the work in these systems, although there is some nitrogen uptake by the plants. The surface water system most closely approximates a natural wetland. Typically, these systems are long, narrow basins, with depths of less than 0.6 m, that are planted with aquatic vegetation such as bulrush (*Scirpus* spp.) or cattails (*Typha* spp.). The shallow groundwater systems use a gravel or sand medium, approximately 0.45 m deep, which provides a rooting medium for the aquatic plants and through which the wastewater flows.

Aquaculture systems are distinguished by the type of plants grown in the wastewater holding basins. These plants are commonly water hyacinth (*Eichhornia crassipes*) or duckweed (*Lemna* spp.). These systems are basically shallow ponds covered with floating plants that detain wastewater at least one week. The main purpose of the plants in these systems is to provide a suitable habitat for bacteria which remove the vast majority of dissolved nutrients. The design features of such systems are summarized in Table 5.5.

Table 5.5 Typical Design Features for Constructed Wetlands

Design factor	Surface water flow	Subsurface water flow
Minimum surface area	2.43-12.15 ha	0.24-4.86 ha
Maximum water depth	Relatively shallow	Water level below ground surface
Bed depth	Not applicable	12.30 m
Minimum hydraulic residence time	7 days	7 days
Maximum hydraulic loading rate	8.15-40.74 lpd/m^2	20.37-407.41 lpd/m^2
Minimum pretreatment	Primary (secondary optional)	Primary
Range of organic loading as BOD	10.23-20.45 kg/ha/d	2.05-159.09 kg/ha/d

Source: http://www.oas.org/usde/publications/Unit/oea59e/ch25.htm].

Sand filters have been used for wastewater treatment purposes for at least a century in Latin America and the Caribbean. Two types of sand filters are commonly used: intermittent and recirculating. They differ mainly in the method of application of the wastewater. Intermittent filters are flooded with wastewater and then allowed to drain completely before the next application of wastewater. In contrast, recirculating filters use a pump to recirculate the effluent to the filter in a ratio of 3 to 5 parts filter effluent to 1 part raw wastewater. Both types of filters use a sand layer, 0.6 to 0.9 m thick, underlain by a collection system of perforated or open joint pipes enclosed within graded gravel. Water is treated biologically by the epiphytic flora associated with the sand and gravel particles, although some physical filtration of suspended solids by the sand grains and some chemical adsorption onto the surface of the sand grains play a role in the treatment process.

• **Terrestrial Treatment Technologies**

Terrestrial treatment systems include slow-rate overland flow, slow-rate subsurface infiltration, and rapid infiltration methods. In addition to wastewater treatment and low maintenance costs, these systems may yield additional benefits by providing water for groundwater recharge, reforestation, agriculture, and/or livestock pasturage. They depend upon physical, chemical, and biological reactions on and within the soil. Slow-rate overland flow systems require vegetation, both to take up nutrients and other contaminants and to slow the passage of the effluent across the land surface to ensure maximum contact times between the effluents and the plants/soils. Slow-rate subsurface infiltration systems and rapid infiltration systems are 'zero discharge' systems that rarely discharge effluents directly to streams or other surface waters. Each system has different constraints regarding soil permeability.

Although slow-rate overland flow systems are the most costly of the natural systems to implement, their advantage is their positive impact on sustainable development practices. In addition to treating wastewater, they provide an economic return from the reuse of water and nutrients to produce marketable crops or other agriculture products and/or water and fodder for livestock. The water may also be used to support reforestation projects in water-poor areas. In slow-rate systems, either primary or secondary wastewater is applied at a controlled rate, either by sprinklers or by flooding of furrows, to a vegetated land surface of moderate to low permeability. The wastewater is treated as it passes through the soil by filtration, adsorption, ion exchange, precipitation, microbial action, and plant uptake. Vegetation is a critical component of the process and serves to extract nutrients, reduce erosion, and maintain soil permeability.

Overland flow systems are a land application treatment method in which treated effluents are eventually discharged to surface water. The main benefits of these systems are their low maintenance and low technical manpower requirements. Wastewater is applied intermittently across the tops of terraces constructed on soils of very low permeability and allowed to sheet-flow across the vegetated surface to the runoff collection channel. Treatment, including nitrogen removal, is achieved primarily through sedimentation, filtration, and biochemical activity as the wastewater flows across the vegetated surface of the terraced slope. Loading rates and application cycles are designed to maintain active microorganism growth in the soil. The rate and length of application are controlled to minimize the occurrence of severe anaerobic conditions, and a rest period between applications is needed. The rest period should be long enough to prevent surface ponding, yet short enough to keep the microorganisms active. Site constraints relating to land application technologies are shown in Table 5.6.

Table 5.6   Site Constraints for Land Application Technologies

Feature	Slow rate	Rapid infiltration	Subsurface infiltration	Overland flow
Soil texture	Sandy loam to clay loam	Sand and sandy loam	Sand to clayey loam	Silty loam and clayey loam
Depth to groundwater	0.9 m	0.9 m	0.9 m	Not critical
Vegetation	Required	Optional	Not applicable	Required
Climatic restrictions	Growing season	None	None	Growing season
Slope	<20%, cultivated land < 40%, uncultivated land	Not critical	Not applicable	2%–8% finished slopes

Source: http://www.oas.org/usde/publications/Unit/oea59e/ch25.htm].

In rapid infiltration systems, most of the applied wastewater percolates through the soil, and the treated effluent drains naturally to surface waters or recharges the groundwater. Their cost and manpower requirements are low. Wastewater is applied to soils that are moderately or highly permeable by spreading in basins or by sprinkling. Vegetation is not necessary, but it does not cause a problem if present. The major treatment goal is to convert ammonia nitrogen in the water to nitrate nitrogen before discharging to the receiving water.

Subsurface infiltration systems are designed for municipalities of less than 2,500 people. They are usually designed for individual homes (septic tanks), but they can be designed for clusters of homes. Although they do require specific site conditions, they can be low-cost methods of wastewater disposal.

5.3.4.2 Extent of Use

These treatment technologies are widely used in Latin America and the Caribbean. Combinations of some of them with wastewater reuse technologies have been tested in several countries. Colombia has extensively tested aerobic and anaerobic mechanical treatment systems. Chile, Colombia, and Barbados have used activated sludge plants, while Brazil has utilized vertical reactor plants. Argentina, Bolivia, Colombia, Guatemala, Brazil, Chile, Curaçao, Mexico, Jamaica, and Saint Lucia have successfully experimented with different kinds of terrestrial and aquatic treatment systems for the treatment of wastewaters. Curaçao, Mexico, and Jamaica have used stabilization or facultative lagoons and oxidation ponds; their experience has been that aquatic treatment technologies require extensive land areas and relatively long retention times, on the order of 7 to 10 days, to adequately treat wastewater. An emerging technology, being tested in a number of different countries, is a hybrid aquatic-terrestrial treatment system that uses wastewaters for *hydroponic cultivation* (i.e., plant cultivation without soil). However, most of the applications of this hybrid technology to date have been limited to the experimental treatment of small volumes of wastewater.

5.3.4.3   *Operation and Maintenance*

Operation and maintenance requirements vary depending on the particular technology used. In mechanical activated-sludge plants, maintenance requirements consist of periodically activating the sludge pumps, inspecting the system to ensure that are no blockages or leakages in the system, and checking BOD and suspended solids concentrations in the plant effluent to ensure efficient operation.

In the case of aquatic treatment systems using anaerobic reactors and facultative lagoons for primary wastewater treatment, the following operational guidelines should be followed:

- Periodically clean the sand removal system (usually every 5 days in dry weather, and every 2 to 3 days in wet weather).
- Daily remove any oily material that accumulates in the anaerobic reactor.
- Daily remove accumulated algae in the facultative lagoons.
- Open the sludge valves to send the sludge to the drying beds.
- Establish an exotic aquatic plant removal programme (aquatic plant growth can hamper the treatment capacity of the lagoons).
- Properly dispose of the materials removed, including dried sludge.

A preventive maintenance programme should also be established to increase the efficiency of the treatment systems and prolong their lifespan. When using terrestrial treatment systems or hybrid hydroponic cultivation systems for wastewater treatment, it is advisable to have two parallel systems, and to alternate applications of wastewater to these systems every 12 hours in order to facilitate aeration and to avoid damage to the system. Care is required to avoid hydraulic overload in these systems, as the irrigated plant communities could be damaged and the degree of treatment provided negated. Periodic removal of sediments accumulated in the soil is also required to improve the soil-plant interaction and to avoid soil compaction/subsidence.

5.3.4.4 Level of Involvement

Government involvement is essential in the implementation of most of the wastewater treatment technologies. The private sector, particularly the tourism industry, has successfully installed "packaged" or small-scale, self-contained sewage treatment plants at individual sites. In some cases, the installation of these plants has been combined with the reuse of the effluent for watering golf courses, lawns, and similar areas. The selection and construction of the appropriate wastewater treatment technology is generally initiated and financed, at least partially, by the government, with the subsequent operation and maintenance of the facility being a responsibility of the local community. Nevertheless, despite the large number of well-known and well-tested methods for wastewater treatment, there still exist a significant number of local communities in Latin America which discharge wastewater directly into lakes, rivers, estuaries, and oceans without treatment. As a result, surface water degradation, which also affects the availability of freshwater resources, is more widespread than is desirable within this region.

5.3.4.5 Costs

Operation and maintenance as well as capital costs for wastewater treatment systems with a capacity of 0.1 to 1 million gallons per day (mgd) are summarized in Figures 5.24 and 5.25. Most of the cost data come from systems implemented in the United States. Similar systems in Latin America might be less expensive, in some cases, owing to lower labor costs and price differentials in construction materials. Nevertheless, the relative cost comparison among technologies is likely to be applicable to all countries.

Figure 5.24 compares the operating and maintenance costs (labor, energy, chemicals, and materials such as replacement equipment and parts) of the various systems of 0.1 to 1 mgd (i.e., 3,785 m^3/d) treatment capacity. All costs are presented in dollars per million gallons of wastewater treated. The cost for mechanical systems is significantly larger than for any of the other systems, particularly at smaller flows. The cost of harvesting plants from

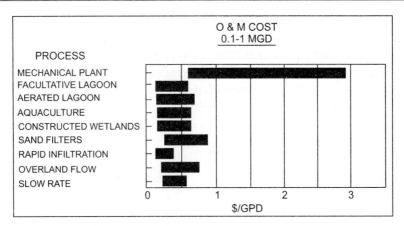

Figure 5.24 Comparative Operation and Maintenance Cost of Wastewater Treatment Technologies

[Source: http://www.oas.org/usde/publications/Unit/oea59e/p162a.GIF].

aquaculture systems is not included; this could be a significant amount for some systems.

Figure 5.25 compares the capital cost of the wastewater treatment processes. All natural systems are assumed to have a facultative lagoon as the primary treatment unit. The cost of chlorination/disinfection is included for all systems except the slow rate and rapid infiltration systems. The cost of land is excluded in all cases, as is the cost of liners for the aquatic treatment systems. The mechanical treatment plant cost was derived as the cost of an oxidation ditch treatment system, and includes the cost of a clarifier, oxidation ditch, pumps, building, laboratory, and sludge drying beds. These costs also include the cost of engineering and construction management, in addition to the costs for piping, electrical systems, instrumentation, and site preparation. All costs are in March 1993, in dollars.

5.3.4.6 Effectiveness of the Technology

Natural treatment systems are capable of producing an effluent quality equal to that of mechanical treatment systems. Figure 5.26 summarizes the treatment performance of each of the systems. All can meet the limits generally established for secondary treatment, defined as biological oxygen demand (BOD) and total suspended solids (TSS) concentrations of less than 30 mg/l. All except the lagoon systems can also produce effluents that meet the criteria generally categorized as advanced treatment, defined as BOD and TSS concentrations of less than 20 mg/l. The results of a project conducted in Bogota, Colombia, to compare the performance of different sewage treatment processes are summarized in Table 5.7.

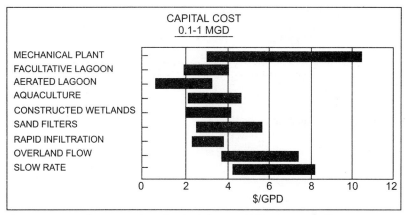

Figure 5.25 Comparative Capital Cost of Wastewater Treatment Technologies

Source: http://www.oas.org/usde/publications/Unit/oea59e/p162b.GIF].

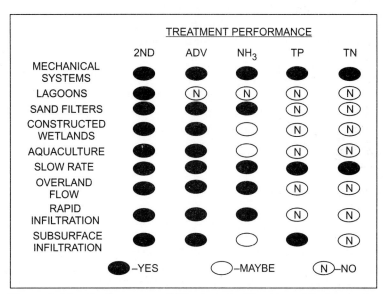

Figure 5.26 Performance of Wastewater Treatment Technologies

[Source: http://www.oas.org/usde/publications/Unit/oea59e/p164.GIF].

* 2ND = secondary limits of treatment for BOD and suspended solids < 30 mg/l.
* ADV = advanced treatment limits for BOD and total suspended solids < 20 mg/l.
* NH_3 = Ammonia = 2 mg/l, TP = Total phosphates < 2 mg/l, TN = Total nitrates < 2 mg/l.

Table 5.7 Comparative Performance of Sewage Treatment Systems

Process	Oxygen supply	Reactor volume	Retention time	Removal efficiency
Activated sludge	Pressurized air	10 m³	4–6 hr	90%–95% organic matter 90%–95% suspended solids
Biologic rotary discs	Air	1 m³	1–3 hr	90%–95% organic matter
Ascendant flow	Anaerobic	2 m³	24 hr	50%–60% organic matter; 57% suspended solids
Anaerobic filtration	Anaerobic	2 m³	36 hr	40%–50% organic matter; 52% suspended solids
Septic tank	Anaerobic	2 m³	36 hr	25% organic matter
Hydroponic cultivation	Aerobic/anaerobic	6 m³	12 hr	65%–75% organic matter

Source: http://www.oas.org/usde/publications/Unit/oea59e/ch25.htm].

5.3.4.7 Suitability

Mechanical systems are more suitable for places where land availability is a concern, such as hotels and residential areas. Mechanical plants are the least land intensive of the wastewater treatment methods based on natural processes.

Lagoon and oxidation pond technologies are suitable where there is plenty of land available. Slow-rate systems require as much as 304 hectares. Hybrid hydroponic cultivation techniques, using aquatic and terrestrial plants for the treatment of wastewater, also require relatively large amounts of land, and are best suited to regions where suitable aquatic plants can grow naturally.

5.3.4.8 Advantages

Table 5.8 summarizes the advantages of the various wastewater treatment technologies. In general, the advantages of using natural biological processes relate to their "low-tech/no-tech" nature, which means that these systems are relatively easy to construct and operate, and to their low cost, which makes them attractive to communities with limited budgets. However, their simplicity and low cost may be deceptive in that the systems require frequent inspections and constant maintenance to ensure smooth operation. Concerns include hydraulic overloading, excessive plant growth, and loss of exotic plants to natural watercourses. For this reason, and also because of the land requirements for biologically based technologies, many communities prefer mechanically-based technologies, which tend to require less land and permit better control of the operation. However, these systems generally have a high cost and require more skilled personnel to operate them.

Table 5.8 Advantages and Disadvantages of Conventional and Non-conventional Wastewater Treatment Technologies

Treatment type	Advantages	Disadvantages
Aquatic Systems		
Stabilization lagoons	Low capital cost Low operation and maintenance costs Low technical manpower requirement	Requires a large area of land May produce undesirable odors
Aerated lagoons	Requires relatively little land area Produces few undesirable odors	Requires mechanical devices to aerate the basins Produces effluents with a high suspended solids concentration
Terrestrial Systems		
Septic tanks	Can be used by individual households Easy to operate and maintain Can be built in rural areas	Provides a low treatment efficiency Must be pumped occasionally Requires a landfill for periodic disposal of sludge and septage
Constructed wetlands	Removes up to 70% of solids and bacteria Minimal capital cost Low operation and maintenance requirements and costs	Remains largely experimental Requires periodic removal of excess plant material Best used in areas where suitable native plants are available
Mechanical Systems		
Filtration systems	Minimal land requirements; can be used for household-scale treatment Relatively low cost Easy to operate	Requires mechanical devices
Vertical biological reactors	Highly efficient treatment method Requires little land area Applicable to small communities for local-scale treatment and to big cities for regional-scale treatment	High cost Complex technology Requires technically skilled manpower for operation and maintenance Needs spare-parts availability Has a high energy requirement
Activated sludge	Highly efficient treatment method Requires little land area Applicable to small communities for local-scale treatment and to big cities for regional-scale treatment	High cost Requires sludge disposal area (sludge is usually land-spread) Requires technically skilled manpower for operation and maintenance

Source: http://www.oas.org/usde/publications/Unit/oea59e/ch25.htm].

5.3.4.9 Disadvantages

Table 5.8 also summarizes the disadvantages of the various wastewater treatment technologies. These generally relate to the cost of construction and ease of operation. Mechanical systems can be costly to build and operate as they require specialized personnel. Nevertheless, they do offer a more controlled environment which produces a more consistent quality of effluent. Natural biological systems, on the other hand, are more land-intensive, require less skilled operators, and can produce effluents of variable quality depending on time of year, type of plants, and volume of wastewater loading. Generally, the complexity and cost of wastewater treatment technologies increase with the quality of the effluent produced.

5.3.4.10 Cultural Acceptability

Governments and the private sector in many Latin American countries fail to fully recognize the necessity of wastewater treatment and the importance of water quality in improving the quality of life of existing and future generations. The contamination of natural resources is a major impediment to achieving the objective of environmentally sustainable economic growth and development.

5.3.4.11 Further Development of the Technology

The cost-effectiveness of all wastewater treatment technologies needs to be improved. New designs of mechanical systems—which address this concern, are being introduced by the treatment plant manufacturing industry. The use of vertical reactors with an activated-sludge system, being tested in Brazil in order to acquire data for future improvement of this technology, is one example of the innovation going on in the industry. Similar product development is occurring in the use of aquatic and terrestrial plants and hybrid hydroponic systems, as a means of wastewater treatment; however, these technologies are still in an experimental phase and will require more testing and research prior to being accepted as standard treatment technologies. In addition, education to create an awareness of the need for wastewater treatment remains a critical need at all levels of government and non-governmental organizations (NGOs).

5.3.5 Wastewater Reuse

Once freshwater has been used for an economic or beneficial purpose, it is generally discarded as waste. In many countries, these wastewaters are discharged, either as untreated waste or as treated effluent, into natural watercourses, from which they are abstracted for further use after undergoing 'self-purification' within the stream. Through this system of indirect reuse, wastewater may be reused up to a dozen times or more before being discharged to the sea. Such indirect reuse is common in the larger river systems of Latin America. However, more direct reuse is also possible: the technology to reclaim wastewaters as potable or process waters is a technically feasible option for agricultural and some industrial purposes (such as for cooling water or sanitary flushing), and a largely experimental option for the supply of domestic water. Wastewater reuse for drinking raises public health, and possibly religious, concerns among consumers. The adoption of wastewater treatment and subsequent reuse as a means of supplying freshwater is also determined by economic factors.

In many countries, water quality standards have been developed governing the discharge of wastewater into the environment. Wastewater, in this context, includes sewage effluent, stormwater runoff, and industrial discharges. The necessity to protect the natural environment from wastewater-related pollution has led to much improved treatment techniques. Extending these technologies to the treatment of wastewaters to potable standards was a logical extension of this protection and augmentation process.

5.3.5.1 Technical Description

One of the most critical steps in any reuse programme is to protect the public health. To this end, it is most important to neutralize or eliminate any infectious agents or pathogenic organisms that may be present in the wastewater. For some reuse applications, such as irrigation of non-food crop plants, secondary treatment may be acceptable. For other applications, further disinfection, by such methods as chlorination or ozonation, may be necessary. Table 5.9 presents a range of typical survival times for potential pathogens in water and other media.

Table 5.9 Typical Pathogen Survival Times at 20–30°C (in days)

Pathogen	Freshwater and sewage	Crops	Soil
Viruses	< 120 but usually < 50	< 60 but usually < 15	< 100 but usually < 20
Bacteria	< 60 but usually < 30	< 30 but usually < 15	< 70 but usually < 20
Protozoa	< 30 but usually < 15	< 10 but usually < 2	< 70 but usually < 20
Helminths	Many months	< 60 but usually < 30	Many months

Source: http://www.oas.org/usde/publications/Unit/oea59e/ch25.htm].

A typical example of wastewater reuse is the system at the Sam Lords Castle Hotel in Barbados. Effluent consisting of kitchen, laundry, and domestic sewage ('grey water') is collected in a sump, from which it is pumped, through a comminutor, to an aeration chamber. No primary sedimentation is provided in this system, although it is often desirable to do so. The aerated mixed liquor flows out of the aeration chamber to a clarifier for gravity separation. The effluent from the clarifier is then passed through a 4.8 m deep chlorine disinfection chamber before it is pumped to an automatic sprinkler irrigation system. The irrigated areas are divided into sixteen zones; each zone has twelve sprinklers. Some areas are also provided with a drip irrigation system. Sludge from the clarifier is pumped, without thickening, as a slurry to suckwells, where it is disposed of. Previously the sludge was pumped out and sent to the Bridgetown Sewage Treatment Plant for further treatment and additional desludging.

5.3.5.2 Extent of Use

For health and aesthetic reasons, reuse of treated sewage effluent is presently limited to non-potable applications such as irrigation of non-food crops and provision of industrial cooling water. There are no known direct reuse schemes using treated wastewater from sewerage systems for drinking. Indeed, the only known systems of this type are experimental in nature, although in some cases treated wastewater is reused indirectly, as a source of aquifer recharge. Table 5.10 presents some guidelines for the utilization of wastewater, indicating the treatment required and other details. In general, wastewater reuse is a technology that has some use, primarily in small-scale projects in the region, owing to concerns about potential public health hazards.

Table 5.10 Guidelines for Wastewater Reuse

Type of reuse	Treatment required	Reclaimed water quality	Recommended monitoring	Setback distances
AGRICULTURAL Food crops commercially processed Orchards and Vinyards	Secondary Disinfection	pH = 6-9 BOD $\leq$ 30 mg/l SS = 30 mg/l FC $\leq$ 200/100 ml Cl_2 residual = 1 mg/l minimum	pH weekly BOD weekly SS daily FC daily Cl_2 residual continuous	90 m from potable water supply wells 30 m from areas accessible to public
PASTURAGE Pasture for milking animals Pasture for livestock	Secondary Disinfection	pH = 6-9 BOD $\leq$ 30 mg/l SS $\leq$ 30 mg/l FC $\leq$ 200/100 ml Cl_2 residual = 1 mg/l minimum	pH weekly BOD weekly SS daily FC daily Cl_2 residual continuous	90 m from potable water supply wells 30 m from areas accessible to public

Contd.

Table 5.10 Contd.

FORESTATION	Secondary Disinfection	pH = 6-9 BOD ≤ 30 mg/l SS ≤ 30 mg/l FC ≤ 200/100 ml Cl₂ residual = 1 mg/l min.	pH weekly BOD weekly SS daily FC daily Cl₂ residual continuous	90 m from potable water supply wells 30 m from areas accessible to the public
AGRICULTURAL Food crops not commercially processed	Secondary Filtration Disinfection	pH = 6-9 BOD ≤ 30 mg/l Turbidity ≤ 1 NTU FC = 0/100 ml Cl₂ residual = 1 mg/l minimum	pH weekly BOD weekly Turbidity daily FC daily Cl₂ residual continuous	15 m from potable water supply wells
GROUNDWATER RECHARGE	Site-specific and use-dependent	Site-specific and use-dependent	Depends on treatment and use	Site-specific

Source: http://www.oas.org/usde/publications/Unit/oea59e/ch25.htm].

Wastewater reuse in the Caribbean is primarily in the form of irrigation water. In Jamaica, some hotels have used wastewater treatment effluent for golf course irrigation, while the major industrial water users, the bauxite/alumina companies, engage in extensive recycling of their process waters. In Barbados, effluent from an extended aeration sewage treatment plant is used for lawn irrigation. Similar use of wastewater occurs on Curaçao.

In Latin America, treated wastewater is used in small-scale agricultural projects and, particularly by hotels, for lawn irrigation. In Chile, up to 220 lit/s of wastewater is used for irrigation purposes in the desert region of Antofagasta. In Brazil, wastewater has been extensively reused for agriculture. Treated wastewaters have also been used for human consumption after proper disinfection, for industrial processes as a source of cooling water, and for aquaculture. Wastewater reuse for aquacultural and agricultural irrigation purposes is also practiced in Lima, Peru. In Argentina, natural systems are used for wastewater treatment. In such cases, there is an economic incentive for reusing wastewater for reforestation, agricultural, pasturage, and water conservation purposes, where sufficient land is available to do so. Perhaps the most extensive reuse of wastewater occurs in Mexico, where there is large-scale use of raw sewage for the irrigation of parks and the creation of recreational lakes.

In the United States, the use of reclaimed water for irrigation of food crops is prohibited in some states, while others allow it only if the crop is to be processed and not eaten raw. Some states may hold, for example, that if a food crop is irrigated in such a way that there is no contact between the edible portion and the reclaimed water, a disinfected, secondary-treated effluent is acceptable. For crops that are eaten raw and not commercially processed, wastewater reuse is more restricted and less economically attractive. Less stringent requirements are set for irrigation of non-food crops.

International water quality guidelines for wastewater reuse have been issued by the World Health Organization (WHO). Guidelines should also be established at national level and at the local/project level, taking into account the international guidelines. Some national standards that have been developed are more stringent than the WHO guidelines. In general, however, wastewater reuse regulations should be strict enough to permit irrigation use without undue health risks, but not so strict as to prevent its use. When using treated wastewater for irrigation, for example, regulations should be written so that attention is paid to the interaction between the effluent, the soil, and the topography of the receiving area, particularly if there are aquifers nearby.

5.3.5.3   Operation and Maintenance

The operation and maintenance required in the implementation of this technology is related to the previously discussed operation and maintenance of the wastewater treatment processes, and to the chlorination and disinfection technologies used to ensure that pathogenic organisms will not present a health hazard to humans. Additional maintenance includes the periodic cleaning of the water distribution system conveying the effluent from the treatment plant to the area of reuse; periodic cleaning of pipes, pumps, and filters to avoid the deposition of solids that can reduce the distribution efficiency; and inspection of pipes to avoid clogging throughout the collection, treatment, and distribution system, which can be a potential problem. Further, it must be emphasized that, in order for a water reuse programme to be successful, stringent regulations, monitoring, and control of water quality must be exercised in order to protect both workers and the consumers.

5.3.5.4   Level of Involvement

The private sector, particularly the hotel industry and the agricultural sector, are becoming involved in wastewater treatment and reuse. However, to ensure the public health and protect the environment, governments need to exercise oversight of projects in order to minimize the deleterious impacts of wastewater discharges. One element of this oversight should include the sharing of information on the effectiveness of wastewater reuse. Government oversight also includes licensing and monitoring the performance of the wastewater treatment plants to ensure that the effluent does not create environmental or health problems.

5.3.5.5 Costs

Cost data for this technology are very limited. Most of the data relate to the cost of treating the wastewater prior to reuse. Additional costs are associated with the construction of a dual or parallel distribution system. In many cases, these costs can be recovered out of the savings derived from the reduced use of potable freshwater (i.e., from not having to treat raw water to potable standards when the intended use does not require such extensive treatment). The feasibility of wastewater reuse ultimately depends on the cost of recycled or reclaimed water relative to alternative supplies of potable water, and on public acceptance of the reclaimed water. Costs of effluent treatment vary widely according to location and level of treatment (see the previous section on wastewater treatment technologies). The degree of public acceptance also varies widely depending on water availability, religious and cultural beliefs, and previous experience with the reuse of wastewaters.

5.3.5.6 Effectiveness of the Technology

The effectiveness of the technology, while difficult to quantify, is seen in terms of the diminished demand for potable-quality freshwater and, in the Caribbean islands, in the diminished degree of degradation of water quality in the near-shore coastal marine environment, the area where untreated and unreclaimed wastewaters were previously disposed. The analysis of beach waters in Jamaica indicates that the water quality is better near the hotels with wastewater reuse projects than in beach areas where reuse is not practiced: Beach #1 in Table 5.11 is near a hotel with a wastewater reuse project, while Beach #2 is not. From an aesthetic point of view, also, the presence of lush vegetation in the areas where lawns and plants are irrigated with reclaimed wastewater is further evidence of the effectiveness of this technology.

Table 5.11 Water Quality of Beach Water in Wastewater Reuse Project in Jamaica

Site	BOD	TC	FC	NO$_3$
Beach # 1	0.30	<2	<2	0.01
Beach # 2	1.10	2.400.00	280.00	0.01

Source: http://www.oas.org/usde/publications/Unit/oea59e/ch25.htm].
TC : Total Chlorides
FC : Fecal Coliform

5.3.5.7 Suitability

This technology has generally been applied to a small-scale projects, primarily in areas where there is a shortage of water for supply purposes. However, this technology can be applied to larger-scale projects. In many developing countries, especially where there is a water deficit for several months of the year, implementation of wastewater recycling or reuse by industries can reduce demands for water of potable quality, and also reduce impacts on the environment.

Large-scale wastewater reuse can only be contemplated in areas where there are reticulated sewerage and/or stormwater systems. (Micro-scale wastewater reuse at the household or farmstead level is a traditional practice in many agricultural communities that use night soils and manures as fertilizers.) Urban areas generally have sewerage systems, and, while not all have stormwater systems, those that do are ideal localities for wastewater reuse schemes. Wastewater for reuse must be adequately treated, biologically and chemically, to ensure the public health and environmental safety. The primary concerns associated with the use of sewage effluents in reuse schemes are the presence of pathogenic bacteria and viruses, parasite eggs, worms, and helminths (all biological concerns) and of nitrates, phosphates, salts, and toxic chemicals, including heavy metals (all chemical concerns) in the water destined for reuse.

5.3.5.8 Advantages

- This technology reduces the demands on potable sources of freshwater.
- It may reduce the need for large wastewater treatment systems, if significant portions of the waste stream are reused or recycled.
- The technology may diminish the volume of wastewater discharged, resulting in a beneficial impact on the aquatic environment.
- Capital costs are low to medium, for most systems, and are recoverable in a very short time; this excludes systems designed for direct reuse of sewage water.
- Operation and maintenance are relatively simple except in direct reuse systems, where more extensive technology and quality control are required.
- Provision of nutrient-rich wastewaters can increase agricultural production in water-poor areas.
- Pollution of seawater, rivers, and groundwaters may be reduced.
- Lawn maintenance and golf course irrigation is facilitated in resort areas.

- In most cases, the quality of the wastewater, as an irrigation water supply, is superior to that of well water.

5.3.5.9 Disadvantages

- If implemented on a large scale, revenues to water supply and wastewater utilities may fall as the demand for potable water for non-potable uses and the discharge of wastewaters is reduced.
- Reuse of wastewater may be seasonal in nature, resulting in the overloading of treatment and disposal facilities during the rainy season; if the wet season is of long duration and/or high intensity, the seasonal discharge of raw wastewaters may occur.
- Health problems, such as water-borne diseases and skin irritations, may occur in people coming into direct contact with reused wastewater.
- Gases, such as sulfuric acid, produced during the treatment process can result in chronic health problems.
- In some cases, reuse of wastewater is not economically feasible because of the requirement for an additional distribution system.
- Application of untreated wastewater as irrigation water or as injected recharge water may result in groundwater contamination.

5.3.5.10 Cultural Acceptability

A large percentage of domestic water users are afraid to use this technology to supply of potable water (direct reuse) because of the potential presence of pathogenic organisms. However, most people are willing to accept reused wastewater for golf course and lawn irrigation and for cooling purposes in industrial processes. On the household scale, reuse of wastewaters and manures as fertilizer is a traditional technology.

5.3.5.11 Further Development of the Technology

Expansion of this technology to large-scale applications should be encouraged. Cities and towns that now use mechanical treatment plants that are difficult to operate, expensive to maintain, and require a high skill level can replace these plants with the simpler systems; treated wastewater can be reused to irrigate crops, pastures, and lawns. In new buildings, plumbing fixtures can be designed to reuse wastewater, as in the case of using gray water from washing machines and kitchen sinks to flush toilets and irrigate lawns. Improved public education to ensure awareness of the technology and its benefits, both environmental and economic, is recommended.

5.3.6 Other Water Conservation Practices

The importance of water conservation and water loss reduction should always be an integral part of the management of freshwater resources and needs to be given prominence in freshwater resources planning. As is suggested by the three interlinking arrows in the recyclable materials symbol, reduction of waste is the first of the several means of resource conservation (the other means being reuse and recycling). An excellent reference book is *Efficient Water Use*, edited by Hector Garduño and Felipe Arreguín-Cortés.

For water management purposes, the community can be divided into two basic groups: system users (such as households, industry, and agriculture) and system operators (such as municipal, state, and local governments and privately owned suppliers). These users have a choice of a number of different practices, which promote or enhance the efficiency of their use. These practices fall into two basic categories: *engineering practices*, based on modifications to hardware (e.g., plumbing and fixtures) and/or water supply operational procedures, and *behavioral practices*, based on changing water use habits.

Engineering practices are generally technical or regulatory measures, while behavioral practices typically involve market-oriented measures. Collectively, these measures, which affect water use and reduce waste and loss from the source, are known as 'demand management' measures. Such measures include leak detection; waste reduction (encouraging consumers to cut out wasteful uses); investment in appliances, processes, and technologies that reduce water input without reducing consumer satisfaction and/or output; treatment of industrial effluents and wastewaters to a standard suitable for recycling and reuse; and reallocation of freshwater resources to the area of greatest social good. The policies that encourage demand management include pricing water at an economic rate, charging for pollution or community-based pollution control practices, regulating and restricting specific water uses, exhorting and informing the consumer of the ways and means of use reduction and recycling, and encouraging water trading among and between users.

5.3.6.1 Technical Description

Water conservation practices can be followed by residential users, industrial and commercial users, and agricultural users. They can also be followed by local utilities and/or regional water supply plants. Table 5.12 shows some of the more common practices recommended for use by the

Table 5.12 Recommended Water Conservation Practices

User group	Engineering practices	Behavioral practices
Residential	Plumbing changes	Changing water use habits
	Low-flush toilets	Pricing
	Toilet tank volume displacement devices	Public information and education
	Low-flow showerheads	Lawn irrigation scheduling
	Faucet aerators	Drought management practices
	Pressure reduction devices	
	Gray Water reuse landscaping	
	Drought-tolerant plants	
	Xeriscaped landscapes	
Agricultural	Irrigation	Irrigation scheduling
	Low volume irrigation technologies	
	Wastewater reuse and recycling	
	Soil management	
Industrial and commercial	Water reuse and recycling	Monitoring water use
	Cooling water recirculation	Enforcing water use practices
	Wash water recycling	Educational programs on water
	Landscape irrigation	

Source: http://www.oas.org/usde/publications/Unit/oea59e/ch31.htm]

different user groups. A brief description of the most common conservation practices follows.

  i) *Residential Water Conservation Measures:* Low-flow plumbing fixtures and retrofit programmes are permanent, one-time conservation measures that can be implemented with little or no additional cost over the lifetime of the fixtures. In some cases, these fixtures can even save the residents money over the long term. The most commonly recommended low-flow plumbing fixtures are pressure reduction devices, faucet aerators, toilet displacement devices, low-flush toilets, low-flow showerheads, and plumbing modifications for grey water reuse. A typical breakdown of residential water use is shown in Figure 5.27.

Pressure Reduction. Homeowners can reduce the water pressure in a home by installing pressure reducing valves. A reduction in water pressure can save water in other ways: it can reduce the likelihood of leaking water pipes, leaking water heaters, and dripping faucets.

Faucet Aerators. Faucet aerators, which break the flowing water into fine droplets and entrain air while maintaining wetting effectiveness, are inexpensive devices that can be installed in sinks to reduce the volume of water used. Aerators are easily installed and can reduce the volume of water

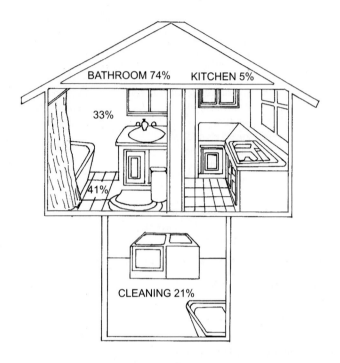

Figure 5.27 Typical Breakdown of Interior Water Use
Source: http://www.oas.org/usde/publications/Unit/oea59e/ch31.htm]

use at a faucet by as much as 60% while still maintaining a strong flow. More efficient kitchen and bathroom faucets that use only 7.5 lit/min, in contrast to standard faucets, which use 12 to 20 lit/min, are also available.

Toilet Displacement Devices. Non-toxic bricks or plastic containers (e.g., milk jugs filled with water or pebbles) can be placed in a toilet tank to reduce the amount of water used per flush. By placing between one and three such containers in the tank, more than 4 liters of water can be saved per flush. A toilet dam, which holds back a reservoir of water when the toilet is flushed, can also be used instead of the displacement device to save water.

Low-flush Toilets. Conventional toilets use 15 to 20 liters of water per flush, but low-flush toilets use only 6 liters of water or less. Since low-flush toilets use less water, they also reduce the volume of wastewater produced. A schematic representation of a low-flush toilet is shown in Figure 5.28. Even in existing residences, replacement of conventional toilets with low-flush toilets is a practical and economical water-saving alternative.

Figure 5.28   A schematic Representation of a low-flush toilet

Source: http://www.oas.org/usde/publications/Unit/oea59e/ch31.htm].

1. The 6-liter flush design of this gravity toilet has a different flush mechanism.
2. Steep bowl sides and a narrow trapway to allow the siphoned water to gain velocity for more effective removal of waste.
3. This is where the water pushes waste into the trapway.
4. Stored water flows into the bowl.

Composting Toilets. A composting toilet is a well-ventilated container that provides the optimum environment for unsaturated, but moist, human excrement for biological and physical decomposition under sanitary, controlled aerobic conditions.

The main components of a composting toilet (see Figure 5.29) are:

1) A composting reactor connected to a dry or micro-flush toilet(s).
2) A screened air inlet and an exhaust system (often fan-forced) to remove odors and heat, carbon dioxide, water vapor, and the byproducts of aerobic decomposition.
3) A mechanism to provide the necessary ventilation to support the aerobic organisms in the composter.
4) A means of draining and managing excess liquid and leachate (optional).
5) Process controls to optimize and facilitate management of the processes.
6) An access door for removal of the end-product.

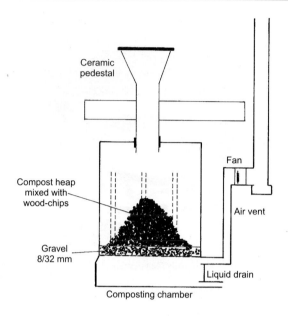

Figure 5.29   A Typical Composting Toilet
[Source: Kahlon, 2004]

7) Additives such as sawdust or wood-chips are required to be added at
    regular intervals of use.

The composting unit must be constructed to separate the solid fraction
from the liquid fraction and produce a stable, humus material with less than
200 MPN per gram of fecal coliform. Once the leachate has been drained or
evaporated out of the unit, the moist, unsaturated solids are decomposed by
aerobic organisms using molecular oxygen. Bulking agents can be added to
provide spaces for aeration and microbial colonization. The compost
chamber in some composting toilets is solar or electrically heated to provide
and maintain optimum temperature requirements for year-round usage.

Originally commercialized in Sweden, composting toilets have been an
established technology for more than 30 years, and perhaps longer in site-
built forms. As they require little to no water, composting toilet systems can
provide a solution to sanitation and environmental problems in non-
sewered, rural, and suburban areas and in both developed and
underdeveloped countries. A composting (or biological) toilet system
contains and processes excrement, toilet paper, carbon additive, and
sometimes, food waste. Unlike a septic system, a composting toilet system
relies on unsaturated conditions where aerobic bacteria break down waste.

This process is similar to a yard waste composter. If sized and maintained properly, a composting toilet breaks down waste to 10% to 30% of its original volume. Public health professionals are beginning to recognize the need for environmentally sound human waste treatment and recycling methods. The composting toilet is a non-water-carriage system that is well-suited for (but is not limited to) remote areas where water is scarce, or areas with low percolation, high water tables, shallow soil, or rough terrain. Because composting toilets eliminate the need for flush toilets, this significantly reduces water use and allows for the recycling of valuable plant nutrients in the form of compost.

The primary objective of composting toilet systems is to contain, immobilize, or destroy pathogens, thereby reducing the risk of human infection to acceptable levels without contaminating the environment or negatively affecting the life of its inhabitants. This should be accomplished in a manner that is consistent with good sanitation (minimizing the availability of excrement to disease vectors, such as flies, and minimizing human contact with unprocessed excrement), thus producing an inoffensive and reasonably dry end-product that can be handled with minimum risk.

Although there are many different composting toilet designs that continue to evolve, the basic concept of composting remains the same. In the composting process, organic matter is transformed by naturally occurring bacteria and fungi that break down the excrement into an oxidized, humus-like end product. These organisms thrive by aeration, without the need for water or chemicals. Various parameters control and manage environmental factors—air, heat, moisture—to optimize the process.

ADVANTAGES OF COMPOSTING TOILETS

- Composting toilet systems do not require water for flushing, and thus, reduce domestic water consumption.
- These systems reduce the quantity and strength of wastewater to be disposed of on-site.
- They are especially suited for new construction at remote sites where conventional onsite systems are not feasible.
- Composting toilet systems have low power consumption
- Self-contained systems eliminate the need for transportation of wastes for treatment/disposal.
- Composting human waste and burying it around tree roots and non-edible plants keeps organic wastes productively cycling in the environment

- Composting toilet systems can accept kitchen wastes, thus reducing household garbage.
- In many states, installing a composting toilet system allows the property owner to install a reduced-size leach field, minimizing costs and disruption of landscapes.
- Composting toilet systems divert nutrient and pathogen containing effluent from soil, surface water, and groundwater.

DISADVANTAGES OF COMPOSTING TOILETS

- Maintenance of composting toilet systems requires more responsibility and commitment by users and owners than conventional wastewater systems.
- Using a composting toilet requires electricity and hence in regions where Power supply is not stable some difficulty might arise over maintaining aerobic conditions unless back-up power is supplied.
- Composting toilet systems must be used in conjunction with a grey water system in most circumstances.
- Improper maintenance makes cleaning difficult and may lead to health hazards and odor problems.
- Using an inadequately treated end product as a soil amendment may have possible health consequences.
- There may be aesthetic issues because the excrement in some systems may be in sight.
- Too much liquid residual (leachate) in the composter can disrupt the process if it is not drained and properly managed.
- Use of power source increases running costs.
- Improperly installed or maintained systems can produce odors and unprocessed material.

ii) Landscaping Water Conservation Practices

Drought-tolerant Plants. Water conservation in landscaping can be accomplished through the use of plants that need little water, thereby saving not only water but labor and fertilizer as well. Careful landscape design can significantly reduce water use; it can also take advantage of native plants which have evolved water-saving or water-tolerant characteristics ideally suited for the local climatic conditions. Use of native plants can also help to minimize the spread of exotic plant species that disrupt local ecosystems. In addition to the selection of the plant species to be used in landscaping, practices such as the use of low precipitation rate sprinklers that have better distribution uniformity, bubbler/soaker systems, and/or drip or point irrigation systems can also conserve water used for landscaping purposes.

iii) Agricultural Water Conservation Practices

Water saving irrigation practices fall into three categories: field practices, management strategies, and system modifications. Examples of these practices include, respectively, the chisel plow aeration of extremely compacted soils, furrow diking to prevent uncontrolled runoff, and leveling of the land surface to distribute water more evenly. A number of such practices have been previously detailed in Section 2.3 of Chapter 2.

Irrigation Scheduling. Improved irrigation scheduling can reduce the amount of water required to irrigate a crop effectively by reducing evaporative losses, supplying water when most needed by the irrigated plants, and applying the water in a manner best suited to the plants being irrigated. A careful choice of the rate and timing of irrigation can help farmers to maintain yields with less water. In making scheduling decisions, irrigators should consider:

- The uncertainty of rainfall and the timing of crop water demands.
- The limited water storage capacity of many irrigated soils.
- The finite pumping capacity of most irrigation systems.
- The price of water and changes in water prices as additional operators increase water demand.

Irrigation Management. Management strategies involve monitoring soil and water conditions and collecting information on water use and efficiency. The methods include measuring rainfall, determining soil moisture levels, monitoring pumping plant efficiency, and scheduling irrigation. Typical system modifications include the addition of drop tubes to a center pivot irrigation system, retrofitting wells with smaller pumps, installing a surge or demand irrigation system, and/or constructing a tailwater or return flow recovery system.

iv) Industrial and Commercial Water Conservation Practices

Water recycling is the reuse of water for the same application for which it was originally used. Recycled water might require treatment before it can be reused. Cooling water recirculation and washwater recycling are the most widely used water recycling practices. The following guidelines should be used when considering water reuse and recycling in industrial and commercial applications:

- Identification of water reuse opportunities: Are there areas within the factory or in the production process that currently use water only once that would be amenable to reuse?

- Determination of the minimum water quantity needed for the given use: Are there areas within the factory or in the production process where more water is being supplied than is needed to accomplish the purpose?
- Identification of wastewater sources that satisfy the water quality requirements: Does the process require potable water or water of a lesser standard? Can the same result be achieved with lower-quality water?
- Determination of how the water can be transported to the new use: What modifications, if any, in the process or factory may be needed to permit recovery and recirculation/recycling of the water currently sent to waste? What additional treatment may be necessary to reuse this water? What is the relative cost of the required modifications versus the cost of the raw water over the life of the modifications?

Cooling Water Recirculation: Recycling water within a recirculating cooling system can greatly reduce water consumption by using the same water to perform several cooling operations. The water savings are generally sufficiently substantial to result in an overall cost saving to industry. Such savings can be even greater if the waste heat is used as a heat source elsewhere in the manufacturing process. Three cooling water conservation approaches are typically used to reduce water consumption: evaporative cooling, ozonation, and heat exchange.

Washwater Recycling: Another common use of water by industry is in the use of fresh or deionized water for removing contaminants from products and equipment. Deionized water can generally be recycled after its first use, although the reclamation treatment cost of recycling this water may be as great as or greater than the cost of purchasing raw water from a producer and treating it. The same processes required to produce deionized water from municipal water can be used to produce deionized water from used washwater. It is also possible to blend used washwater with raw water, which also would result in an overall water saving. The reuse of once-used deionized water for a different application within the same factory should also be considered as a water conservation option. For example, used washwater may be perfectly acceptable for washing vehicles or the factory premises.

v) Water Conservation Practices for Water Utilities

Common practices used by water supply utilities include metering, leak detection, repairing water lines, well capping, retrofitting programmes, pricing, wastewater reuse, and developing public education programmes and drought management plans.

Metering. The measurement of water use with a meter provides essential data for charging fees based on actual customer use. Submetering may also be used in multiple-unit operations such as apartment buildings, apartments, and mobile homes to indicate water use by individual units within a complex. In such cases, the entire complex of units might be metered by the main supplier, while the individual units might be monitored by either the owner or the water utility.

Leak Detection. It has been estimated that in many distribution systems up to half of the water supplied by the water treatment plant is lost to leakage; even more may be lost due to unauthorized abstraction. One way to detect leaks and identify unauthorized connections is to use listening equipment to survey the distribution system, identify leak sounds, and pinpoint the locations of hidden underground leaks. Metering can also be used to help detect leaks in a system. It is not unusual for unaccounted water losses to drop by up to 36% after the introduction of metering and leak detection programmes.

Water Distribution Network Rehabilitation. A water utility can improve the management and rehabilitation of its water distribution network by a well-planned preventive maintenance programme based on a sound knowledge of the distribution network. This knowledge is often embodied in a distribution system database that includes the following data:

- An inventory of the characteristics of the system components, including information on their location, size, and age and the construction material(s) used in the network.
- A record of regular inspections of the network, including an evaluation of the condition of mains and degree of corrosion (if any).
- An inventory of soil conditions and types, including the chemical characteristics of the soils.
- A record of the quality of the product water in the system.
- A record of any high or low pressure problems in the network.
- Operating records, such as of pump and valve operations, failures, or leaks, and of maintenance and rehabilitation costs.
- A file of customer complaints.
- Metering data.

Through the monitoring of such records, advance warning of possible problems can be achieved. For example, excessive water use, or numerous complaints or demands for spare parts, could be early warning signs of an impending breakdown in the system. This system should also include a regular programme of preventive maintenance to minimize the possibility of system failures.

Well Capping. Well capping is the sealing of abandoned wells. In the case of artesian wells, rusted casings can spill water in a constant flow into drainage ditches, resulting in evaporative loss or runoff losses. In non-artesian wells, uncapped abandoned wells form points of entry for contaminants into the groundwater system.

Pricing. Placing an economic value on freshwater is the principal means of achieving water conservation. Pricing provides a financial incentive to conserve water. Rate structures may be variable and/or graduated, with prices fixed on the basis of class of service (residential versus industrial or agricultural, for example) and quantity used (for example, the unit price for quantities below 400 lit/day might be significantly lower than for quantities which exceed that amount for a single-family residence). Pricing has the advantage of minimizing the costs of overt regulation, restrictions, and policing, while providing a high degree of freedom of choice for individual water customers.

Retrofit Programmes. Retrofitting involves the replacement of existing plumbing fixtures with equipment that uses less water. The most successful water-saving fixtures are those which operate in the same manner as the fixtures being replaced; for example, toilet tank inserts, faucet aerators, and low-flow showerheads do not significantly change the operation of the systems into or onto which they are placed, but they do result in substantial water savings.

Water Audit Programmes. Various types of audits can be undertaken. For example, residential water audits may involve sending trained water auditors into participating households, free of charge, to encourage water conservation efforts, or providing them with record sheets to note down their water use for external analysis. Water audits may also be undertaken in commercial and industrial facilities, and may be combined with an assessment of the potential for implementing water reuse and recycling programmes. A pre-implementation and post-implementation water audit in factories adopting a reuse and recycling programme would be a valuable means of demonstrating and quantifying the water savings achieved.

Public Information and Education. Public information and education programmes can be undertaken to inform the public about the basics of water use and conservation. Programmes should be developed for specific applications and may be targeted at specific user groups or age groups; for example, at housekeepers, to encourage domestic water conservation, or at schoolchildren, to provide information on the wider implications of water conservation for future consumption, the environment and other uses. Basic information should include the following:

- How water is delivered and how wastewater is disposed of.
- The costs of water and water supply services.
- Why water conservation is important.

The programmes should provide guidance on how the user groups and individuals can participate in conservation efforts. It should be noted that there is a large body of public information and education materials available, particularly in the United States, which may be obtained from a variety of public agencies and NGOs at little or no cost and form the basis of a local public awareness initiative.

Demand Management. Demand management is closely linked with water conservation practices. Table 5.13 shows a summary of short-term measures that can be used to reduce demand during periods of drought and the expected levels of reduction. These measures may also be considered in concert with other conservation measures noted above.

Table 5.13   Short-term Measures to Reduce Water Demand and their Effectiveness

Creation of public awareness: 0-15%	Voluntary measures: 15-25%	Mandatory measures (after a drought determination): 25-39%
Explain water conservation practices	Encourage voluntary restrictions on use	Adopt regulatory measures
Implement a public information programme	Conduct water audits of water-intensive customers	Develop water rationing, with penalties
Intensify conservation efforts	Implement conservation-related rate structures	Restrict annexations and new connections

Source: http://www.oas.org/usde/publications/Unit/oea59e/ch31.htm]

5.3.6.2   Extent of Use

Water conservation measures have been practiced primarily in the United States, although some Latin American countries have implemented specific measures. For example, in Brazil, the pharmaceutical, food processing, and dairy industries were required to pay effluent charges that contributed to reductions in water use and wastewater production of between 42% and 62%. In Mexico, increased water prices contributed to an increase in wastewater reuse and the recycling of cooling water.

Chile is the only country in the region with a comprehensive water law that has encouraged the development of water markets. The 1981 National Water Law in Chile established secure, tradable, and transferable water use rights for both surface and groundwaters. As a result, during periods of low rainfall, farmers shift from the production of water-intensive crops, such as corn and oilseeds, to higher-valued and less water-intensive crops, such as fruits and vegetables.

Water recycling is used at a Container Corporation of America Mill in Santa Clara, California (U.S.A), which manufactures paperboard from the recycled fibers of newspapers, corrugated cardboard clippings, and ledger paper. Historically, water has been used in this process for a variety of purposes. In recent years, however, the mill has begun recycling water used in its rinsing processes after clarification. The mill has also installed a closed loop cooling tower, which has resulted in an additional reduction in water use. These water conservation and use efficiency practices have resulted in an estimated saving of approximately 2,800 m^3/day, compared to its 1980 water use rates. These water reductions amount to approximately 900,000 m^3/year and saved the company approximately $348,200 per year.

5.3.6.3 Operation and Maintenance

Given the variety of measures that might be undertaken to address conservation needs within a specific geographic area, of which a number are mechanical but many may be technological or informational, it is difficult to identify specific operational requirements. However, some of the more obvious requirements include the following: low-flow water conservation devices require periodic maintenance and repair; leak detection equipment and meters require periodic testing and repair; drought and water conservation management strategies, such as pricing and user charges, require monitoring and enforcement; and well-capping programmes require monitoring and trained personnel in order to be effective. Maintenance requirements range from regular inspections of mechanical devices to the review of legislation and conservation plans to ensure their continued relevance.

5.3.6.4 Level of Involvement

The installation and maintenance of low-flow household and irrigation devices may require governmental incentives in order to be accepted. In some cases, employees of the water utility may install and maintain these systems at little or no charge in order to effect the desired water savings. Alternatively, government regulations may be necessary to provide incentives for the implementation of industrial and agricultural water conservation measures. Government action is required in the promulgation of plumbing codes for new construction that will contribute to the adoption of residential water conservation measures. Government or utility involvement is also needed for leakage detection and the repair of distribution systems. Metering, in addition to requiring technical personnel and equipment to be effective, generally requires governmental action to implement and government authority to establish or regulate water tariffs. However, community participation and voluntary conservation are a key element if this technology is to be effective.

5.3.6.5 Costs

The cost of water conservation measures varies with the cost of any equipment required and with size and location. The cost of replacing a conventional toilet with a low-flush toilet is about $250 per unit. Low-flow showerheads, in contrast, cost about $5 each. Meter installation costs range from about $200 for interior meters to $500 for external meters. Leak control has been estimated at $40/million liters.

Costs associated with water conservation are often offset by cost savings incurred after implementation. For example, the use of treated wastewater for cooling at an industrial plant in California, U.S.A., resulted in a saving of $150,000 in 1989, while modifications to the sinks in a computer manufacturing plant in Denver, Colorado, resulted in a saving of $81,000 also in 1989. Close monitoring of water use in a packing facility in Santa Clara, California, produced an annual saving of $40,000. Elsewhere, the introduction of water markets in Chile in 1993 increased agricultural profits by $1.5 billion.

5.3.6.6 Effectiveness of the Technology

Water conservation measures are highly effective. However, this technology may not be too popular with consumers, who may be asked to pay a higher price for the water they consume, and can be, politically, very unpalatable. Nevertheless, studies carried out in Seattle, Washington, U.S.A., reported the following results from water conservation measures:

- According to detailed data on the performance of low-flow water devices in 308 single-family residences, indoor per capita water use dropped 6.4% after low-flow showerheads were installed.
- Easily installed aerators reduced water use at a faucet by as much as 60% while still maintaining a strong flow.
- A reduction in water pressure from 690 kPa to 345 kPa at an outlet resulted in a water flow reduction of about one-third of the pre-existing use.

- Gray water reuse saved a volume of water equivalent to that needed to supply more than 7,000 residences and businesses.
- Outdoor water use was reduced by restricting watering times to the early morning or late evening; watering on cooler days, when possible, also reduced outdoor water use. All these measures contributed to reduced evaporative losses.
- As many as 600 liters of water were saved when washing a car by turning the hose off between rinses; additional benefits and water savings were achieved by washing the car on the lawn, which both watered the lawn and reduced runoff.
- Sweeping sidewalks and driveways, instead of hosing them down, saved about 200 liters of water every 5 minutes.

In other studies, such as an industrial water conservation project in California, the conversion of an industrial process from a single-pass freshwater cooling system to a closed-loop cooling system, with circulating chilled water, has saved an estimated 20,000 to 28,000 lit/day, while cities in the hemisphere that have large, old, deteriorating systems, leak detection programmes have been especially efficient in minimizing water losses.

5.3.6.7 Suitability

Water conservation measures are suitable and recommended for all public water supply systems, industries with high water use, agricultural enterprises, and individual residential users in Latin American countries and the Caribbean islands.

5.3.6.8 Advantages

Residential water users
- Low-flow devices result in water use savings of 20% to 40%.
- Pressure reductions save up to 33% of the water normally consumed.
- Conservation-based landscape irrigation practices also produce significant water use savings.

Industrial/commercial users
- Water recycling greatly reduces water use.
- Deionized water can be recycled after its first use at little or no additional cost, using the same equipment used to produce the deionized water from the municipal supply.
- Proper scheduling of landscape irrigation optimizes water use by minimizing evaporative losses.

Agricultural users
- Water savings can be achieved through a combination of field practices, monitoring, and system modifications.
- Wastewater reuse can produce significant water savings.

Water supply plants
- Widespread leakages and illegal connections may account for 30% to 50% of the water loss in a distribution system.
- Metering allows for greater accountability and assists in the development of a pricing structure that is fair and appropriate to the individual water supply system and that provides incentives for conservation.
- Equipment repairs to water mains and valves, and capping unused wells, can reduce unnecessary water loss, and prevent contamination of both piped water and groundwater.
- Retrofit programmes can produce long-term savings of water and money.

5.3.6.9 Disadvantages

Residential users:
- The initial cost of low-flow devices can be high.
- Changes or modifications in water use habits are not readily accepted.

Agricultural users:
- Low-volume irrigation systems may be costly in some cases.
- The use of wastewater for irrigation may pose potential health risks.

Industrial/commercial users:
- Modifications to manufacturing processes may be required in some cases, incurring an initial capital charge to the user.
- Changes in the piping system within a plant can be costly.

Water supply plants:
- Implementation of leak detection, control and metering is costly.
- Meters and leak detection devices require regular maintenance.

5.3.6.10 Cultural Acceptability

Most conservation measures have been applied in response to government regulations or conservation programmes. As was noted above, public acceptance is limited despite the economic benefits.

5.3.6.11 Further Development of the Technology

Improved equipment for use in leak detection and metering is required. Such devices need to be more robust and less costly. Meters should be able to withstand tampering. It would also be desirable for low-flow plumbing devices to be more cost effective so as to be more attractive to consumers. Implementation of educational programming on the necessity and the economic and environmental benefits of water conservation is also likely to lower consumer resistance to water conservation technologies.

5.3.7 Groundwater Conservation and Recharge

Beneath the surface of the earth—from a few metres to hundreds of metres down—water slowly seeps through the pores, cracks, and fractures of rocks, and in the spaces between sand and gravel. When these spaces are filled or saturated with water, that water is called *groundwater*. If there is enough groundwater to supply a well or spring, the zone of saturation is considered an aquifer. For example, in Arizona, USA most of the water that seeps into aquifers (recharge) comes from streams fed by mountain runoff. Additional recharge occurs when crops and turf are irrigated and from storm water and municipal discharge of treated sewage. Typically, aquifers in central and southern Arizona are made of sands and gravels and yield considerable groundwater. Northern Arizona aquifers are characterized by fractured rock; in many places, wells must be drilled to great depths, and may yield little or no water.

To tap groundwater, a well must be drilled deep enough to penetrate the water table, which is the surface of the zone of saturation. When pumping occurs, the water table near the well drops, creating a cone of depression. If more water is pumped from an aquifer than is recharged, overdraft occurs and the water table drops. In many parts of the world, overdraft is a serious problem that has lead to land settling and cracking (subsidence), lower well yields, quality degradation and the drying up of streams and rivers. In many countries and provinces, laws have been enacted to regulate pumping in areas with substantial overdraft and to bring pumping and recharge into balance.

Although in many areas groundwater supply is large, it has accumulated over thousands of years. Much of it is might not be of good quality or cannot easily be pumped. Efforts have been initiated globally to conserve groundwater by artificially recharging the aquifers so as to store water for future use and reduce overdraft. This can be done by artificial structures like percolation pit, dug-cum-recharge bore well as shown in Section 4.6 of Chapter 4. Eventually it would lead to an increased groundwater recharge than the natural groundwater recharge capability, as explained in that Section.

5.3.7.1 Artificial Recharge of Groundwater

The increasing demand for water has increased awareness towards the use of artificial recharge to augment groundwater supplies. Stated simply, artificial recharge is a process by which excess surface water is directed into the ground—either by spreading on the surface, by using recharge wells, or by altering natural conditions to increase infiltration—to replenish an aquifer. It refers to the movement of water through man-made systems from the surface of the earth to underground water-bearing strata where it may be stored for future use. Artificial recharge (sometimes called planned recharge) is a way to store water underground in times of water surplus to meet demand in times of shortage.

Purposes: One may wonder why one should pour water into a well only to haul it up again. Instead why not keep the water on the surface itself to begin with, if he has much water to spare? This is not the true picture of what really happens.

Water used for recharge purposes is always surplus water that has already been used for some purpose, water from high flow periods of a river, which is not needed for use exactly at that season. Such water, if not stored in surface or underground water reservoirs, would just run down to the sea and be lost for beneficial use by man.

Artificial recharge techniques have been used throughout the world for more than 200 years for a variety of purposes. Occasions arise where it is desirable to artificially recharge water underground to achieve one or more objectives, which may include:

(a) To stem the decline of water levels.
(b) To supplement existing supplies.
(c) To remove suspended solids by infiltration through the soil.
(d) To store cyclic water surpluses for use in dry periods.
(e) To prevent loss of land surface resulting from excessive groundwater or oil exploration.
(f) To use the aquifer as a distributory system.
(g) To decrease the size of areas required for water supply system.
(h) To increase stream flow.

5.3.7.2 Methods of Artificial Groundwater Recharge

The methods of artificial groundwater can be broadly grouped into direct and indirect methods as described in the following sections.

i) Direct Methods of Artificial Groundwater Recharge

Spreading basins: This method involves surface spreading of water in basins that are excavated in the existing terrain. For effective artificial recharge highly permeable soils are suitable and maintenance of a layer of water over the highly permeable soils is necessary. When direct discharge is practiced the amount of water entering the aquifer depends on three factors—the infiltration rate, the percolation rate, and the capacity for horizontal water movement. In a homogenous aquifer the infiltration rate is equal to the percolation rate. At the surface of the aquifer however, clogging occurs by deposition of particles carried by water in suspension or in solution, by algal growth, colloidal swelling and soil dispersion, microbial activity etc. Recharge by spreading basins is most effective where there are no impending layers between the land surface and the aquifer and where clear water is available for recharge; however, more turbid water can be tolerated than with well recharge. The common problem in recharging by surface spreading is clogging of the surface material by suspended sediment in the recharge water or by microbial growth. In coarse-grained materials removal of fine suspended sediment is difficult. Playa lakes or wet weather lakes are depressions that collect water after rainfall or periods of snowmelt. Playa lakes in Texas, New Mexico and Colorado have been used in artificial recharge projects (O'Hare et al., 1986). Many Playa lakes have tight clay deposits that restrict leakage of water. Most of the water is lost by evaporation or by non-beneficial growth of vegetation in the lake. Heavy clay soils can be broken up and the lake bottom regraded for maximum recharge. In a demonstration project near Lubbock, Texas, playa lakes were modified by excavating concentration pits and using the excavated soil to raise the elevation of some of the previously flooded lands.

Recharge pits and shafts: Conditions that permit surface spreading methods for artificial recharge are relatively rare. Often lenses of low permeability lie between the land surface and water table. In such situations artificial recharge systems such as pits and shafts could be effective in penetrating the less permeable strata in order to access the dewatered aquifer. The rate of recharge has been found to increase as the side slopes of the pits increased.

Unfiltered runoff waters leave a thin film of sediment on the sides and bottom of the pits, which require maintenance in order to sustain the high recharge rates. Shafts may be circular, rectangular, or of square cross-section and may be backfilled with porous material. Excavation may terminate above the water table level or may be hydraulic connectors and extend below the water table. Recharge rates in both shafts and pits may decrease with time due to accumulation of fine-grained materials and the plugging effect brought about by microbial activity.

Ditches: A ditch could be described as a long narrow trench, with its bottom width less than its depth. A ditch system can be designed to suit the topographic and geologic conditions that exist at a given site. A layout for a ditch and a flooding recharge project could include a series of ditches trending down the topographic slope. The ditches could terminate in a collection ditch designed to carry away the water that does not infiltrate in order to avoid ponding and to reduce the accumulation of fine material (O'Hare et al., 1986).

Recharge wells: Recharge or injection wells (Figure 5.30) are used to directly recharge water into deep water-bearing zones. Recharge wells could be cased through the material overlying the aquifer and if the earth materials are unconsolidated, a screen can be placed in the well in the zone of injection. In some cases, several recharge wells may be installed in the same borehole. Recharge wells are suitable only in areas where a thick impervious layer exists between the surface of the soil and the aquifer to be replenished.

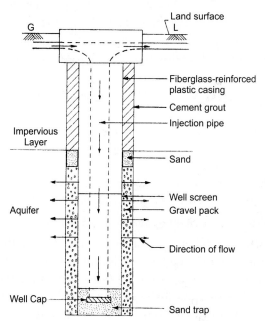

Figure 5.30   A Typical Recharge Well

Source: http://www.cee.vt.edu/program_areas/environmental/teach/gwprimer/recharge/recharge.html#Methods]

They are also advantageous in areas where land is scarce. A relatively high rate of recharge can be attained by this method. Clogging of the well screen or aquifer may lead to excessive buildup of water levels in the recharge well. In ideal conditions a well will accept recharge water at least as readily as it will yield water by pumping. Factors that cause the build up of water levels in a recharge well to be greater than the corresponding drawdown in a discharging well may include the following.

- Suspended sediment in the recharge water, including organic and inorganic matter.
- Entrained air in the recharge water.
- Microbial growth in the well.
- Chemical reactions between the recharge water and the native groundwater, the aquifer material, or both.
- Ionic reactions that result in dispersion of clay particles and swelling of colloids in a sand-and-gravel aquifer.
- Iron precipitation.
- Biochemical changes in recharge water and the groundwater involving iron-reducing bacteria or sulfate-reducing organisms.
- Differences in temperature between recharge and aquifer water.

Factors that cause the build up of water levels in a recharge well to be less than the corresponding drawdown in a discharging well may include the following.

- Recharge water is warmer than native groundwater and therefore, less viscous.
- Increase in the saturated thickness and transmissitivity of the aquifer due to the higher water levels that result when a water table aquifer is recharged.

Recharge water that is unsaturated with respect to calcium carbonate. Such water may dissolve parts of a carbonate aquifer (O'Hare et al, 1986).

ii) Indirect Methods of Artificial Groundwater Recharge

Enhanced streambed infiltration (induced infiltration): This method of induced recharge consists of setting a gallery or a line of wells parallel to the bank of a river and at a short distance from it. Without the wells there would be unimpended outflow of groundwater to the river. When small amounts of groundwater are withdrawn from the gallery parallel to the river, the amount of groundwater discharged into the river decreases. The water recovered by the gallery consists wholly of natural groundwater. Each groundwater withdrawal is accompanied by a drawdown in the water table. For high recovery rates this drawdown tends to lower the groundwater table at the shoreline below that at the river. Thus, surface water from the river will be induced to enter the aquifer and to flow into the gallery. In areas where the stream is separated from the aquifer by materials of low permeability, leakage from the stream may be so small that the system is not feasible (O'Hare et al., 1986)

Conjunctive wells: A conjunctive well is one that is screened in both a shallow confined aquifer and a deeper artesian aquifer. Water is pumped from the deeper aquifer and if its potentiometric surface is lowered below the shallow water table, water from the shallow aquifer drains directly into the deeper aquifer. Water augmentation by conjunctive wells has the advantage of utilizing sediment-free groundwater, which greatly reduces the damage of clogging well screens.

Other benefits are:

- It reduces the amount of evapotranspiration water loss from the shallow water table.
- Reduces flooding effects in some places.

Environmental effects from the conjunctive well method must be carefully studied to assure that unwanted de-watering of wetlands or reduction of base flow will not occur. The possibility of coagulation due to mixing of chemically different types of groundwater should also be investigated. Table 5.14 lists some of the major factors to be considered in artificial groundwater recharge.

Table 5.14   Some Factors to be Considered in Artificial Groundwater Recharge

1.	Availability of waste water
2.	Quantity of source water available
3.	Quality of source water available
4.	Resulting water quality (reactions with native water and aquifer materials)
5.	Clogging potential
6.	Underground storage space available
7.	Depth to underground storage space
8.	Transmission characteristics
9.	Topography/applicable methods (injection or infiltration)
10.	Legal/institutional constraints
11.	Costs
12.	Cultural/social considerations

Source: http://www.cee.vt.edu/program_areas/environmental/teach/gwprimer/recharge/recharge.html#Methods]

5.3.7.3 Advantages of Artificial Groundwater Recharge

Artificial recharge has several potential advantages, as stated below:

- The use of aquifers for storage and distribution of water and removal of contaminants by natural cleaning processes which occur as polluted rain and surface water infiltrate the soil and percolate down through the various geological formations.
- The technology is appropriate and generally well understood by both the technicians and the general population.
- Very few special tools are needed to dig drainage wells.
- In rock formations with high, structural integrity few additional materials may be required (concrete, soft stone or coral rock blocks, metal rods) to construct the wells.
- Groundwater recharge stores water during the wet season for use in the dry season, when demand is highest.
- Aquifer water can be improved by recharging with high quality injected water.
- Recharge can significantly increase the sustainable yield of an aquifer.
- Recharge methods are environmentally attractive, particularly in arid regions.
- Most aquifer recharge systems are easy to operate.
- In many river basins, control of surface water runoff to provide aquifer recharge reduces sedimentation problems.
- Recharge with less-saline surface waters or treated effluents improves the quality of saline aquifers, facilitating the use of the water for agriculture and livestock.

5.3.7.4 Disadvantages of Artificial Groundwater Recharge

Artificial groundwater recharge has some disadvantages too, as listed below:

- In the absence of financial incentives, laws, or other regulations to encourage landowners to maintain drainage wells adequately, the wells may fall into disrepair and ultimately become sources of groundwater contamination.
- There is a potential for contamination of the groundwater from injected surface water runoff, especially from agricultural fields and roads surfaces. In most cases, the surface water runoff is not pre-treated before injection.
- Recharge can degrade the aquifer unless quality control of the injected water is adequate.

- Unless significant volumes can be injected into an aquifer, groundwater recharge may not be economically feasible.
- The hydrogeology of an aquifer should be investigated and understood before any future full-scale recharge project is implemented. In karstic terrain, dye tracer studies can assist in acquiring this knowledge.
- During the construction of water traps, disturbances of soil and vegetation cover may cause environmental damage to the project area.

5.3.7.5 Costs of Artificial Groundwater Recharge

The cost of treating wastewater to potable standards is generally prohibitive. Therefore, it may be appropriate to consider artificial groundwater recharge without adversely affecting water quality or public health.

The estimated cost of infiltration of surface water in Argentina, using basins and canals, is $0.20/m^3$. The basins and canals used in the 1977 experiment in the San Juan River basin incurred a capital cost of $31,300. The comparable cost of water traps in Argentina was estimated to be between $133 and $167. The capital cost of a 5,700 m^3 cutwater, equipped with a 14 m extraction well, was estimated at $6,325. The operation and maintenance cost was estimated at $248 per year. The production costs were estimated to be about $0.30/m for the first five years of operation, $0.17/m for the next five years (five to ten years of operation), and $0.15/m for the following five years (ten to fifteen years of operation).

In Jamaica, the initial capital cost of the sinkhole injection system was estimated at less than $15,000. This cost is primarily related to the construction of the inflow settling basin and channels conveying the runoff water to the sinkholes. Maintenance costs were low, less than $5,000 for the 18-month project (or under $3,500/year) [O'Hare et al., 1986].

5.4 SUSTAINABLE DEVELOPMENT THROUGH INTEGRATED WATER MANAGEMENT

Sustainable development has already been discussed in Section 5.1 of this Chapter. The meaning of sustainable development essentially includes achieving development by conserving all natural resources such as air, water, land etc., qualitatively and quantitatively. Although a perfect conservation may not be possible in a real life situation always, the deviations in the qualitative/quantitative parameters of the natural resource are maintained at acceptable minimum levels, so that eventually the natural resource will get preserved in an overall sense.

Integrated water management involves the management of water resources in an integrated manner so that the water resources are conserved qualitatively and quantitatively. Therefore, integrated water management is a pre-requisite for sustainable development. While planning for sustainable development, it is necessary to ensure the conservation of natural resources at all levels including the highest micro level. Accordingly the concept of a *sustainable house* at the individual house level in Australia, underground dams at the river basin level in Brazil, desalination for freshwater augmentation at the regional level across the globe, integrated groundwater management at a regional level in Israel and sustainable water management at the national level in Germany are elaborated here in this Section.

5.4.1 *Sustainable House* in Sydney, Australia

A 'sustainable house' [Mobbs, 1998] is essentially an individual house that
- creates its own potable water,
- will recycle and/reuse all its wastewater (sewage or stormwater) and
- generates its own power through renewable sources to meet its requirements.

In 1996, a Sydney family set out to renovate their 100-year-old terrace house in the inner-city suburb of Chippendale. With a bit of vision, some common sense, and a lot of tenacity, they built what most of us would think impossible—a house in the middle of Australia's biggest city that produces its own power and water, and reuses its sewage on site.

Mike Mobbs, a Sydney environmental lawyer, and his lawyer wife Heather Armstrong wanted to renovate their house—to expand their kitchen and make a bit more living space for their two kids. But when they sat down to plan the job they decided to build a house that would be less of a drain on the planet's resources. Despite living in one of the most densely populated suburbs in Sydney's inner-west, they dreamed-up a house that would:

• Collect all its drinking water from the roof
• Generate all of its electricity from the sun
• Process all of its wastewater, including sewage, on site

These might seem hopelessly optimistic objectives for a house on a block 35 m long and 5 m wide, but what began as a private experiment a few years ago is today an example of how a little ingenuity and perseverance can change a house from a waste-belching brick box into a self-contained, sustainable home, without huge sacrifices in lifestyle, routine or comfort.

Mike and Heather's plan was more than a hippie pipe-dream, and environmental impact wasn't their only concern. They had a budget to work and the house might be needed to be resold one day and they couldn't afford to reduce its market value. It needed to be safe and hygienic for themselves and their two children, and they weren't prepared to have weird bits of equipment constantly breaking down and needing repair all the gadgets had to be off-the-shelf, thoroughly tested technology. Finally, to top-off their wish list, they wanted to carry out the renovations without using rainforest timber or materials that could release toxic chemicals into the house.

Mike wanted to collect all of the water the family needed off their roof. They had never been big water users despite running the washing machine every day, showering and doing all the normal things a family of four does. Even before they renovated, they used just 350 liters of water a day, or about half that of the average Sydney household. But over a year, this still adds up to around 100,000 liters of water.

The biggest issue in collecting rainwater is keeping it free of muck such as leaves, bird droppings and dead animals, and avoiding contamination with pollutants like heavy metals and dust. The Chippendale house has four simple but clever adaptions to get around this.

- the enclosed gutter on the roof excludes leaves and bird droppings but lets water in through special sinks
- the downpipe contains a sloping mesh trap to exclude leaves and debris without blocking water flow
- a simple diversion system in the downpipe directs the first 6-10 liters of rainwater (carrying dirt from the roof) away from the water tank and into the garden, and
- a sump with a fine mesh excludes all the heavy sediments before the water enters the tank

The guttering system feeds into a concrete tank hidden below the house's back deck that holds about 8,500 liters. When it is full the water overflows into a mini-wetland to reduce stormwater runoff from the block.

5.4.1.1 *All the Potable Water from Rainwater*

The house is less than 2 km from Sydney's Central Business District (CBD), sandwiched between two busy streets, choked frequently with buses and cars. So with two young kids, Mike and Heather were initially concerned about the quality of the water they'd collect off their roof. They were pleasantly surprised. Today, their drinking water is cleaner than that of most of the households. Test carried out by the University of Technology, Sydney, demonstrate consistently low turbidity and faecal coliform counts and, importantly, the highest level of lead ever measured recorded in the tank water was 0.03 mg/l—below the safety threshold of 0.05 mg/l recommended by the National Health and Medical Research Council in 1991.

The rainwater tank system had its problems. The family has run out of rainwater four times, which meant popping across to the neighbours to borrow 1,500 liters of water occasionally. Installing a bigger tank would have alleviated this problem, but not solved it—there simply isn't enough rainfall on their tiny roof to sustain the family's needs. It would be enough for two. These shortfalls aside, the system has served them well and means that an extra 100,000 liters of fresh water stays in the local dam each year.

5.4.1.2 All the Non-potable Water from Wastewater

A key to the house's astonishing water efficiency and waste disposal is a nifty low-maintenance recycling system that lives under the back deck. All the dirty water from the house (toilet, shower, bath, dishwasher, washing machine, sinks and tubs) drains into a single sewer pipe that empties into an underground concrete tank. Food scraps and other biodegradable waste (both theirs and the neighbours) are added through a hatch in the deck.

Inside the tank are three filter bedlayers of sand and peat packed with worms and bugs of all sizes and tastes. As the sewage works its way through this organic industrial disposal unit (Figure 5.31), it gets chewed up, spat out and generally cleaned by the organisms, in much the same way that rainwater is naturally filtered by soil before entering river systems.

The water exiting in the other end of the tank is clean enough to be reused in the house as grey water to flush toilets, wash clothes and water the garden and any excess overflows into a dry reed bed (Figure 5.32).

A reed bed is basically a hole in the ground, lined to stop water loss, filled with gravel and wastewater, and planted with reeds (which grow hydroponically). Inlet and outlet wastewater pipes are positioned below the gravel surface, so the standing wastewater level is always below the gravel. This stops mosquitoes and children accessing the wastewater, and stops odors escaping the reedbed. As untreated wastewater enters the reedbed, it pushes treated wastewater out the other end.

Beneficial micro-organisms (bugs) provide most of the wastewater treatment in a reedbed. Aerobic bugs breathe oxygen supplied by reeds. Reeds pump oxygen to their roots, surrounding them with a thin layer of air to stop them drowning. Aerobic bugs live here and capture passing pollutants.

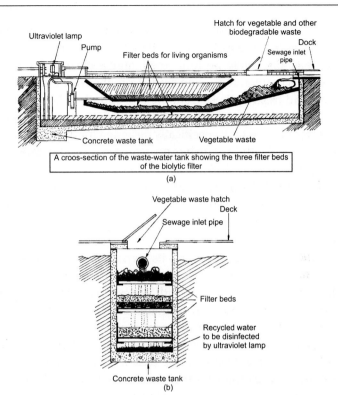

Figure 5.31 is shown with labels:
Ultraviolet lamp, Pump, Filter beds for living organisms, Hatch for vegetable and other biodegradable waste, Dock, Sewage inlet pipe, Concrete waste tank, Vegetable waste

A croos-section of the waste-water tank showing the three filter beds of the biolytic filter

(a)

Vegetable waste hatch
Deck
Sewage inlet pipe
Filter beds
Recycled water to be disinfected by ultraviolet lamp
Concrete waste tank
(b)

Figure 5.31 Wastewater Treatment and Reuse System in the 'Sustainable House', Sydney, Australia

[Source: http://www.abc.net.au/science/planet/house/special.htm]

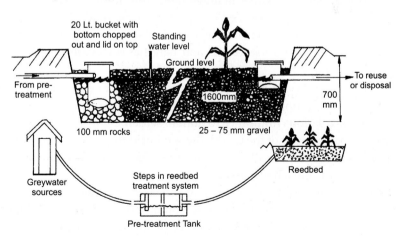

20 Lt. bucket with bottom chopped out and lid on top
Standing water level
Ground level
From pre-treatment
To reuse or disposal
1600mm
700 mm
100 mm rocks
25 – 75 mm gravel
Reedbed
Greywater sources
Steps in reedbed treatment system
Pre-treatment Tank

Figure 5.32 A Profile of a Reedbed for Domestic Wastewater Treatment Locally

[Source: Kahlon, 2004]

Any residual nasty bacteria or viruses are killed off as the water leaves the tank with a UV radiation zapper (powered by a small solar panel). The water looks clear, has absolutely no foul smell, and is clean enough to drink (although not recommended)... and shirts from the washing machine look just as white and fresh as those from the best suburban laundromats. The system did suffer three breakdowns during its first year, the result of the unusually long, narrow tank specifically designed for the small yard, but it now functions perfectly, requiring no maintenance and producing no offensive smells.

The water recycling and sewerage disposal systems in the Chippendale house processes around 100,000 liters of sewage each year, preventing it from entering the Pacific Ocean. The organic composting also cuts the local council's waste by several tonnes.

The 'Sustainable House' generates its own electricity through eighteen number of 120 W solar panels installed on the rooftop. During daytime, excess power is fed into the New South Wales power grid. There is also a solar hot water system installed in the house. Thus the 'sustainable house' is essentially self-sustaining in its water and power requirements most of the time. Successful examples of sustainable non-residential houses like 'RETREAT' near New Delhi, India and 'Green Business Centre' in Hyderabad, India have been already mentioned in Section 3.7.1 of Chapter 3.

5.4.2   Underground Dams in Brazil

To achieve integrated water management, the amount and depth of groundwater has to be conserved irrespective of the spatial and temporal variation in rainfall. The underground dams in Brazil conserve groundwater, similar to the johads and bandharas described in Section 5.2.1.1 of this Chapter.

Climatic instability in north east Brazil has more to do with irregular rainfall than with drought. The lack of a reliable water supply to meet even basic needs is a serious hindrance to human settlement in rural areas. Like other semi-arid regions of the world, the semi-arid tropical region of Brazil has shallow, rocky soils with low water retention capacity, a low organic material content, and a high susceptibility to erosion.

There are various options for creating and tapping water reserves in this region. Surface reservoirs are the most commonly used, since geological conditions are conducive to a high degree of surface drainage. However, evaporative losses are high. Another option is to make use of groundwater. However, the underlying crystalline bedrock lacks the porous structure necessary to store a large volume of water and maintain a high rate of extraction. To overcome these shortcomings, a further option, that of creating artificial aquifers using underground dams has been devised as a means of storing large quantities of good quality water for family or community needs, for use by animals, or even for small-scale irrigation. Under semi-arid tropical conditions, alluvial pools are a widespread phenomenon. This natural pooling of water, very common in watersheds with crystalline bedrock, lends itself to the building of underground dams in the alluvium. Such dams have the advantages of being able to store larger volumes of water than the natural aquifers in this area, and of being less susceptible to evaporative losses as the water is stored underground. The use of these underground dams also takes advantage of the naturally occurring alluvium.

5.4.2.1 Technical Description

An underground dam is any structure designed to contain underground flow, from a natural aquifer or from an artificial one, built with an impermeable barrier. Two major types have been distinguished in the literature: underground or submersible dams and submerged dams, both shown in Figures 5.33 and 5.34. Underground or submersible dams are defined as dams with walls that begin at the impermeable layer and extend above the surface of the alluvium, causing pools to form upstream during rainy periods. Water is stored both above and below the alluvial surface. The wall of a submerged dam, on the other hand, is entirely enclosed in the alluvium, and water is stored in the saturated soil. These types of dams have been built in north east Brazil since the turn of the century to augment rural water supplies.

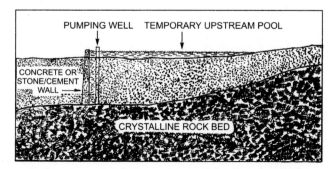

Figure 5.33   Cross-section of an Underground or Submersible Dam
Source: http://www.oas.org/usde/publications/Unit/oea59e/ch34.htm]

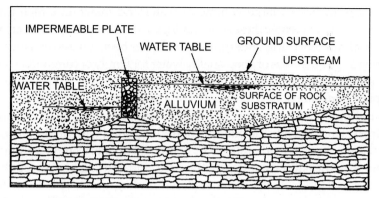

Figure 5.34 Cross-section of a Submerged Dam
Source: http://www.oas.org/usde/publications/Unit/oea59e/ch34.htm].

The dam wall, also called an impermeable plate, intercepts the flow of underground and/or surface waters, creating and/or raising the water table and pool elevation within an alluvial area. The dam wall is the main component of this technology. It extends from the bedrock or other subsurface impermeable layer up to, or beyond, the surface of the alluvial soil. It can be built of various materials, such as layers of compressed clay; packed mud; masonry; polyethylene or PVC plastic canvas; concrete; or a combination of materials.

• **Construction of an Underground Dam**

Site selection. The first step is site selection. Information on the soil distributions in an area is used to identify the best site. Sites with alluvial soils no more than 3 m to 4 m deep, of medium to coarse texture, and having a gradient of no more than 5% are preferred. Such sites may coincide with natural drainage routes, known as creeks, which carry large amounts of rainwater runoff in the region. In order to make optimal use of creek beds, a knowledge of the soil profile, and hence the depth to the impermeable layer, is necessary. Once a group of sites has been identified on the basis of topography, a further selection should be made on the basis of the salt content of the surface water and the average annual flow rates. Sites that have high salinities and high flow rates that could jeopardize the dam structure should be excluded from further consideration.

Topographic survey. Once a site has been selected on the basis of the topographic, salinity, and flow rate criteria, an on-site topographic survey should be performed, using 20 m × 20 m quadrants, to better determine the situation of the components. In systems that do not include a natural watercourse, this determination should include the delineation of the catchment area and location of the wall. For schemes that are being built for agricultural purposes, it is also necessary to locate the planting area to be served from the dam.

Construction of the wall. In the area chosen for the dam wall, a gutter, or cut-off trench, is dug across, or perpendicular to, the bed of the river or drainage route, down to the impermeable layer; its width depends upon the depth of the impermeable layer, the type of soil, and the material to be used in building the wall. In very dry, sandy alluvia, banks with low cohesion may constantly collapse, making excavation difficult and requiring the use of a trenching shield or other type of support to prevent slumping of the trench walls. Nevertheless, areas of sandy alluvium are desirable as dam sites because the water table is easily found there. It may be necessary to control the level of the water table by pumping so that excavation down to the impermeable layer may proceed. Some materials that can be used in constructing the wall include the following:

- *Layers of clay.* The clay should be deposited in the trench in uniform, 10 cm thick layers, moistened, and compressed to about half that thickness (5 cm). This is usually done by hand using wooden blocks. Multiple layers are placed and compressed, until the clay layers reach the surface of the soil.

- *Packed mud.* The mud, called 'ambor' by farmers in the western part of Rio Grande do Norte, is a mixture of mud and water, similar to that used in rural areas to build mud huts, which is deposited evenly in the trench up to the surface of the soil.

- *Masonry.* A double row of bricks, joined with a cement-and-sand mortar (1:4 ratio), is used to form a vertical wall. The space between the wall and the downstream slope of the cut should be filled. The upstream side of the wall should be plastered with cement-and-sand mortar (1:3 ratio) and sealant diluted with water (1:15 ratio). The bricks in this wall must be well-baked and salt-free to minimize the risk of dam failure or seepage.

- *Stone.* In very rocky areas, masonry bricks can be replaced with stones joined with cement-and-sand mortar (1:4 ratio). The stones should be properly set in the mortar, leaving no crevices where seepage could occur. It is recommended that the wall be plastered with cement-and-sand mortar (1:3 ratio) and sealant diluted with water (1:15 ratio). Because the stones are less regular than bricks, use of this material normally requires more skilled manpower to ensure the integrity of the structure.

- *Plastic canvas.* It is also possible to use an artificial fabric core in this technology. When doing so, however, it is recommended that a mud-and-water plaster be used on the downstream side of the trench to smooth the slope cut and to prevent sharp stones, roots, etc., from puncturing the fabric. At the bottom of the trench, on the upstream side, a small gutter (20 cm × 20 cm) should be dug in the impermeable layer, and a similar gutter dug in the soil surface at the top of the trench, on the downstream side. These gutters are used to secure and seal the ends of the plastic canvas, using the same mud mortar as in the plaster. Care should be taken, when laying the canvas, to avoid stretching it; to lay it in low wind and non-extreme temperature conditions, so as to minimize expansion and contraction effects; and to protect it from sharp surfaces. If the canvas is pierced, it should be patched with a piece of plastic and an adhesive appropriate for the material.

• Management of Underground Dams

Soil and water management in underground dams has been the subject of much discussion, primarily as it relates to the potential for salination. In order to avoid salinity problems, a discharge pipe should be placed at the bottom of the dam, on top of the impermeable layer. At the upstream end, this pipe passes through the wall parallel to the trench floor and to the water course. At the downstream end, the pipe connects to a vertical pipe, which functions as the abstraction point as well as an overflow/outlet. Water can be pumped or drained through the vertical pipe and discharged for use or onto the soil as waste. The pipe allows an annual drawdown of the dam as a means of removing dissolved salts.

5.4.2.2 Extent of Use

Underground dams are an option for rural areas that lack more traditional sources of water for agricultural and other uses. They are widely used in the semi-arid region of Brazil, and may be used in other semi-arid regions where similar conditions occur.

5.4.2.3 Operation, Maintenance and Level of Involvement

While an underground dam is a simple technology—which does not require any particular level of training to operate or maintain, it does require some degree of care in siting and construction. Certain factors must be taken into account when building underground dams, including the average rainfall in the region, the average rates of flow of rivers/streams or drainage lines, the porosity and texture of the soil in the area, the salinity of the water, the aquifer storage capacity, and the depth of the impermeable layer.

Farmers in the western region of Brazil are generally satisfied with the operation of the underground dams. Problems that have arisen have generally done so in other areas of Brazil and primarily relate to aspects of dam construction. Some of the problems have had to do with water loss by seepage through the dam wall, which is likely to be caused by the dam wall's not extending to the impermeable layer. Other construction-related problems have to do with the drainage ditches that provide water to underground dam sites not located on natural watercourses. Where the ditches have not been adequately sized to cope with high flows, problems such as erosion and contamination of the artificial aquifer during the rainy season may result. Generally, these problems can be solved by rural extension technicians. As with any technology, the users must be familiar with its operating principles to take full advantage of it.

Underground dams are under construction throughout the semi-arid region of Brazil, with funding from state and municipal governments and from farmers.

5.4.2.4 Costs

The costs involved in building underground dams vary depending on such factors as length of the wall, materials used, depth of the impermeable layer, and availability of manpower. An underground dam with a drainage area of 1.0 ha, built with a polyethylene plastic canvas wall, costs an average of $500. If 4 mm PVC canvas is used for the wall instead, the dam will cost about $1,700.

5.4.2.5 Effectiveness of the Technology

Although simple to build, underground dams must be constructed with considerable care if they are to work effectively. For example, the dam wall should extend all the way down to the impermeable layer to prevent seepage; when plastic canvas is used for the wall, every effort should be made to prevent punctures, and, should they occur, the canvas should be patched with a piece of the same plastic and an appropriate glue. The canvas should never be left uncovered and exposed to direct sunlight, as it easily dries out and may split. A drainage ditch should also be provided as a means of managing the salinity of the impounded water.

5.4.2.6 Suitability

Underground dams can be introduced throughout the semi-arid region. Given the agro-ecological and socioeconomic conditions that inhibit agricultural development in the area, this technology has the potential to take maximum advantage of the available water. Underground dams have been accepted throughout the semi-arid northeast region of Brazil because of their benefit to users. Their use is primarily by farmers, owing to the relatively high cost of building them.

5.4.2.7 Advantages

- Underground dams are based on a simple technology, are inexpensive to build, and can make use of locally available materials and manpower.
- Once water has been stored in the alluvial soils, they have low evaporation rates compared to surface water reservoirs.
- They can be combined with other technologies, such as soil and water conservation techniques, and dug wells upstream.

5.4.2.8 Disadvantages

- Because underground dams store water within the alluvial soil profile, their capacities are low compared with those of conventional dams.
- Given the socioeconomic circumstances of farmers in the semi-arid tropical region of Brazil, the cost of building these dams is a real obstacle to the widespread adoption of this technology.

5.4.2.9 Further Development of the Technology

In order to make the technology more acceptable for farmers and other users, certain matters must be addressed, such as the development of alternative construction materials having a lower capital cost, the provision of training programmes for farmers in the proper management of soil and water resources, and the introduction of selection criteria for appropriate crops to grow with water supplied from underground dams.

5.4.3 Seawater/Brackish Water Desalination by Reverse Osmosis

Many coastal areas as well as areas subjected to waterlogging and salinity in the world face the problem of scarcity of freshwater. For such areas desalination of brackish water is essential for freshwater augmentation. This is generally achieved either by distillation or by reverse osmosis. Distillation can be done either by solar heat or by any other source of heat. In this Section, the reverse osmosis technique for desalination is described.

Desalination is a separation process used to reduce the dissolved salt content of saline water to a usable level. All desalination processes involve three liquid streams: the saline feedwater (brackish water or seawater), low-salinity product water, and very saline concentrate (brine or reject water).

The saline feedwater is drawn from oceanic or underground sources. It is separated by desalination process into two output streams: the low-salinity product water and very saline concentrate streams. The use of desalination overcomes the paradox faced by many coastal communities, that of having access to a practically inexhaustible supply of saline water but having no way to use it. Although some substances dissolved in water, such as calcium carbonate, can be removed by chemical treatment, other common constituents, like sodium chloride, require more technically sophisticated methods, collectively known as desalination. In the past, the difficulty and expense of removing various dissolved salts from water made saline waters an impractical source of potable water. However, starting in the 1950s, desalination began to appear to be economically practical for ordinary use, under certain circumstances.

The product water of the desalination process is generally water with less than 500 mg/l dissolved solids, which is suitable for most domestic, industrial, and agricultural uses. A by-product of desalination is brine. Brine is a concentrated salt solution (with more than 35 000 mg/l dissolved solids) that must be disposed of, generally by discharge into deep saline aquifers or surface waters with a higher salt content. Brine can also be diluted with treated effluent and disposed of by spraying on golf courses and/or other open space areas.

5.4.3.1 Technical Description

There are two types of membrane process used for desalination: reverse osmosis (RO) and electrodialysis (ED). The latter is not generally used in Latin America and the Caribbean. In the RO process, water from a pressurized saline solution is separated from the dissolved salts by flowing through a water-permeable membrane. The permeate (the liquid flowing through the membrane) is encouraged to flow through the membrane by the pressure differential created between the pressurized feedwater and the product water, which is at near-atmospheric pressure. The remaining feedwater continues through the pressurized side of the reactor as brine. No heating or phase change takes place. The major energy requirement is for the initial pressurization of the feedwater. For brackish water desalination the operating pressures range from 1,724 to 2,758 kPa, and for seawater desalination from 5,516 to 6,895 kPa.

In practice, the feedwater is pumped into a closed container, against the membrane, to pressurize it. As the product water passes through the membrane, the remaining feedwater and brine solution becomes more and more concentrated. To reduce the concentration of dissolved salts remaining, a portion of this concentrated feedwater-brine solution is withdrawn from the container. Without this discharge, the concentration of dissolved salts in the feedwater would continue to increase, requiring ever-increasing energy inputs to overcome the naturally increased osmotic pressure.

A reverse osmosis system consists of four major components/processes: (1) pretreatment, (2) pressurization, (3) membrane separation, and (4) post-treatment stabilization. Figure 5.35 illustrates the basic components of a reverse osmosis system.

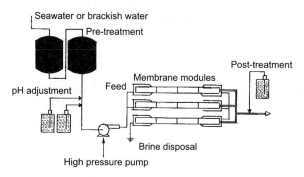

Figure 5.35 Elements of the Reverse Osmosis Desalination Process
Source: http://www.oas.org/usde/publications/Unit/oea59e/ch20.htm]

Pretreatment: The incoming feedwater is pretreated to be compatible with the membranes by removing suspended solids, adjusting the pH, and adding a threshold inhibitor to control scaling caused by constituents such as calcium sulfate.

Pressurization: The pump raises the pressure of the pretreated feedwater to an operating pressure appropriate for the membrane and the salinity of the feedwater.

Separation: The permeable membranes inhibit the passage of dissolved salts while permitting the desalinated product water to pass through. Applying feedwater to the membrane assembly results in a freshwater product stream and a concentrated brine reject stream. Because no membrane is perfect in its rejection of dissolved salts, a small percentage of salt passes through the membrane and remains in the product water. Reverse osmosis membranes come in a variety of configurations. Two of the most popular are spiral wound and hollow fine fiber membranes (see Figure 5.36). They are generally made of cellulose acetate, aromatic polyamides, or, nowadays, thin film polymer composites. Both types are used for brackish water and

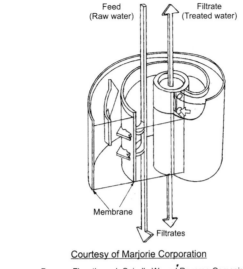

Courtesy of Marjorie Corporation

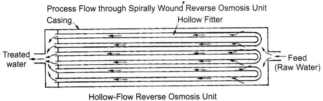

Process Flow through Spirally Wound Reverse Osmosis Unit

Hollow-Flow Reverse Osmosis Unit

Figure 5.36   Two Types of Reverse Osmosis Membranes

Source: http://www.oas.org/usde/publications/Unit/oea59e/ch20.htm]

seawater desalination, although the specific membrane and the construction of the pressure vessel vary according to the different operating pressures used for the two types of feedwater.

Stabilization: The product water from the membrane assembly usually requires pH adjustment and degasification before being transferred to the distribution system for use as drinking water. The product passes through an aeration column in which the pH is elevated from a value of approximately 5 to a value close to 7. In many cases, this water is discharged to a storage cistern for later use.

5.4.3.2 *Extent of Use*

The capacity of reverse osmosis desalination plants sold or installed during the 20-year period between 1960 and 1980 was 1,050 600 m^3/day. Since then, this capacity has continued to increase as a result of cost reductions and technological advances. RO-desalinated water has been used as potable water and for industrial and agricultural purposes.

Potable Water Use: RO technology is currently being used in Argentina and the northeast region of Brazil to desalinate groundwater. New membranes are being designed to operate at higher pressures (7 to 8.5 atm) and with greater efficiencies (removing 60% to 75% of the salt plus nearly all organics, viruses, bacteria, and other chemical pollutants).

Industrial Use: Industrial applications that require pure water, such as the manufacture of electronic parts, speciality foods, and pharmaceuticals, use reverse osmosis as an element of the production process, where the concentration and/or fractionating of a wet process stream is needed.

Agricultural Use: Greenhouse and hydroponic farmers are beginning to use reverse osmosis to desalinate and purify irrigation water for greenhouse use (the RO product water tends to be lower in bacteria and nematodes, which also helps to control plant diseases). Reverse osmosis technology has been used for this type of application by a farmer in the State of Florida, U.S.A., whose production of European cucumbers in an 8.8 Ha. greenhouse increased from about 4,000 dozen cucumbers/day to 7,000 dozen when the farmer changed the irrigation water supply from a contaminated surface water canal source to an RO-desalinated brackish groundwater source. A 300 lit/d reverse osmosis system, producing water with less than 15 mg/l of sodium, was used.

In some Caribbean islands like Antigua, the Bahamas, and the British Virgin Islands, reverse osmosis technology has been used to provide public water supplies with moderate success.

In Antigua there are five reverse osmosis units, which provide water to the Antigua Public Utilities Authority, Water Division. Each RO unit has a capacity of 750,000 lit/d. During the eighteen-month period between January 1994 and June 1995, the Antigua plant produced between 6.1 million lit/d and 9.7 million lit/d. In addition, the major resort hotels and a bottling company have desalination plants.

In the British Virgin Islands, all water used on the island of Tortola, and approximately 90% of the water used on the island of Virgin Gorda, is supplied by desalination. On Tortola, there are about 4,000 water connections serving a population of 13,500 year-round residents and approximately 256,000 visitors annually. In 1994, the government water utility bought 950 million liters of desalinated water for distribution on Tortola. On Virgin Gorda, there are two seawater desalination plants. Both have open seawater intakes extending about 450 m offshore. These plants serve a population of 2,500 year-round residents and a visitor population of 49,000, annually. There are 675 connections to the public water system on

Virgin Gorda. In 1994, the government water utility purchased 80 million liters of water for distribution on Virgin Gorda.

In South America, particularly in the rural areas of Argentina, Brazil, and northern Chile, reverse osmosis desalination has been used on a smaller scale.

In southern India, the Bhabha Atomic Research Centre (BARC) has established the world's largest desalination plant at the Kalpakkam Atomic Power Station near Chennai. The plant produces 1.8 million liters of potable water per day through reverse osmosis since 2002. The second section was expected to be functional by 2004 and produce 4.5 million liters using the thermal process. Water is supplied to people living nearby at less than what tanker water costs in neighboring Chennai.

Experts opine that a desalination plant for Chennai could produce 135 million liters/day at Rupees 4 billions (US $ 87 millions approximately). A private company has set up at least 40 such plants under the National Drinking Water Mission in the southern state of Tamil Nadu, Lakshadweep islands and elsewhere.

The cost of desalinated water can't be more than 70 Rupees (US $ 1.5 approximately) per m^3. However, the daunting cost of fuel to run the plants could be a negative point. Alternate sources like waste heat and solar power could provide a solution.

5.4.3.3 *Operation and Maintenance*

Operating experience with reverse osmosis technology has improved over the past 15 years. Fewer plants have had long-term operational problems. Assuming that a properly designed and constructed unit is installed, the major operational elements associated with the use of RO technology will be the day-to-day monitoring of the system and a systematic programme of preventive maintenance. Preventive maintenance includes instrument calibration, pump adjustment, chemical feed inspection and adjustment, leak detection and repair, and structural repair of the system on a planned schedule.

The main operational concern related to the use of reverse osmosis units is fouling. Fouling is caused when membrane pores are clogged by salts or obstructed by suspended particulates. It limits the amount of water that can be treated before cleaning is required. Membrane fouling can be corrected by backwashing or cleaning (about every 4 months), and by replacement of the cartridge filter elements (about every 8 weeks). The lifetime of a membrane in Argentina has been reported to be 2 to 3 years, although, higher life spans have been reported.

Operation, maintenance, and monitoring of RO plants require trained engineering staff. Staffing levels are approximately one person for a 200 m^3/day plant, increasing to three persons for a 4 000 m^3/day plant.

5.4.3.4 Level of Involvement

The cost and scale of RO plants are so large that only public water supply companies with a large number of consumers, and industries or resort hotels, have considered this technology as an option. Small RO plants have been built in rural areas where there is no other water supply option. In some cases, such as the British Virgin Islands, the government provides the land and tax and customs exemptions, pays for the bulk water received, and monitors the product quality. The government also distributes water and in some cases provides assistance for the operation of the plants.

5.4.3.5 Costs

The most significant costs associated with reverse osmosis plants, aside from the capital cost, are the costs of electricity, membrane replacement, and labor. All desalination techniques are energy-intensive relative to conventional technologies. Table 5.15 presents generalized capital and operation and maintenance costs for a 18,925 m^3/d reverse osmosis desalination in the United States. Reported cost estimates for RO installations in Latin American and the Caribbean are shown in Table 5.16. The variation in these costs reflects site-specific factors such as plant capacity and the salt content of the feedwater.

The International Desalination Association (IDA) has designed a Seawater Desalting Costs Software Program to provide the mathematical tools necessary to estimate comparative capital and total costs for each of the seawater desalination processes.

5.4.3.6 Suitability

This technology is suitable for use in regions where seawater or brackish groundwater is readily available.

Table 5.15 U.S. Army Corps of Engineers Cost Estimates for RO Desalination Plants in Florida

Feedwater type	Capital cost per unit of daily capacity ($/m^3/day)	Operation and maintenance cost per unit of production ($/m^3)
Brackish water	380 – 562	0.28 – 0.41
Seawater	1341 – 2379	1.02 – 1.54

Source: http://www.oas.org/usde/publications/Unit/oea59e/ch20.htm]

Table 5.16   Comparative Costs of RO Desalination for Several Latin American and Caribbean Developing Countries

Country	Capital cost ($/m³/day)	Operation and maintenance ($/m³)	Production cost* ($/m³)[a]
Antigua	264 – 528	0.79 – 1.59	
Argentina		3.25	
Bahamas			4.60 – 5.10
Brazil	1454 – 4483		0.12 – 0.37
British Virgin Islands	1190 – 2642		[b]3.40 – 4.30
Chile	1300		1.00

Source: http://www.oas.org/usde/publications/Unit/oea59e/ch20.htm]

[a]Includes amortization of capital, operation and maintenance, and membrane replacement.

[b]Values of $2.30 – $3.60 were reported in February 1994.

5.4.3.7   Effectiveness of the Technology

Since 1970s, researchers were struggling to separate product waters from 90% of the salt in feedwater at total dissolved solids (TDS) levels of 1 500 mg/l, using pressures of 4,137 kPa and a flux through the membrane of 18 l/m²/day. Today, typical brackish installations can separate 98% of the salt from feedwater at TDS levels of 2 500 to 3 000 mg/l, using pressures of 13.6 to 17 atm. and a flux of 24 lit/m²/day and guaranteeing to do it for 5 years without having to replace the membrane. The state-of-the-art technology uses thin film composite membranes in place of the older cellulose acetate and polyamide membranes. The composite membranes work over a wider range of pH, at higher temperatures, and within broader chemical limits, enabling them to withstand more operational abuse and conditions more commonly found in most industrial applications. In general, the recovery efficiency of RO desalination plants increases with time as long as there is no fouling of the membrane.

5.4.3.8   Advantages

- The processing system is simple; the only complicating factor is finding or producing a clean supply of feedwater to minimize the need for frequent cleaning of the membrane.
- Systems may be assembled from prepackaged modules to produce a supply of product water ranging from a few liters per day to 750,000 lit/day for brackish water, and to 400,000 lit/day for seawater; the modular system allows for high mobility, making RO plants ideal for emergency water supply use.

- Installation costs are low.
- RO plants have a very high space/production capacity ratio, ranging from 25,000 to 60,000 lit/day/m^2.
- Low maintenance, nonmetallic materials are used in construction.
- Energy use to process brackish water ranges from 1 to 3 kWh per m^3 of product water.
- RO technologies can make use of an almost unlimited and reliable water source, the sea.
- RO technologies can be used to remove organic and inorganic contaminants.
- Aside from the need to dispose of the brine, RO has a negligible environmental impact.
- The technology makes minimal use of chemicals.

5.4.3.9 Disadvantages

- The membranes are sensitive to abuse.
- The feedwater usually needs to be pretreated to remove particulates (in order to prolong membrane life).
- There may be interruptions of service during stormy weather (which may increase particulate resuspension and the amount of suspended solids in the feedwater) for plants that use seawater.
- Operation of an RO plant requires a high quality standard for materials and equipment.
- There is often a need for foreign assistance to design, construct, and operate plants.
- An extensive spare parts inventory must be maintained, especially if the plants are of foreign manufacture.
- Brine must be carefully disposed of to avoid deleterious environmental impacts.
- RO technologies require a reliable energy source.
- There is a risk of bacterial contamination of the membranes; while bacteria are retained in the brine stream, bacterial growth on the membrane itself can introduce tastes and odors into the product water.
- Desalination technologies have a high cost when compared to other methods, such as groundwater extraction or rainwater harvesting.

5.4.3.10 Cultural Acceptability

RO technologies are perceived to be expensive and complex, a perception that restricts them to high-value coastal areas and limited use in areas with saline groundwater that lack access to more conventional technologies. At this time, use of RO technologies is not widespread.

5.4.3.11 Further Development of the Technology

The seawater and brackish water reverse osmosis process would be further improved with the following advances:

- Development of membranes that are less prone to fouling, operate at lower pressures, and require less pretreatment of the feedwater.
- Development of more energy-efficient technologies that are simpler to operate than the existing technology; alternatively, development of energy recovery methodologies that will make better use of the energy inputs to the systems.
- Commercialization of the prototype centrifugal reverse osmosis desalination plant developed by the Canadian Department of National Defense; this process appears to be more reliable and efficient than existing technologies and to be economically attractive.

5.4.4 Integrated Groundwater Management in Israel

Although, Israel is well-known for its sustainable agriculture through water conservation techniques like drip irrigation, there is a need to extend this concept of sustainability over time and space. In addition to this, it is essential to prevent hazard of groundwater pollution due to natural and artificial reasons. In order to mitigate adverse aquifer behavior in groundwater quality and attain sustainable development, groundwater remedial activities must be implemented. This can be achieved with maximal efficiency by simultaneously addressing environmental and social concerns, implementing land use and water management guidelines, and coping with public needs and constraints as regards various disciplines.

5.4.4.1 Pyramid Analogous to Maslow's Pyramid of Human Needs

Around the year 1940, Maslow conceptualized a hierarchy of five levels of human needs as shown in the triangle BCD of Figure 5.37. The Maslow's Thesis stipulated that until the lowest levels of needs are met, a person could not go to higher levels. Similar to human needs, proper groundwater management can also be perceived as a hierarchy of needs. As regards land-use, only when the local land-use needs have been fulfilled will the society be interested in improving national and international infrastructure.

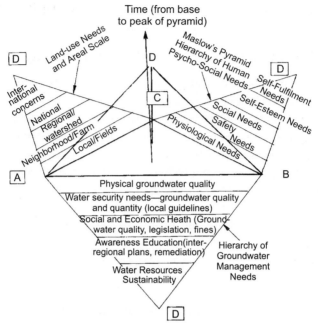

Figure 5.37 Integrated Three-Dimensional Pyramid of Needs.
ABD=Groundwater Management; ADC=Land-Use
DBC=Psycho-social Needs
[Courtesy: Melloul and Collin, 2001]

Management decision-making whether for water resources, land use, or human welfare management, must therefore relate to a logical process of continuous transition of levels of user needs and benefits, from the base of the higher and more sophisticated levels. An integrated hierarchy of groundwater and land-use can take the form as Maslow's three-dimensional pyramid as shown in Figure 5.37.

In this Figure, the hierarchies of groundwater management needs, land-use needs and psycho-social needs are represented by the three inclined surfaces ABD, ACD and BCD of the triangular pyramid with base ABC and vertex D. These three inclined surfaces are unfolded into three triangles, with five levels of development in each of them. The lowest levels in each of these triangles represent the initial basic level of development. Moving towards vertex D through corresponding next higher levels of development in the three triangles, their topmost levels representing the ultimate level of sustainable development can be reached eventually.

5.4.4.2 Case Study

The case study considered here is the northern portion of Israel's Coastal aquifer. Its groundwater basin can be considered a lens of fresh water impinged upon laterally from the west by seawater, from the east and below by saline, and/or salty brines, and from above by percolating anthropogenic pollution from agriculture, industry and domestic solid and liquid waste. The lens of fresh groundwater is a major supplier of drinking water and can be differentiated in two sub-regions: the seashore and the inland sub-region. Seawater intrusion endangers remaining fresh groundwater resources in the Seashore sub-region, whilst inland; the degree of pollution has risen to new heights.

There are thus two major concerns as follows:

(i) Near the coast most observation wells are more than 40 years old, sharply reducing their effectiveness as an early warning of seawater intrusion, mostly due to corrosion and clogging (Melloul and Dax, 1990). On the other hand, imported freshwater and treated water from other aquifers has been recharged into wells. An adequate water quality-monitoring network must yet be developed.

(ii) Inland, potential pollution hazards and salinization from anthropogenic contamination are on the increase. Remediation efforts have been undertaken in this sub-region, including recharge of imported freshwater and treated effluents. However, not all of these remedial operations have attained a level of resultant groundwater quality required for drinking water usage. Additionally, these operations lose efficiency owing to lack of adequate legislation and amendments against environmental contaminators, as well as insufficient public involvement and participation in this effort. Much recharged water is thus immediately consumed by unconstrained pumpage or polluted by anthropogenic activities.

With regard to social concerns, people having a wide range of income populate the aquifer's ground surface in this region. They live in densely urban as well as rural situations, with significant variations in educational backgrounds. However, some groups are still striving economically to achieve a secure future. These socio-economic data indicate the quality of life in the study area to be below level 3 on Maslow's pyramid.

Hydrological and environmental land-use data relating to the pyramid of groundwater management needs imply certain operational measures being carried out in the study area involve concerns at level 4 (national). The area would thus be found in a lower situation from the social standpoint of

Maslow's pyramid than on the corresponding hierarchy of groundwater management needs. This could indicate that society is not yet prepared for requisite groundwater remediation measures, and certainly not envisioning the sacrifice required by the sustainability management.

The pyramid shown in Figure 5.37 shows that, higher levels of groundwater management and needs cannot be attained until lower level needs have been fulfilled. Groundwater management measures must be synchronized with society's land use needs, its educational, economic, and political concerns. Sustainable groundwater management cannot be achieved without overcoming such unsustainability factors as uncontrolled anthropogenic pollution and salinization sources, unchecked population growth, and low levels of education and salaries. Concern over future water resources needs must appear to the resident population not as a luxury but a necessary condition in order maintain groundwater reservoir and storage levels for time of drought, and to fulfill their children's needs in the future.

In the case study discussed above, near the coast sustainable groundwater management will involve appropriate regulation of pumpage and the building of hydrological barriers to mitigate seawater intrusion. Inland, domestic, industrial, and agricultural effluents must have regional treatment, and there must be an ecologically sound long-term urban and rural land-use planning framework. Aquifer remediation of inland areas can be effected by importing and injecting fresh water into these reservoirs where salinity trends are still relatively low and in which any remedial measure can still be efficient. But these efforts must be bolstered by adequate public support to maintain the gains afforded by these measures. Sustainable water resource management will only be achievable when decision makers have social backing and readiness on the part of their population to pay the price for long-term initiatives. Such initiatives would include improved monitoring networks along the coast and inland, which would lead to better control of water resource quality and quantity trends.

To achieve sustainable groundwater resources, local as well as global monitoring and remediation measures with regional, national, and international impact are required.

5.4.5 Sustainable Water Management in Germany

Similar to the elite group of eight industrialized nations of the world (G-8), Flavin (1997) has proposed another group of eight nations which shape the global environmental trends. This group is called the E-8 and it includes the following nations:

1. China, the country with the largest population,
2. USA, the largest economy and carbon emissions in the world,
3. Brazil, with the richest array of biodiversity,
4. Russia, the nation with largest share of forest areas,
5. Japan, the nation with the second largest share of the gross world product,
6. Germany, the nation with the third largest economy and a great public awareness,
7. India, the nation with second largest population and
8. Indonesia, the nation with a high share in the biodiversity of the world.

Among these nations, Germany is credited with a great public awareness in general and environmental awareness in particular. Therefore in the following sections, sustainable water management in Germany is discussed.

Recognizing that problems concerning water quality and quantity tend to be of a regional rather than a global nature, the German Federal Environmental Agency has commissioned a study to identify sustainable and non-sustainable trends in water management [Bismuth, 1997].

The study bases on interviews with German water management experts, who were requested to give their opinion on the concept of sustainability, to supply their personal definition and give (counter-) examples for sustainability. Common perceptions as well as conflicting ideas were then identified and again discussed with the experts previously asked. They were further elaborated in an international seminar and in a national workshop with the aim to formulate a common national definition of sustainable water management.

5.4.5.1 Definitions

A definition for sustainable water management should respect the following requirements:

- Water management action must be subject to democratic control and local co-determination.
- The basic economic functions of the local community must be supported.
- The basic functions of water management, water supply and sewage disposal and the maintenance of recreation areas must be guaranteed for all citizens.

Based on these requirements an extended definition had to be developed and discussed with the German water experts:

'Sustainable Water Management' means the integrated management of all artificial or natural water cycles in accordance with the following aims:

- The protection of water as a natural life habitat or as a central element for life habitats respectively.
- The preservation of water in its various facets as a resource for the existing as well as for the future generations.
- The evolution of options which assure the natural basis for life for a long-term economic and social development compatible with nature.

The achievement of the aims respects all requirements of other sectors for a sustainable development.

5.4.5.2 Principles

Considering the above quoted definition, principles of sustainable water management were developed and discussed in regard to German Water Management:

- The regional principle
- The integration principle
- The polluters pay principle
- The co-operation and participation principle
- The resource minimizing principle
- The precautionary principle
- The sources reduction principle
- The reversibility principle
- The intergeneration principle

The Regional Principle The regional principle demands that each region avoids to externalize its water related problems to other regions and protects its own local resources. A region should be based on a river basin. The implementation of technical or organizational solutions will not be possible, if they are not adapted to the regional needs and do not respect the natural and social conditions.

The Integration Principle The integration principle calls for integration of environmental exigencies in other politic sectors. In German water management a fragmentation of the resources water for different uses and qualities has to be noted. So water is separated into drinking water and used water (sewage). The cycles which connects both fractions are usually not registered. The separation of water in different fragments has only been possible by the establishment of complex technical structures. Together with the specialization of the technical systems also a specialization of the controlling administration and management units occurred. The economic and ecological costs are rising with the fragmentation: Many rivers were optimized in view to navigation but ecological aspects were neglected with consequences for the flood prevention. To master such problems or better not to let them evolve, the different users, servers, administration units etc. have to develop an integrated perception of the water resource.

The Polluters Pay Principle Several damages to the environment and also damages to the social system are not calculated and compensated by the responsible parties. Instead there are many costs incurred by the state and the social community. Partly takes a temporary or spatial transfer place, with the result that future generations or foreign countries have to pay for the damages of recent uses. This does not conform with sustainable development especially in its social dimension.

The demand for an adequate assignment of costs and responsibilities is the central statement of the polluters pay principle. The polluters pay principle includes also the resource-user-pay-principle. To assure the water resource for future generations as well as an ecological habitat the polluters pay principle has to be set into practice in all is aspects. Special attention has to be given to non-point sources.

The Co-operation and Participation Principle The involvement of the people in decision making process is one of the key postulates of the Agenda 21 of the United Nations Commission on Sustainable Development (UNCSD). The co-operation and participation principle demands that before decisions are taken all interests are considered and that concerned parties are embedded in the decision making process.

The implementation of this principle means for the water management to take active part in the dialogue with other social and economic groups. An open dialogue implies transparency in the decision making process.

Generally has the broad public only restricted access to information and to public bodies. There is the urgent need to improve transparency and involvement of the public in the decision making process. The water and soil boards as well as the river basin boards should participate with environmental associations. The several technical-scientific associations in Germany which are responsible for numerous technical guidelines, regarded as technical standards, should involve the public in their decision making process, to assure environmental protection as a public matter.

The Resource Minimizing Principle Water management should try to minimize the use of resources and energy. Until now only few measurements have been undertaken to implement this principle. Resource management is not only a task of the water management but of the whole market itself. The conditions of the local and world market have therefore to be changed to minimize the use of resources.

The Precautionary Principle According to this principle no actions should be undertaken which would cause severe damages or even if there is only a slight possibility of damage. This is also due to measurements which have no clearly defined risk potential.

The large amounts of substances and their ubiquity appearance in the water bodies make the fulfilment of this principle very difficult. As a solution the concept of critical loads or carrying capacity was proposed. But in reality it is not possible to predict the effects of a substance in the water bodies, especially if one considers the synergetic effects and reactions with other substance. There is also the danger that "carrying capacity" could be misunderstood as "ideal load" which is perfectly reasonable to reach. The principle of prevention prescribed in the Federal Water Act and national provisions could be watered down.

As a consequence the establishment of quality objectives should be limited to those particular substances whose use leads to increased concentrations in the environment despite the implementation of precautionary measures at the source.

The Source Reduction Principle This principle demands to stop emission at their source. It further demands an integrated view of production processes but also of consumption processes with the aim to modify them in such a way, that no or only very few environmental damages occur. On the long-term should all dangerous substances be substituted or only be used in closed production cycles.

The Reversibility Principle This principle postulates on the one hand that all measurements of water management should be reversible on the other hand it postulates that already during the planning processes of measurements their possible adjustment to changing needs should be considered. This principle demands from water management to respect the protection of species and ecological habitats. It prohibits further the overuse of non-renewable water sources.

The Intergeneration Principle The intergeneration principle demands to respect the interests of future generations. To implement this principle long-term plans and advanced prognostic methods are needed.

5.4.5.3 State of German Water Management in the Year 1997

By subject the discussion was divided into the following issues of water management:

- quantitative water management,
- the morphology of surface waters and landscape,
- water quality,
- sewage-water management,
- administrative and legislative aspects.

Quantitative Water Management In erstwhile West Germany water resources far exceed present and future water use. Nevertheless, water supply problems arise due to the irregular regional distribution of usable water resources and the demand for water as well as the poor quality of many sources. Thus it has been necessary to supply water in particular densely populated regions with water using water pipelines (long distance water supply) and reservoirs. But there is an inherent danger with long distance water supplies that people lose touch with their own region. Difficulties that occur in the water supply areas then only constitute secondary problems for the water consumers.

The Morphology of Surface waters and Landscape The pressures exerted on waters by users (shipping, hydroelectric power, flood control etc.) in the form of technical measures negatively effect their morphological state. Such pressures reduce the natural diversity of different environments and the dynamics of water, interrupt water flow (dams, barrage weirs), and disrupt the unity of rivers and floodplains. This leads to a reduction in natural species diversity and a displacement of the spectrum of river fishes.

It was generally accepted that waterways have to be restored to a near-natural state in the future and the remaining relatively intact floodplains have to be protected from environmental pressures, such as lock construction for shipping. This will be one of the main coming tasks in German water policies but also one of the main conflict fields with the traffic sector.

The trend towards ever-bigger ships and the necessity for widening the waterways runs counter to any improvement in the morphological situation of the waters of the erstwhile Federal Republic of Germany. Other solutions, such as flat-bottomed boats or changing transport systems, must be found to accommodate the demands of users.

Water Quality Up to now, water quality objectives in the sense of quality targets have been set by the International Commission for the Protection of the Rhine. The quality of German surface waters has improved during the last decade even though small surface waters still show deficits in respect to their water quality. This is mainly due to the inputs of nutrients and pesticides from agricultural sources. Also for the Rhine, the quality targets for Nitrate (25 mg/l), Ammonium (0.2-0.4mg/l) and Phosphorus (0.16-0.2 mg/l total) have been exceeded. From a scientific point of view, even significantly lower concentration must be achieved in lakes in order to prevent unwanted eutrophication effects. The insufficient reduction in the quantity of nutrients also constitutes a key problem for coastal waters. The quality targets for two third of the substances named in the Rhine Action Programme have already been attained. For nine substances however the targets were not attained (mercury, cadmium, copper, zinc, HCH, trichloromethane, HCB, PCB and ammonium). Due to the large share of diffuse sources concerning a number of the substances listed, it will probably not be possible to achieve the quality targets within the next years.

Erstwhile West Germany satisfied approximately 70% of its drinking water requirements from the groundwater. The most important problems for the groundwater quality are the nitrate values and the crop protection products found. An additional factor which promotes high nitrate values in groundwater is the atmospheric nitrogen deposition, expressly nitrogen oxides from burning processes (traffic) and ammonia from agriculture with high livestock rates. A problem which has yet not been focused are the synergetic effects between different micro pollutants, among others pesticides.

One approach that can be adopted to protect groundwater resources is the installation of water protection zones, where restrictions on the use of water and precautionary measures against a potential impairment of the groundwater become progressively more stringent as one approaches the water withdrawal point. This frequently involves conflicts between water management interests and the interest of other users of the area in question and particularly those of agriculture and of local authority residential and industrial developments. Additional problems arise with surveillance and enforcement here, particularly in the case of application over wide areas.

Sewage-Water Management From 1970 to 1991 only in West Germany 120 millions of DM were invested in to canalization systems and waste water treatments plants. In 1991, 90.6% of the population were connected to a central sewage system.

A decentralized treatment of sewage waters has been favored by the representatives of the NGOs. They noted that a decentralized treatment of waste water, which favors anaerobic processes, does not only lead to considerable reductions but also to higher energy efficiency and would therefore, be more sustainable. Agreement was reached that decentralized waste water treatment could be installed in sparcely populated areas.

Administrative and Legislative Aspects of German Water Management The positive results achieved over the past years are taken as a sign that the administration can be reduced. At the same time the increased costs in the field of water supply and sewage disposal put the administration under strong pressure to justify any further measures that it takes. This also holds true for a wide range of generally small and highly diverse organisations involved in water management: municipal companies, private companies, special-purpose associations, water and land associations and statutory associations, etc. A further problem at present is the poor integration of all the areas of policy that are dependent on the land. Area development, water management, nature conservation and agriculture are nearly always accommodated in separate authorities, making it virtually impossible for a coherent policy to be developed.

Citizen participation is guaranteed up to a certain extent by the municipal self-administration. On the point to what degree and how other forms of citizen participation could be implemented, a controversial discussion started between members of administration and representatives of the NGOs. The NGOs suggested the installation of local water councils with the participation of all interested stakeholders. The administration mentioned that it is in some cases the last instance to prevent the violation of environmental laws (e.g., construction of buildings in flood areas). It seems that this debate which does not only cover administrative issues but also constitutional and democratic issues is only at its starting point.

5.4.5.4 Concluding Remarks on Sustainable Water Management in Germany

If a number of aspects of water management in the erstwhile Federal Republic of Germany are rated as non-sustainable, then this should not be interpreted as meaning that the existing water management system has proved to be inefficient. The efficiency of a system can only be judged on the basis of the requirements that are placed on it.

The institutional debate has to be continued also in view of the proposed European water framework directive. It is an undeniable fact that German water management will have to change if the concept of sustainability is implemented into practice. Actions of priority will be the restoration of the morphology of the river and flood plains, and the reduction of non-point sources, especially from agricultural sources.

5.5 REFERENCES

Acosta Baladon, A.N., (1995), Agricultural uses of occult precipitation. Agrometeorological Applications Associates, Ornex, France.

Agarwal, A. and Narain, S., (1997), *Dying Wisdom, Rise, Fall and Potential of India's Traditional Water Harvesting Systems, State of India's Environment 4, A Citizens' Report,* Centre for Science and Environment, New Delhi, India.

Agarwal, A. and Narain, S., (2001), *Making Water everybody's Business, Practice and Policy of Water Harvesting,* Centre for Science and Environment, New Delhi, India.

Ahlbck, A.J., (1997), Management planning of industrial plantations in Tanzania—principles and efforts. Paper to Topic 12, 11th World Forestry Congress, 13-22 Oct., Antalya, Turkey.

Apsey, T.M. and Reed, F.L.C. (1996), World timber resources outlook: current perceptions with implications for policy and practice. Commonwealth Forestry Review 75: 155-160.

ASAE, (1980), *Design and Operation of Farm Irrigation Systems.* St. Joseph, Missouri, USA.

AWWA, (1994), *Water Wiser: The Water Efficiency Clearinghouse.* Denver, Colorado, USA.

Barbier, E.B., (1987), 'The concept of sustainable economic development', *Environmental Conservation,* 14: 101-110.

Beecher, J.A., and Laubach, A.P., (1989), *Compendium on Water Supply, Drought, and Conservation.* National Regulatory Research Institute, (Report No. NRRI 89-15), Columbus, Ohio, USA

Billings, R.B., and Day, W.M., (1989), 'Demand Management Factors in Residential Water Use: The Southern Arizona Experience', *Journal of the American Water Works Association,* 81(3): 58-64.

Bhatia, R., Cestti, R., and Winpenny, J., (1995), *Water Conservation and Reallocation: Best Practice Cases in Improving Economic Efficiency and Environmental Quality.* World Bank, Washington, D.C. (A World Bank-ODI Joint Study), USA.

Birkett, J.D. (1987). 'Factors Influencing the Economics of Desalination', *Non-conventional Water Resources Use in Developing Countries,* United Nations, pp. 89-102. (Natural Resources/Water Series No. 22), New York, USA.

Bismuth, C., (1998), 'Sustainable Water Management in Germany', INBO Workshop, *International Conference on Water and Sustainable Development,* March 19-21, Paris, France; also available at http://www.oieau.fr/ciedd/contributions/atriob/contribution/bismuth.htm

Bosch, D.J., and Ross, B.B., (1990), 'Improving Irrigation Schedules to Increase Returns and Reduce Water Use in Humid Regions', *Journal of Soil and Water Conservation,* 45(4): 485-488.

Brown, J.W., and Hurst, W.F., (1990), 'Opportunities for Water Marketing and Conservation in California', *Water World News,* 6(6): 10-13.

Brundtland Report, (1988), *Our Common Future*, The World Commission on Environment and Development, Alianza Publications, Madrid, Spain; also available at http://www.brundtlandnet.com/brundtlandreport.htm

Burley, J. and Adlard, P.G., (1992), 'Plantation silvicultural research and genetic improvement', *Proc. IUFRO Centennial Meeting*, Berlin, September. p. 13, IUFRO, Vienna, Austria.

Buros, O.K., (1987), 'An Introduction to Desalination', *Non-conventional Water Resources Use in Developing Countries*, United Nations, pp. 37-53. (Natural Resources/Water Series No. 22), New York, USA.

Bush, D.B., (1988), 'Dealing for Water in the West Water Rights as Commodities', *Journal of the American Water Works Association*, 80(3): 30-37.

California Department of Water Resources, (1992), *Water Audit and Leak Detection Guide-Book*. State of California Department of Water Resources, Water Conservation of Office, and AWWA, California/Nevada Section, Sacramento, CA, USA.

Cant, R.V., (1980), 'Summary of Comments on R.A. Tidball's 'Lake Killarney Reverse Osmosis Plant', P. Hadwen (ed.). *Proceedings of the United Nations Seminar on Small Island Water Problems*, Barbados, UNDP, pp. 552-554, New York, USA.

Cavalcanti, C., (1996), 'Brazil's new forests bring profit and pain', *People and the Planet*, 5(4): 14-16.

Cavelier, J., and Goldstein, G., (1989), 'Mist and Fog Interception in Elfin Cloud Forests in Colombia and Venezuela', *Journal of Tropical Ecology*, 5: 309-322.

Cereceda, P., and Larraín, H., (1981), *Pure Water Flowing from the Clouds*. UNESCO/ROSTLAC, Montevideo, Uruguay.

Cereceda, P., Schemenauer, R.S., and Suit, M., (1992), 'Alternative Water Supply for Chilean Coastal Desert Villages', *Water Resources Development*, 8(1): 53-59.

Childs, W.D., and Dabiri, A.E., (1992), 'Desalination Cost Savings or VARI-RO', *Pumping Technology*, 87: 109-135.

Chritchley, W., Reij, C. and Seznec, A., (1992), Water Harvesting for Plant Production. Vol 2. Case Studies and Conclusions from Sub-Saharan Africa. World Bank Techn. Paper 157.

Chritchley, W., Reij, C. and Turner, S.D., (1992), Soil and Water Conservation in Sub-Saharan Africa: towards sustainable production by the rural poor. IFAD, Rome, Italy and CDCS, Amsterdam, The Netherlands.

Collinge, R.A., (1992), 'Revenue Neutral Water Conservation: Marginal Cost Pricing With Discount Coupons', *Water Resources Research*, 28(3): 617-622.

Coughlan, B. and Singleton, J.A. (1989), *Opportunities to Protest Instream Flows and Wetland Uses of Water in Kentucky*. U.S. Department of the Interior, Fish and Wildlife Service, Biological Report No. 89(9), Washington, D.C., USA.

Countryside and Forestry Commissions, (1991). 'Forests for the community', *Countryside Commission*, p. 4, London, UK.

Crook J., Asano, T., and Nellor, M., (1990), 'Groundwater Recharge with Reclaimed Water in California', *Municipal Wastewater Reuse*. USEPA, pp. 67-74, Report No. EPA-43 0/09-91-022, Washington, D.C., USA.

Cuthbert, W.R., (1989), 'Effectiveness of Conservation Oriented Water Rates in Tucson', *Journal of the American Water Works Association*, 81(3): 65-73.

de Gunzbourg, J., and Froment, T., (1987), 'Construction of a Solar Desalination Plant (40 cum/day) for a Caribbean Island', *Desalination*, 67: 53-58.

Dyballa, C., and Connelly, C., (1991), 'State Programs Incorporating Water Conservation Resources', In: *Engineering and Operations for the New Decade* (1991), Annual Conference Proceedings, AWWA, Philadelphia, PA., U.S.A.). Denver, CO, USA.

Eddy, N., (1993), 'Water Conservation Program Provides Interim Relief for Native American Wastewater Woes,' *Small Flows,* 7(2): 15.

Eger H., (1988). *Runoff Agriculture.* Reichert, Wiesbaden, Germany.

Eisenberg, T.N., and Middlebrooks, E.J., (1992), 'A Survey of Problems with Reverse Osmosis Water Treatment', *American Water Works Association Journal,* 76(8): 44.

Evans, J., (1992), *Plantation Forestry in the Tropics,* 2nd ed., 403 pp., Clarendon Press, Oxford, UK.

Evans, J., (1997), 'The sustainability of wood production in plantation forestry',. Paper to Topic 12, 11th World Forestry Congress, 13-22 Oct., Antalya, Turkey.

Evenari, M., Shanan, L. and Tadmor, N., (1971), The Negev: The Challenge of a Desert. Harvard University Press, Cambridge, Massachusetts, USA.

FAO, (1985), *Tree growing by rural people.* FAO Forestry Paper 64, 130 pp., FAO, Rome, Italy.

FAO, (1993), *Forest resources assessment 1990—Tropical countries.* FAO Forestry Paper 112, 61 pp + appendices, FAO, Rome, Italy.

Flavin, C., (1997), 'Rio and After', *State of the World,* W.W. Norton, New York, USA.

Frederiksen, H.D., (1992), *Drought Planning and Water Efficiency Implications in Water Resources Management,* World Bank Technical Paper No. 185, Washington, D.C., USA .

Furukawa, D.H., and Milton, G., (1977), 'High Recovery Reverse Osmosis with Strontium and Barium Sulfate in a Brackish Wellwater Source', *Desalination,* 22(1,2,3): 345.

Garduno, H. and F. Arreguin-Cortes, eds., (1991), *Efficient Water Use.* International Seminar on Efficient Water Use, October 21-25, 375, Mexico City, Mexico.

Gilbertson D.D. (1986). 'Runoff (floodwater) farming and rural water supply in arid lands', *Applied Geography,* 6: 5-11.

Gilmour, D.A., King, G.C., Applegate, G.B., and Mohns, B., (1990), 'Silviculture of plantation forests in central Nepal to maximise community benefits', *Forest Ecology and Management,* 32: 173-186.

Gischler, C., (1977), *Camanchaca as a Potential Renewable Water Resource for Vegetation and Coastal Springs along the Pacific in South America.* UNESCO/ROSTAS, Cairo, Egypt.

Gischler, C., (1991), *The Missing Link in a Product! on Chain: Vertical Obstacles to Catch Camanchaca,* UNESCO/ROSTLAC, Montevideo, Uruguay.

Gollnitz, W.D., (1988), 'Source Protection and the Small Utility', *Journal of the American Water Works Association,* 80(8): 52-57.

Gomez, Evencio G., (1979), 'Ten Years Operation Experience at 7.5 Mgd Desalination Plant', *Desalination,* 31(1): 77-90, Rosarito, B. Cafa, Mexico.

Grisham, A., and Fleming, W.M., (1989), 'Long Term Options for Municipal Water Conservation', *Journal of the American Water Works Association,* 81(3): 33.

Gould, J.E., (1992), 'Rainwater Catchment Systems for Household Water Supply', *Environmental Sanitation Reviews,* No. 32, ENSIC, Asian Institute of Technology, Bangkok, Thailand.

Gould, J.E. and McPherson, H.J., (1987), 'Bacteriological Quality of Rainwater in Roof and Groundwater Catchment Systems in Botswana', *Water International,* 12: 135-138.

Habibian, A., (1992), 'Developing and Utilizing Data Bases for Water Main Rehabilitation', *Journal of the American Water Works Association,* 84(7): 75-79.

Haines, R. J., (1994), *Biotechnology in forest tree improvement*, FAO Forestry Paper 118, 230 pp., Rome, Italy.

Hall, W.A., (1980), 'Desalination: Solution or New Problem for Island Water Supplies', P. Hadwen (ed.), *Proceedings of the United Nations Seminar on Small Island Water Problems*, Barbados, UNDP, pp. 542-543, New York, USA.

http://www.cee.vt.edu/program_areas/environmental/teach/gwprimer/recharge/recharge.html#Methods

http://www.montecitowater.com/tips.htm

http://www.oas.org/usde/publications/Unit/oea59e/ch20.htm

http://www.oas.org/usde/publications/Unit/oea59e/ch25.htm

http://www.oas.org/usde/publications/Unit/oea59e/ch26.htm

http://www.oas.org/usde/publications/Unit/oea59e/ch31.htm

http://www.oas.org/usde/publications/Unit/oea59e/ch34.htm

http://timesofindia.indiatimes.com/articleshow/735144.cms

http://www.unep.or.jp/ietc/Publications/TechPublications/TechPub-8e/rainwater1.asp

http://www.unep.or.jp/ietc/Publications/TechPublications/TechPub-8e/rainwater2.asp

http://www.unesco.org/mab/capacity/ucep/etp5e.htm

http://www.unesco.org/iau/sd/definitions.html

IDA, (1988), *Worldwide Inventory of Desalination Plants*, Topsfield, Massachusetts, USA.

Inions, G., (1995), 'Lessons from farm forestry in Western Australia', *Outlook 95*, 1: 416-422, ABARE, Canberra, Australia.

Jarett, A.R., Fritton, D.D., and Sharpe, W.E., (1985), *Renovation of Failing Absorption Fields by Water Conservation and Resting*, ASAE, Paper No. 85-2630, St. Joseph, Missouri, USA.

Jensen, R., (1991), 'Indoor Water Conservation', *"Texas Water Resources*, 17(4) Austin, Texas, USA.

Kahlon, K.S., (2004), *Sustainable Water and Wastewater Management for Residential Buildings*, Unpublished Bachelor of Technology Thesis, Deptt. of Civil Engineering, Indian Institute of Technology, Kharagpur, India.

Kanowski, P.J., (1997), Afforestation and Plantation Forestry For The 21st Century, 11th World Forestry Congress, 13-22 Oct., Antalya, Turkey; also available at http://www.fao.org/montes/foda/wforcong/PUBLI/V3/T12E/1-7.HTM#TOP

Kanowski, P.J. and Savill, P.S., (1992), 'Forest plantations: towards sustainable practice', pp. 121-151, Chapter 6, C. Sargent and S. Bass (Eds). In: *Plantation Politics: Forest Plantations in Development*, Earthscan, London, UK.

Keh, S.K., (1997), 'Whither goest Myanmar teak plantation establishment?' Paper to Topic 12, 11th World Forestry Congress, 13-22 Oct., Antalya, Turkey.

Kolarkar A.S., Murthy K. and Singh N., (1983), 'Khadin—a method of harvesting water for agriculture in the Thar desert', *Arid Environment* 6: 59-66.

King, W.D., Parkin, D.A., and Handsworth, R.J., (1978), 'A Hot-Wire Water Device Having Fully Calculable Response Characteristics', *J. Appl. Meteor.*, 17: 1809-1813.

Knollenberg, R.G., (1972), 'Comparative Liquid Water Content Measurements of Conventional Instruments with an Optical Array Spectrometer', *J. Appl. Meteor.*, 11: 501-508.

Kraft, H., (1995), *Draft Preliminary Report on Design, Construction and Management of a Root Zone Waste Water Treatment Plant at Hurricane Hole Hotel,* Castries, CARICOM/GTZ Environmental Health Improvement Project, Marigot Bay, Saint Lucia.

Kranzer, B.S., (1988), *Determinants of Residential Water Conservation Behavior. An Investigation of Socio-economic and Psycho-dynamic Factors,* Unpublished Ph.D. dissertation, Southern Illinois University, Carbondale, IL, USA.

Kromm, D.E., and White, S.E. (1990), 'Adoption of Water Saving Practices by Irrigation in the High Plains', *Water Resources Bulletin,* 26(6): 999-1012.

Kutsch H., (1982), 'Principal Features of a Form of Water-concentrating Culture on Small-Holdings With Special Reference to the Anti-Atlas', *Trierer Geography.* Isernhagen, Germany

Lawand, T.A., (1987), 'Desalination With Renewable Energy Sources', *Non-conventional Water Resources Use in Developing Countries,* United Nations, pp. 66-86. (Natural Resources/Water Series No. 22), New York, USA.

Libert, J.J., (1982), 'Desalination and Energy', *Desalination,* 40: 401-406.

Martin, W.E., and S. Kulakowski, (1991), 'Water Price as a Policy Variable in Managing Urban Water Uses: Tucson, Arizona', *Water Resources Research,* 27(2): 157-166.

Mather, A.S., (1990), *Global forest resources,* 341 pp., Bellhaven Press, London, UK.

Mather, A. S. (Ed)., (1993), *Afforestation: Policies, Planning and Progress,* 223 pp., Bellhaven Press, London, UK.

Melloul, A.J. and Collon, M.L., (2001), 'A Hierarchy of Groundwater Management, Land-use, and Social Needs Integrated for Sustainable Resource Development', *Environment, Development and Sustainability,* 3: 45-59, Kluwer, Netherlands.

Melloul, A.J. and Dax, A., (1990), 'Monitoring the Deterioration of Small-diameter Observation Wells', *Water Resources Management,* 4: 135-153.

Mobbs, M., (1998), *Sustainable House,* Choice Books, Marrickville, New South Wales, Australia; also available at http://www.abc.net.au/science/planet/house/special.htm.

Montgomery, J.M. Consulting Engineers, (1991), *Water Conservation Analyses, Evaluation, and Long Range Planning Study,* Draft Final Report to City of San Jose, Walnut Creek, CA, USA.

Mooney, M.J., (1995), 'Come the Camanchaca', *Américas,* 47(4): 30-37.

Nagel, J.F., (1959), 'Fog Precipitation on Table Mountain', *Quart. J. Roy. Meteor. Soc.,* 82: 452-460.

National Academy of Sciences, (1974), 'More Water for Arid Lands', *Promising Technologies and Research Opportunities,* Washington DC, USA.

National Association of Plumbing, Heating and Cooling Contractors (NAPHCC), (1992), *Low Flow Plumbing Products Fact Sheet,* Alexandria, VA, USA.

Nelson Da Franca Ribeiro Dos Anjos (1998), *Source Book of Alternative Technologies for Freshwater Augmentation in Latin America and the Caribbean,* International Journal of Water Resources Development, Vol. 14, No. 3, pp. 365-398. Also available at http://www.oas.org/usde/publications/Unit/oea59e/begin.htm

NEOS Corporation, (1990), *Technical Assistance for the City of Lompoc: Energy Savings Through Water Conservation: Final Report,* Prepared for Western Area Power Administration, Conservation & Renewable Energy Program, Sacramento Area Office. Lafayette, CA, USA.

Nieswiadomy, M.L., and Molina., D.J., (1989), 'Comparing Residential Water Demand Estimates Under Decreasing and Increasing Block Rates Using Household Data', *Land Economics*, 65(3): 280-289.

Nissen-Petersen, E. (1982). *Rain Catchment and Water Supply in Rural Africa: A Manual.* Hodder and Stoughton Ltd., London, UK.

Oosterbaan R.J. (1983). 'Modern interferences in traditional water resources in Baluchistan', *ILRI Annual Report 1982*, pp.23-34. Wageningen, Germany.

O'Hare, M.P., Fairchild, D.M., Hajali, P.A., and Canter, L.W., (1986), Artificial Recharge Of Groundwater, Lewis Publishers, Boca Raton, FL, USA.

Office for Technology Assessment (OTA), (1988), *Using Desalination Technologies/or Water Treatment,* U.S. Congress, Washington, D.C., USA.

Pacey A. and Cullis A., (1986), *Rainwater Harvesting. The Collection of Rainfall and Runoff in Rural Areas,* IT, London, UK.

Padmanabha, A., (1991), 'Water Conservation Program Combats Increased Wastewater', *Water Environment and Technology,* 3(7): 18-22.

Pandey, D., (1995), 'Forest resources assessment –1990 Tropical forest plantation resources', *FAO Forestry Paper 128,* 81 pp., FAO, Rome, Italy.

Pandey, D., (1997), 'Major issue of tropical forest plantations', Paper to Topic 12, 11th World Forestry Congress, 13-22 Oct., Antalya, Turkey.

Pearson, F.H., (1993), 'Study Documents Water Savings with Ultra Low Flush Toilets', *Small Flows,* 7(2): 8-9, 11.

Prinz, D. and Wolfer, S., (1998), *Opportunities To Ease Water Scarcity (Water Conservation Techniques and Approaches; Added Values and Limits),* University of Karlsruhe, D-76128 Karlsruhe, Germany; also available at www.ubka.uni-karlsruhe.de/indexer-vvv/1998/bau-verm/6

Rathnau, M.M., (1991), 'Submetering—Water Conservation', *Water Engineering and Management,* 138(3): 24-25,37.

Reed, S.C., Middlebrooks, E.J., and Crites, R.W., (1988), *Natural Systems for Waste Management and Treatment,* McGraw-Hill, New York, USA.

Robins, L., McIntrye, K., and Woodhill, J., (1996), *Farm forestry in Australia: integrating commercial and conservation benefits,* 54 pp., Greening Australia, Canberra, Australia.

Roche, M., (1992), 'Privatizing the exotic forest estate: the New Zealand experience', pp. 139-154, J. Dargavel and R. Tucker (eds). *Changing Pacific Forests.* Proc Forest History Society/IUFRO Conference, Forest History Society, Durham, North Carolina, USA.

Sargent, C., (1990), *The Khun Song Plantation Project,* 176 pp., IIED, London, UK.

Savill, P.S. and Evans, J., (1986), *Plantation Silviculture in Temperate Regions,* 246 pp., Clarendon Press, Oxford, UK.

Schemenauer, R.S. and Cereceda, P., (1991), 'Fog-Water Collection in Arid Coastal Locations', *Ambio,* 20(7): 303-308.

Schemenauer, R.S. and Cereceda, P., (1992), 'Water from Fog-covered Mountains', *Waterlines,* 10(4): 10-13.

Schemenauer, R.S. and Cereceda, P. (1992), 'The Quality of Fog Water Collected for Domestic and Agricultural use in Chile', *Journal of Applied Meteorology,* 31(3): 275-290.

Schemenauer, R.S., Cereceda, P., and Carvajal, N., (1987), 'Measurement of Fog Water Deposition and their Relationships to Terrain Features', *Journal of Climate and Applied Meteorology,* 26(9): 1285-1291.

Schiller, E.J. and Latham, B. G., (1987), 'A Comparison of Commonly Used Hydrologic Design Methods for Rainwater Collectors', *Water Resources Development*, 3.

Schlette, T.C., and Kemp, D.C., (1991), 'Setting Rates to Encourage Water Conservation', *Water Engineering and Management*, 138(5): 25-29.

Schoolmaster, F.A., and Fries, T.J., (1990), 'Implementing Agricultural and Urban Water Conservation Programs: A Texas Case Study', *The Environmental Professional*, 12: 229-240.

Sengupta, N., (1993), *User-Friendly Irrigation Designs*, Sage, New Delhi, India.

Shiva, V., (1993), *Monocultures of the Mind*, 184 p., Zed Books, London, UK.

Strauss, S.D., (1991), 'Water Management for Reuse/Recycle', *Power*, 135(5): 13-23.

Sutton, W.R.J., (1991), 'Are we too concerned about wood production?', *New Zealand Forestry* 36(3): 25-28.

Texas Water Development Board, (1986), *A Homeowner's Guide to Water use and Water Conservation*. Austin, TX, USA.

Tobbi B., (1993), *Water Harvesting: Historic, Existing and Potentials in Tunisia*. FAO, Rome, Italy.

Toelkes, W.E., (1987), 'The Ebeye Desalination Project: Total Utilization of Diesel Waste Heat', *Desalination*, 66: 59-66.

Torres, M., Vera, J.A., and Fernandez, F., (1985), '20 Years of Desalination in the Canary Islands, Was It Worth It?', *Aqua*, 3: 151-155.

Troyano, F., (1979), 'Introductory Report (Desalination: Operation and Economic Aspects of Management)', *Proceedings of the Seminar on Selected Water Problems in Islands and Coastal Areas with Special Regard to Desalination of Groundwater, San Anton, Malta*, Pergamon Press, pp. 371-375, New York, USA.

UNEP, (1983), *Rain and Stormwater Harvesting in Rural Areas*, Tycooly, Dublin, Ireland.

UNEP, (1993), 'Re-use of Water: An Overview', Paper for the WHO/FAO/UNCHS/UNEP Workshop on Health, Agriculture, and Environmental Aspects of the Use of Wastewater, Mexico City, Mexico.

UNEP, (1997), *Source Book of Alternative Technologies for Freshwater Augmentation in Latin America and the Caribbean*, UNEP-IETC Joint Technical Publication Series 8c, Unit of Sustainable Development and Environment, General Secretariat, Organization of American States, Washington, D.C., USA; also available at http://www.oas.org/usde/publications/Unit/oea59e/begin.htm#

UNEP, (2000), *Sourcebook of Alternative Technologies for Freshwater Augmentation in Some Countries in Asia*, UNEP-IETC Joint Technical Publication Series 8e; also available at http://www.unep.or.jp/ietc/publications/techpublications/techpub-8e/

UNESCO, (1988), *Final Report: II Meeting of the Major Regional Project (MRP) on Use and Conservation of Water Resources in Rural Areas of Latin America and the Caribbean, La Serena, Chile*. UNESCO/ROSTLAC, Montevideo, Uruguay.

USEPA, (1980), *Process Design Manual: Onsite Wastewater Treatment and Disposal Systems*, (EPA Report No. EPA-625/1-80-012), Cincinnati, Ohio, USA.

USEPA, (1980), *Innovative and Alternative Technology Assessment Manual*, (Report No. PA-430/9-78-009), Washington, D.C., USA.

USEPA, (1980), *Planning Wastewater Management Facilities for Small Communities*, (Report No. EPA-600/8-80-030), Cincinnati, Ohio, USA.

USEPA, (1981), *Process Design Manual: Land Treatment of Municipal Wastewater*, (Report No. EPA-625/1-81-013), Cincinnati, Ohio, USA.

USEPA, (1983), *Process Design Manual: Municipal Wastewater Stabilization Ponds,* (Report No. EPA-625/1-83-015), Cincinnati, Ohio, USA.

USEPA, (1988), *Process Design Manual: Constructed Wetlands and Aquatic Plant Systems,* (Report No. EPA-625/1-88-022), Cincinnati, Ohio, USA.

USEPA, (1990), *State Design Criteria for Wastewater Treatment Systems,* (Report No. EPA-430/9-90-014), Washington, D.C., USA.

USEPA, (1990), *Denver's Water Conservation Program. Compliance Review for 1989 Pursuant to the Foothills Consent Decree,* EPA Region VIII, Denver, CO, USA.

USEPA, (1991a), *Municipal Wastewater Reuse: Selected Readings on Water Reuse,* (Report No. EPA-430/09-91-022), Washington, D.C., USA.

USEPA, (1991b). *Fact Sheet: 21 Water Conservation Measures for Everybody,* (Report No. EPA-570/9-91-100), Washington, D.C., USA.

USEPA, (1992), *Manual: Guidelines for Water Reuse,* (Report No. EPA-625/R-92-104), Washington, D.C., USA.

USEPA, (1992), *Process Design Manual: Wastewater Treatment/Disposal for Small Communities,* (Report No. EPA-625/R-92/005), Cincinnati, Ohio, USA.

USEPA, (1994), *WAVE: Water Alliances for Voluntary Efficiency,* (Report No. EPA-832/F-94-006), Washington, D.C., USA.

USEPA, (1995), *Cleaner Water Through Conservation,* (Report No. EPA- 841/B-95-002), Washington, D.C., USA.

Vickers, A., (1989), 'New Massachusetts Toilet Standard Sets Water Conservation Precedent', *Journal of the American Water Works Association,* 81(3): 51.

Vickers, A., and E.J. Markus, (1992), 'Creating Economic Incentives for Conservation', *Journal of the American Water Works Association,* 84(10): 42-45.

Virginia State Water Control Board, (1979), *Best Management Practices: Agriculture,* (Planning Bulletin 316), Richmond, VA, USA.

Wall, B.H. and McCown, R.L., (1989), 'Designing Roof Catchment Water Supply Systems Using Water Budgeting Methods', *Water Resources Development,* 5: 11-18.

WAHLI and YLBHI, (1992). Mistaking plantations for Indonesia's tropical forest, 70, Wahana Lingkungan Hidup, Jakarta, Indonesia.

Water and Sewage Works, (1988), 'Reverse Osmosis Used for Water Desalination in Sea World', 124(3): 81.

Water Works Journal, (1990), 'Florida Commission Makes Water Conservation Recommendations', 44(6): 7.

Wild, P.M., and G.W. Vickers, (1991), 'The Technical and Economic Benefits of Centrifugal Reverse Osmosis Desalination', *IDA World Conference on Desalination and Water Reuse,* IDP, Topsfield, MA, USA.

Wildenhahn E., (1985), 'Traditional Irrigation Systems in the Southwest of Saudi Arabia', *Traditional Irrigation Schemes and Potential for their Improvement. Irrigation Symposium 1985. DVWK Bulletin 9,* Verlag Paul Parey, Hamburg/Berlin, Germany.

Wilson, R.A., Astorga, R., Gomez, C., and Gonzalez-Rio, F., (1995), 'Papermaking with DNA. ˝Intelligent Fibre˝, B. M. Potts et al (eds). *Eucalypt Plantations: Improving Fibre Yield and Quality.* Proc IUFRO Conference, February, CRC for Temperate Hardwood Forestry, pp. 24-30, Hobart, Tasmania, Australia.

World Resources Institute, (1993), *World Resources 1992-93.* Oxford University Press, New York, USA.

World Water, (1982), 'Desalter Systems for Man-made Islands', July, pp. 39-42.

World Water, (1984), 'RO Renewal Rate Could Be Critical', July, pp. 35-39.

World Water, (1986), 'Reverse Osmosis Still Needs Careful Treatment', December, pp. 33-35.

York, D.W. and Crook, J., (1990), 'Florida's Reuse Program Paves the Way', *Municipal Waste-water Reuse,* USEPA. pp. 67-74. (Report No. EPA-43 0/09-91-422), Washington, D.C., USA.

Appendix

EXAMPLE PROBLEMS WITH SOLUTIONS

Problem 1. A natural channel, which can be approximated with a width of 25 m, has a uniform flow rate of 125 m^3/s. Find the depth of the flow using Newton-Raphson Method, if channel slope is 0.001 m/m and Manning's constant, n=. 015.

Solution: The Manning's formula for flow rate through the natural channel is given by

$$Q = \frac{1}{n}\left[\frac{WY}{W+2Y}\right]^{\frac{2}{3}} AS_0^{\frac{1}{2}}$$

where,

$\quad$ Q = Discharge through the channel

$\quad$ n = Manning's constant

$\quad$ W = Bed width of the channel

$\quad$ A = Area of the approximated equivalent rectangular cross section

$\quad$ S$_0$ = Normal flow in slope of the channel

$\quad$ Y = Depth the channel

Let,

$$f(Y) = Q - \frac{1}{n}\left[\frac{WY}{W+2Y}\right]^{\frac{2}{3}} (WY)_0^{\frac{1}{2}}$$

Differentiating the above function with respect to Y and simplifying,, we get,

$$f'(Y) = -\frac{W^{\frac{5}{3}} Y^{\frac{2}{3}} S_0^{\frac{1}{2}}(5W+6Y)}{3n(W+2Y)^{\frac{5}{3}}}$$

From Newton-Raphson Method, we have the following expression for the successive approximation of Y in the $(n+1)^{th}$ iteration (Y_{n+1}) in terms of the value of Y in n^{th} iteration (Y_n):

$$Y_{n+1} = Y_n - \frac{f(Y_n)}{f'(Y_n)}$$

$$= Y_n - \frac{Q - \frac{1}{n}\left[\frac{WY}{W+2Y_n}\right]^{\frac{2}{3}} A S_0^{\frac{1}{2}}}{-\frac{W^{\frac{5}{3}} Y_n^{\frac{2}{3}} S_0^{\frac{1}{2}} (5W+6Y_n)}{3n(W+2Y_n)^{\frac{5}{3}}}}$$

After simplification and substituting the value of $A = WY_n$,

$$Y_{n+1} = \frac{2W(WY_n)^{\frac{5}{3}} S_0^{\frac{1}{2}} + 3Qn(W+2Y_n)^{\frac{5}{3}}}{W^{\frac{5}{3}} Y_n^{\frac{2}{3}} S_0^{\frac{1}{2}} (5W+6Y_n)}$$

This is the iterative formula. According to the problem,

$$Q = 125 \text{ m}^3/\text{s}$$
$$S_0 = 0.001$$
$$n = 0.015$$
$$W = 25 \text{ m}$$

Substituting these values in the iterative formula, we get,

$$Y_{n+1} = \frac{50(25Y_n)^{\frac{5}{3}} \sqrt{0.001} + 3 \times 125 \times 0.015(25+2Y_n)^{\frac{5}{3}}}{25^{\frac{5}{3}} Y_n \sqrt{0.001}(125+6Y_n)}$$

$$= \frac{338.3265 \, Y_n^{\frac{5}{3}} + 5.625(25+2Y_n)^{\frac{5}{3}}}{6.75927 \, Y_n^{\frac{2}{3}} (125+6Y_n)}$$

By successive iterations the value of depth of channel is obtained as follows-

$$Y_0 = 2$$
$$Y_1 = 1.777406667$$
$$Y_2 = 1.770304684$$
$$Y_3 = 1.770297029$$
$$Y_4 = 1.770297029$$

Hence, the depth of flow is 1.7703 m (approx.)

Problem 2. A discharge of 15 m^3/s occurs in a rectangular channel with a top width of 5m and uniform slope $S_0 = 0.001$. The channel intersects a stream that has a depth of 3 m (as shown in Figure 1). Find the shape of the water surface profile in the channel upstream of its intersection with the stream.

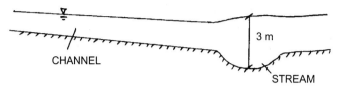

3 m

CHANNEL

STREAM

Figure 1   Channel and Stream Section

Solution:

According to the problem,

$$Q = 15 \text{ m}^3/\text{s}$$
$$W = 5.0 \text{ m}$$
$$S_0 = 0.001$$

It is assumed that

$$n = 0.015$$
$$Y = 3.0 \text{ m}$$

Calculation of normal depth, Y_n,

$$Q = \frac{1}{n}\left[\frac{WY_n}{W+2Y_n}\right]^{\frac{2}{3}} WY_n S_0^{\frac{1}{2}}$$

$$15 = \frac{1}{0.015}\left[\frac{5Y_n}{5+2Y_n}\right]^{\frac{2}{3}} 5Y_n (0.001)^{\frac{1}{2}}$$

$$\frac{Y_n^{\frac{5}{3}}}{(5+2Y_n)^{\frac{2}{3}}} = 0.4867$$

$$Y_n^{\frac{5}{3}} = 0.4867(5+2Y_n)^{\frac{2}{3}}$$

$$Y_n^5 = 0.115287982(5+2Y_n)^{\frac{2}{3}}$$

After solving, by trial & error $Y_n = 1.48989$ m

Calculation of critical depth, Y_c,

$$\frac{Q^2}{g} = \frac{(WY_c)^3}{W}$$

$$\frac{15^2}{9.81} = \frac{(5Y_c)^3}{5}$$

Table 1 Computation of Water Surface Profile in the Solution to Problem 2

Depth (m)	Area (m²)	P (m)	R (m)	V (m/s)	V²/2g (m)	E (m)	S_F (10⁻⁴)	S_F avg. (10⁻⁴)	S_F-S_F avg. (10⁻⁴)	ΔE (m)	Δx (m)	ΣΔx (m)
3	15.0	11.0	1.3636	1.0	0.05099	3.05099	1.4880					0
2.75	13.75	10.5	1.3095	1.0909	0.06068	2.81068	1.8690	1.6785	8.3215	0.24031	288.78	288.78
2.5	12.5	10.0	1.25	1.2	0.07342	2.57342	2.4062	2.1376	7.8624	0.23726	301.76	590.54
2.25	11.25	9.5	1.1842	1.3333	0.09063	2.34063	3.1925	2.7993	7.2007	0.23279	323.29	913.83
2.0	10.0	9.0	1.1111	1.5	0.11472	2.11472	4.3991	3.7958	6.2042	0.22591	364.12	1277.95
1.75	8.75	8.5	1.0294	1.7143	0.14984	1.89984	6.3618	5.3804	4.6196	0.21488	465.15	1743.1
1.5	7.5	8.0	0.9375	2.0	0.20394	1.70394	9.8088	8.0853	1.9147	0.1959	1023.14	2766.24
1.49	7.45	7.98	0.9336	2.0134	0.20668	1.69668	9.9960	9.9024	0.0976	0.00726	743.85	3510.09

$$25Y_c^3 = 22.93578$$

$$Y_c = 0.97168m$$

$\therefore$ Flow is sub-critical. Also, $Y > Y_n > Y_c$.

Hence the surface profile is mild of type, M_1. The computation of surface profile is shown in Table 1 and its plot is shown in Fig. 2.

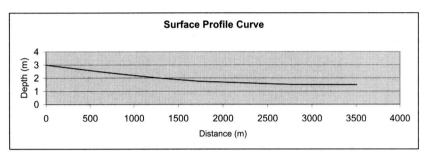

Figure 2 Water Surface Profile in the Solution to Problem 2

Problem 3. A water fishery survey was conducted to estimate wetted usable area in a stream reach as shown in Fig. 3. The wetted usable area is defined as the area multiplied by a preference factor ranging from 0 to 1, depending on factors affecting fish habitat. The preference factor for a particular species is shown in the Figure 4. Find the usable wetted area for the channel where depth varies laterally. Use Manning's formula with n=0.015, and consider bed slope $S_0 = 0.001$.

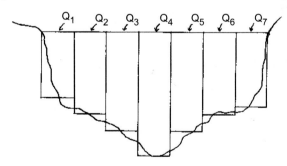

Figure 3 Strip-wise Partitioning of Stream Cross Section

Solution: Since the depths and velocities can vary across a channel cross section, the channel is subdivided laterally and the velocities as well as wetted usable areas are determined for each sub-division which are then added to provide the wetted usable area for the cross section as shown in Table 2, after calculations.

Width of each strip considered = W = 20 m

    Manning's constant = n = 0.015

    Bed slope = S_0 = 0.001

We know,

$$Q = \frac{1}{n}\left[\frac{WY}{W+2Y}\right]^{\frac{2}{3}} AS_0^{\frac{1}{2}}$$

Table 2   Computation of Usable Area

Strip No.	Depth Y (m)	Width W (m)	Area A (m²)	Discharge Q (m³/s)	Velocity V (m/s)	Preference factor, F	Usable area, A_u (m²)
1	8	20	160	911.8156	5.69885	0.86023	137.6368
2	10	20	200	1232.8741	6.16437	0.76713	153.4252
3	12	20	240	1567.8061	6.53253	0.69349	166.4376
•4	15	20	300	2088.3221	6.96107	0.60779	182.3358
5	12	20	240	1567.8061	6.53253	0.69349	166.4376
6	10	20	200	1232.8741	6.16437	0.76713	153.4252
7	9	20	180	1070.3041	5.94613	0.81077	145.9393

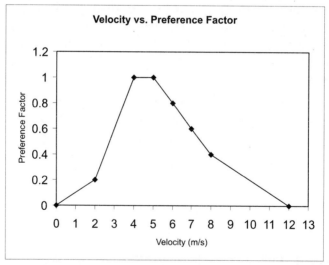

Figure 4   Velocity vs. Preference Factor Plot

Problem 4. A lawn sprinkler (shown in Figure 5) consists of four nozzles of 0.625 cm diameter placed at 30 cm radius from the axis of rotation. It is connected to a water supply line delivering 20 liters/min. of water. The

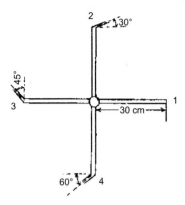

Figure 5   Lawn Sprinkler

nozzles discharge water upward and outward from the plane of rotation. Table 3 gives the angular settings, which are proposed from aesthetic considerations:

Table 3   Angular Settings for the Lawn Sprinkler

Nozzle No.	Outward angle (in deg.) with the tangent to the circumference in the plane of rotation
1	0
2	30
3	45
4	60

Calculate the speed of rotation of sprinkler.

Solution:

Water supply to the sprinkler = 20 lit/min.

$$= \frac{20 \times 10^{-3}}{60} \, m^3/s$$

$$= 3.3 \times 10^{-4} \, m^3/s$$

This water supply will be equally distributed to all the four nozzles.

Hence, water supply to each nozzle $= \dfrac{3.3 \times 10^{-4}}{4}$

$$= 8.33 \times 10^{-5} \, m^3/s$$

For each nozzle,

$$d = 0.625 \text{ cm}$$

$$d = 0.625 \times 10^{-2} \, m$$

$$\text{Area of nozzle} = \frac{\pi d^2}{4} = \frac{\pi \times (0.625 \times 10^{-2})^2}{4}$$

$$= 0.30679 \times 10^{-4} \, m^2$$

Now, the velocity of water coming out of nozzle is given by,

$$V = \frac{Q}{A}$$

$$V = \frac{8.33 \times 10^{-5}}{3.0679 \times 10^{-5}}$$

$$V = 2.7152 \, m/s$$

Component of this velocity in the direction tangential to the plane of rotation will be given by

$$V_{\theta_1} = V \cos\theta_1 = 2.7152 \times \cos 0° = 2.7152 \, m/s$$
$$V_{\theta_2} = V \cos\theta_2 = 2.7152 \times \cos 30° = 2.3514 \, m/s$$
$$V_{\theta_3} = V \cos\theta_3 = 2.7152 \times \cos 45° = 1.9199 \, m/s$$
$$V_{\theta_4} = V \cos\theta_4 = 2.7152 \times \cos 60° = 1.3576 \, m/s$$

Let the angular velocity of the sprinkler is ω rad/sec. Then the absolute velocity of flow from each nozzle will be given by,

For nozzle 1,

$$V_1 = V_{\theta_1} - r\omega$$
$$V_1 = 2.7152 - 0.3\omega$$

Similarly,

$$V_2 = 2.3514 - 0.3\omega$$
$$V_3 = 1.9199 - 0.3\omega$$
$$V_4 = 1.3576 - 0.3\omega$$

Now from the law of conservation of angular momentum, we have,

$$\sum mVr = 0$$

$$mV_1 r + mV_2 r + mV_3 r + mV_4 r = 0$$
$$mr(V_1 + V_2 + V_3 + V_4) = 0$$

As m and r are the mass of water and radius of the system, they cannot be equal to zero.

So,

$$(V_1 + V_2 + V_3 + V_4) = 0$$

$$(2.7152 - 0.3\omega) + (2.3514 - 0.3\omega) + (1.9199 - 0.3\omega) + (1.3576 - 0.3\omega) = 0$$

$$8.3441 - 1.2\omega = 0$$

$$\omega = \frac{8.3441}{1.2}$$

$$\omega = 6.9534 \text{ rad/sec.}$$

$$\omega = \frac{6.9534 \times 60}{2\pi}$$

$$\omega \approx 66 \text{ rpm}$$

Problem 5. Two identical pumps are connected to a pipe system as shown in Figure 6. The head discharge relationship of the pump is given by the following equation:

$$H = 40 + 0.12Q - 0.0012Q^2$$

where,

Q = Pump discharge in lit/sec.

H = Head developed by pump in m.

The pipe XY is 80 m long and of diameter 30 cm. The pipe YZ is 150 m long and 45 cm diameter. Consider f = 0.08 for both the pipes. V is regulatory valve on the pipe YZ. The static lift is 25 m and 20 m. For a given position of valve, V, the pump operating against a static lift of 25 m develops a discharge of 120 lit/sec. Find the discharge in the second pump and the loss coefficient of the valve.

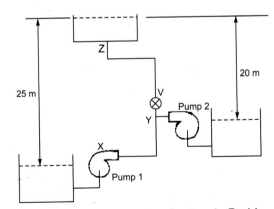

Figure 6 Water Distribution System in Problem 5.

Solution:

For pipe XY,

Length, $\qquad$ l = 80 m

diameter, $\qquad$ d = 30 cm = 0.3 m

$$f = 0.08$$

Head loss in the pipe XY,

$$(h_f)_{XY} = \frac{flV^2}{2gd}$$

$$= 0.08\,(80) \left(\frac{0.12}{\frac{\pi}{4}(0.3)^2} \right)^2 \cdot \frac{1}{2(9.81)0.3}$$

$$= 3.1337 \text{ m}$$

Head developed by pump 1,

$$H_1 = 40 + 0.12\,(120) - 0.0012\,(120^2)$$

$$= 40 + 14.4 - 17.28$$

$$= 37.12 \text{ m}$$

Also,

$$H_1 = (h_f)_{XY} + (h_f)_{YZ} + S_t + h_v$$

where,

$$S_t = \text{Static lift} = 25 \text{ m}$$

$$h_v = \text{Head loss through valve.}$$

$$37.12 = 3.1337 + (h_f)_{YZ} + 25 + h_v$$

$$(h_f)_{YZ} + h_v = 8.9863\text{m} \qquad \qquad \dots(1)$$

Head developed by pump 2,

$$H_2 = 40 + 0.12(1000Q_2) - 0.0012(1000Q_2)^2$$

Also, $\qquad H_2 = 20 + (h_f)_{YZ} + h_v$

$\therefore \qquad (h_f)_{YZ} + h_v = 20 + 0.12(1000Q_2) - 0.0012(1000Q_2)^2 \qquad \dots(2)$

From (1) and (2) we have,

$$20 + 0.12(1000Q_2) - 0.0012(1000Q_2)^2 = 8.9863$$

$$1200Q_2^2 - 120Q_2 - 11.0137 = 0$$

$$Q_2 = 0.15807 \ m^3/s$$

This is the discharge through pump 2.
Now, total discharge,

$$Q = Q_1 + Q_2$$
$$Q = 0.12 + 0.15807$$
$$Q = 0.27807 \ m^3/s$$

For pipe YZ,

$$\text{Length, } l = 150 \ m$$
$$\text{diameter, } d = 45 \ cm = 0.45 \ m$$
$$f = 0.08$$

$$(h_f)_{YZ} = \frac{flV^2}{2gd}$$

$$= 0.08(150) \left[\frac{0.27807}{\frac{\pi}{4}(0.45)^2} \right]^2 \frac{1}{2(9.81)0.45}$$

$$= 4.15478 \ m.$$

From (1) we have,

$$(h_f)_{YZ} + h_v = 8.9863$$
$$h_v = 8.9863 - (h_f)_{Yz}$$
$$h_v = 8.9863 - 4.15478$$
$$h_v = 4.83152 \ m$$

Hence, the co-efficient of loss through valve V,

$$= \frac{headloss}{V^2/2g}$$

$$= 31.01$$

Problem 6. Design a standard Parshall flume for measurement of peak flow of 1.321 m^3/s to control the feed rate of chlorine solution. The dimensions of the channel upstream of the Parshall flume are as follows :

Channel section is rectangular; width of the channel = 2.0 m

Depth of flow at peak design flow, $y_1 = 0.8$ m

Solution:

The slope of the channel is calculated from the Manning's equation, with

$$n = 0.013 \text{ (assumed)}$$

$$Q = A\,(1/n)\,R^{2/3}\,S^{1/2}$$

or,
$$1.321 = (1.60)\,\frac{1}{0.013}\,(0.44)^{2/3}\,S^{1/2}$$

i. e.,
$$S = 0.000344$$

Select the dimensions of the rectangular channel downstream of the Parshall flume,

Width = 2 m, Depth of flow for peak design flow, $y_2 = 1$ m.

The slope is computed by using Manning's formula with n = 0.013

or,
$$1.321 = (2)\,\frac{1}{0.013}\,(0.5)^{2/3}\,S^{1/2}$$

i. e.,
$$S = 0.000186.$$

Select the dimensions of the Parshall flume. The dimensions of the various components of the Parshall flume are shown in Figure 7. Table 4 gives the dimensions of a standard Parshall Flume.

Throat width = 1.22 m, submergence at peak design flow = 70%. At submergence below 70% the flow through a 1.22 m Parshall flume is essentially the same as for free conditions. Q, Discharge, is given by

$$Q = 4WH_a^{1.522\,W^{0.026}}$$

where Q = free flow rate(cfs),
  W = throat width (ft),
  H_a = depth of water at upstream gauge point (ft).
  H_a for design flow of 1.321 m^3/s (= 46.7 cfs) is given by,

$$46.7 = 4\,(4)\,H_a^{1.522\,(4)^{0.026}}$$

or, $H_a = 1.97$ ft (= 0.6 m)
  H_b = depth of water at downstream gaging point = 0.7 (1.97)

at 70% submergence = 1.38 ft = 0.42 m

The head loss through Parshall flume at peak design flow is calculated from Figure 8. At peak discharge of 46.7 cfs (=1.321 m^3/s) and 70% submergence, the head loss is 0.65ft = 0.20 m.

Dimensions as shown in Figure 7

Table 4 Parshall Flume Dimensions (mm) (Courtesy: ILRI)

b	b	A	a	B	C	D	E	L	G	h_1	K	M	N	P	R	X	Y	Z
1 in	25.4	363	242	356	93	167	229	76	203	206	19	–	29	–	–	8	13	3
2 in	50.8	414	276	406	135	214	254	114	254	217	22	–	43	–	–	16	25	6
3 in	76.2	467	311	457	178	259	457	152	305	309	25	–	57	–	–	25	38	13
6 in	152.4	621	414	610	394	397	610	305	610	–	76	305	114	902	406	51	76	–
9 in	228.6	879	587	864	381	575	762	305	457	–	76	305	114	1080	406	51	76	–
1 ft	304.8	1372	914	1343	610	845	914	610	914	–	76	381	229	1492	508	51	76	–
1 ft 6 in	457.2	1448	965	1419	762	1026	914	610	914	–	76	381	229	1676	508	51	76	–
2 ft	609.6	1524	1016	1495	914	1206	914	610	914	–	76	381	229	1854	508	51	76	–
3 ft	914.4	1676	1118	1645	1219	1572	914	610	914	–	76	381	229	2222	508	51	76	–
4 ft	1219.2	1829	1219	1794	1524	1937	914	610	914	–	76	457	229	2711	610	51	76	–
5 ft	1524.0	1981	1321	1943	1829	2302	914	610	914	–	76	457	229	3080	610	51	76	–
6 ft	1828.8	2134	1422	2092	2134	2667	914	610	914	–	76	457	229	3442	610	51	76	–
7 ft	2133.6	2286	1524	2242	2438	3032	914	610	914	–	76	457	229	3810	610	51	76	–
8 ft	2438.4	2438	1626	2391	2743	3397	914	610	914	–	76	457	229	4172	610	51	76	–
10 ft	3048	–	1829	4267	3658	4756	1219	914	1829	–	152	–	343	–	–	305	229	–
12 ft	3658	–	2032	4877	4470	5607	1524	914	2438	–	152	–	343	–	–	305	229	–
15 ft	4572	–	2337	7620	5588	7620	1829	1219	3048	–	229	–	457	–	–	305	229	–
20 ft	6096	–	2845	7620	7315	9144	2134	1829	3658	–	305	–	686	–	–	305	229	–
25 ft	7620	–	3353	7620	8941	10668	2134	1829	3962	–	305	–	686	–	–	305	229	–
30 ft	9144	–	3861	7620	10566	12313	2134	1829	4267	–	305	–	686	–	–	305	229	–
40 ft	12192	–	4877	8230	13818	15481	2134	1829	4877	–	305	–	686	–	–	305	229	–
50 ft	15240	–	5893	8230	17272	18529	2134	1829	6096	–	305	–	686	–	–	305	229	–

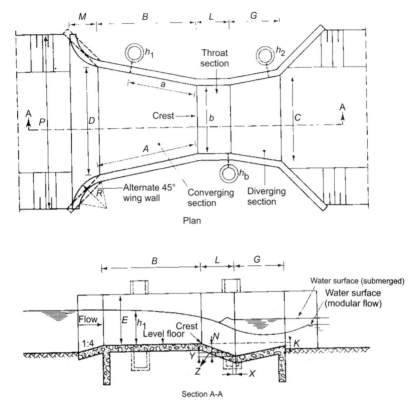

Plan

Section A-A

Figure 7 Geometry of Parshall Flume [Courtesy: ILRI]

The downstream channel bottom depth from the flume crest is calculated as follows. The water surface in the flume at the H_b gauge is essentially level with the surface in the downstream channel.

So $$\Delta = y_2 - H_b = 0.58 \text{ m}.$$

The water surface profile and calibration curve are shown in Figure 8.

Problem 7. A venturi meter is provided in the force main of a sewage pumping station. The force main is 92 cm in diameter. The tube beta ratio (diameter of throat to diameter of the force main) shall be equal to 0.5. The maximum and minimum flow ranges are 1.321 and 0.152 m^3/s respectively. The flow measurement error shall be less than ± 0.75 percent at all flows. The head loss shall not exceed 15 percent of the meter readings at all flows. Develop the calibration curve.

Solution : All differential pressure meters utilize Bernoulli's principle to measure flows. The Bernoulli equation for two sections of a pipe is given by:

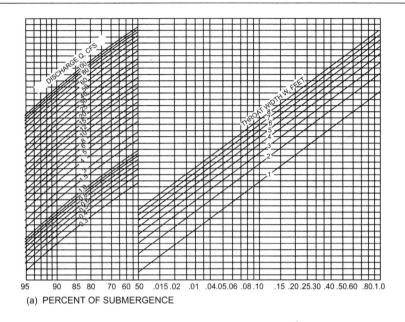

(a) PERCENT OF SUBMERGENCE

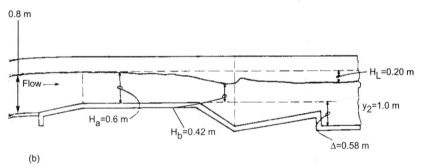

(b)

Figure 8 The Calibration Curve and the Water Surface Profile for Parshall Flume

$$\frac{p_1}{\gamma} + z_1 + \frac{V_1^2}{2g} = \frac{p_2}{\gamma} + z_2 + \frac{V_2^2}{2g} + h_L$$

where p_1, p_2 = internal pressures in the pipe (kPa), γ = Specific Weight of the fluid (kN/m^3), z_1, z_2 = elevations of the center lines of the pipe (m), V_1, V_2 = Velocities of the fluid (m/s), h_L = head loss (m). For horizontal pipe $z_1 = z_2$ and the head loss is negligible.

So,

$$\frac{p_1}{\gamma} - \frac{p_2}{\gamma} = \frac{V_2^2}{2g} - \frac{V_1^2}{2g}$$

replacing $\frac{P_1}{\gamma}, - \frac{P_2}{\gamma}$ by H_1, H_2 and by continuity equation, $Q = V_1 A_1 = V_2 A_2$

So, $$Q = \frac{A_1 A_2 \sqrt{2g(H_1 - H_2)}}{\sqrt{A_1^2 - A_2^2}} = \frac{A_1 A_2 \sqrt{2gh}}{\sqrt{A_1^2 - A_2^2}}$$

where Q = pipe flow (m^3/s), H_1 = upstream piezometric head (m)
H_2 = throat piezometric head (m), A_1 = force main area (m^2),
A_2 = throat area (m^2) $h = (H_1 - H_2)$,
     The above Equation under actual operating conditions including allowance for friction reduces to

$$Q = C_1 K A_2 \sqrt{2g} \sqrt{h}$$

where, C_1 = a combination of velocity, friction or discharge coefficients

$$K = \text{Coefficient} = \frac{1}{\sqrt{1 - (D_2/D_1)^4}}$$

where D_2, D_1, diameters at throat & pipe inlet (m)
for standard venturi meter, the diameter of the throat is 1/3rd to 1/2 of the pipe diameter and the value of K is between 1.0062 and 1.0328. The value of C_1 generally ranges from 0.97 to 0.99 varying with Reynolds number. The value of C_1 is normally provided by the manufacturer.
     Unit sizing and calibration curve
     1. Determine constants; the venturi meter has $D_2/D_1 = 0.5$

$$\text{Throat diameter } D_2 = 46 \text{ cm}, K = \frac{1}{\sqrt{1 - (0.5)^4}} = 1.0328$$

The calibration equation, assuming $C_1 = 0.985$ is

$$Q = (0.985)\, 1.0328 \, \frac{\pi}{4} \, (0.46)^2 \, \sqrt{2(9.8)} \, \sqrt{h}$$

$$= 0.7489 \sqrt{h} \ \ m^3/s$$

     Develop calibration curve: Assuming different values of differential head recorded by the meter the discharge can be obtained vide equation above.
     At maximum and minimum design flows of 1.321 and 0.152 m^3/s, the differential pressure reading will be 3.111 m and 0.041 m respectively. If mercury is used in the manometric glass tube, the differential pressure reading must be adjusted for the specific gravity of mercury = 13.58.

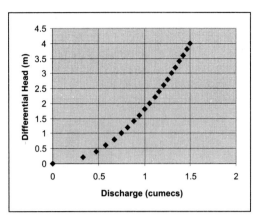

Figure 9   Venturimeter Calibration Curve

Head loss calculations: The head loss in the meter is calculated from $h_L = KV_2^2/(2g)$.

Here, h_L = head loss in the meter (m), K = 0.14 for angle of divergence of 5°. At maximum and minimum flows, the head losses are

$$h_L \text{ at maximum flow} =$$

$$= 0.45 \text{ m.}$$

$$h_L \text{ at minimum flow} = \frac{0.14}{2(9.81)}\left[\frac{0.152}{(\pi/4)(0.46)^2}\right]^2$$

$$= 0.006 \text{ m.}$$

Both these h_L values are within the allowable limit of 15%

Problem 8. Consider the system as shown in Figure 10 with the data given in Table 5 below:

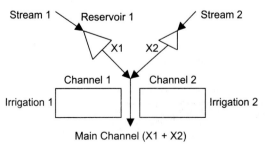

Figure 10   Schematic Diagram of a Reservoir Irrigation System

The capacity of main stream below the confluence is 5 units. If the benefits equal—$2(10)^6$ and $3(10)^6$ per unit of reservoir release from 1 and 2 determine the releases X1 and X2 to maximize the benefits.

Table 5 Details of the Reservoir Irrigation System

Quantity stream	Stream 1 (i = 1)	Stream 2 (i = 2)
Reservoir capacity	19	7
Available release from reservoir	9	6
Capacity of channel below reservoir	4	4
Actual release from reservoir	X1	X2

Solution:

Maximize $Z = 2(10)^6 \, X1 + 3(10)^6 \, X2$,
subject to constraints,

$$X1 + X2 \leq 5$$
$$X1 \leq 4$$
$$X2 \leq 4$$
$$X1 \geq 0$$
$$X2 \geq 0$$

The solution was obtained using Linear Programming software and the result is as follows:

Optimal solution $Z = \$14(10)^6$ with X1 = 1 and X2 = 4.

Problem 9. There are three farms being cultivated by landowner, the output of each being limited by the cultivable area and the availability of irrigation water. The details are as given in Tables 6 and 7.

Table 6 Details of Cultivable Area and Water Available

Farm	Cultivable area (ha)	Availability of water (ha-m)
1	161.87 (=400 ac)	185.02 (=1500 ac-ft.)
2	242.81 (=600 ac)	246.70 (=2000 ac-ft.)
3	121.41 (=300 ac)	111.01(=900 ac-ft.)

Table 7 Details of Crop Irrigation Benefits

Crop	Maximum crop area [Acre]	Water consumption [ft]	Profit per acre [$]
I	700 (=283.28 ha)	5 (=1.52 m)	400
II	800 (=323.74 ha)	4 (=1.22 m)	300
III	300 (=121.41 ha)	3 (=0.91 m)	100

The owner desires to plant three crops differing in their expected profit per acre and consumption of water. It is to be noted that the percentage of cultivable area planted is same at each farm, however any suitable combination of the crops may be grown in any of the farms. The solution to the above problem for maximum profit by linear programming is desired.

Solution:

The decision variables are XI1, XI2, XI3, XII1, XII2, XII3 and XIII1, XIII2, XIII3 respectively representing the acres of land allotted for crop I, II, III in farms 1, 2, and 3.

The objective function to maximize profit is then given as:

$$Z = 400\,(XI1 + XI2 + XI3) + 300\,(XII1 + XII2 + XII3) + 100\,(XIII1 + XIII2 + XIII3)$$

The constraints are respectively as follows:

$$XI1 + XII1 + XIII1 \leq 400$$
$$XI2 + XII2 + XIII2 \leq 600$$
$$XI3 + XII3 + XIII3 \leq 300$$
$$5XI1 + 4XII1 + 3XIII1 \leq 1500$$
$$5XI2 + 4XII2 + 3XIII2 \leq 2000$$
$$5XI3 + 4XII3 + 3XIII3 \leq 900$$
$$XI1 + XI2 + XI3 \leq 700$$
$$XII1 + XII2 + XII3 \leq 800$$
$$XIII1 + XIII2 + XIII3 \leq 300$$

$$\frac{XI1 + XII1 + XIII1}{400}$$
$$= \frac{XI2 + XII2 + XIII2}{600}$$
$$= \frac{XI3 + XII3 + XIII3}{300}$$

Or $\quad 3XI1 + 3XII1 + 3XIII1 - 2XI2 - 2XII2 - 2XIII2 = 0$

And $\quad XI2 + XII2 + XIII2 - 2XI3 - 2XII3 - 2XIII3 = 0$

And $\quad XI1, XI2, XI3, XII1, XII2, XII3, XIII1, XIII2, XIII3 \geq 0$

The optimal solution of the problem using linear programming software comes to

$Z = \$\,120{,}675$ with the variables taking values (in acres) as

XI1 = 300 (=121.41 ha), XI2 = 287.5 (=116.35 ha), XI3 = 112.5 (=45.53 ha), XII1 = 0, XII2 = 75 (=30.35 ha), XII3 = 0, XIII1 = 0, XIII2 = 87.5 (=35.41 ha) and XIII3 = 112.5 (=45.53 ha) respectively.

Problem 10: A schematic diagram of a storm sewer design along with relevant data is shown in Figure 11. Determine the pipe diameters, inlet and outlet invert levels.

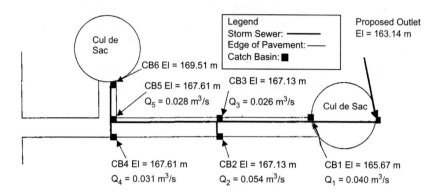

Pipe	Length (m)
Outlet – CB1	68.60
CB1 – CB3	112.20
CB2 – CB3	7.62
CB3 – CB5	93.90
CB4 – CB5	7.62
CB5 – CB6	119.82

Figure 11 Schematic Diagram and Storm Sewer Pipe Details for Problem 10 [Courtesy: ASCE Journal of Hydraulic Engineering, Dec. 2001].

Solution:

Table 7 provides the solution in a tabular form.

Table A1.7   Solution for Problem 10 in a Tabular Form

Outlet CB identity	Inlet CB identity	Flow (cfs)	Calculated pipe diameter (m)	Actual pipe diameter (m)	Flow velocity (m/s)	Pipe length (m)	Inlet invert elevation (m)	Outlet invert elevation (m)	Pipe slope (m/m)	Manning's n	Actual friction slope (m/m)	Outlet ground elevation (m)	EGL Elevation inside outlet CB (m)	Inlet ground elevation (m)	EGL Inlet ground elevation (m)	Elevation inside inlet CB (m)	Check if EGL, is adequate
Outlet	CB1	7.6	0.34	0.38	1.89	68.60	164.27	163.14	0.0164	0.015	0.0184	163.14	163.70	165.64	165.64	165.64	X
CB1	CB3	6.2	0.37	0.38	1.54	112.20	165.85	164.42	0.0128	0.015	0.0123	165.64	165.06	167.13	167.13	166.62	X
CB3	CB2	1.9	—	0.20	1.66	7.62	166.16	166.60	0.0200	0.015	0.0329	167.13	165.06	167.13	167.13	165.52	X
CB5	CB5	4.4	0.36	0.38	0.84	93.90	166.46	166.60	0.0049	0.015	0.0037	167.13	166.62	167.61	167.61	167.02	X
CB5	CB4	1.1	—	0.20	0.96	7.62	166.77	166.62	0.0200	0.015	0.0110	167.611	167.02	167.61	167.61	167.17	X
CB5	CB6	1.3	0.20	0.30	1.14	393	168.29	166.62	0.0140	0.015	0.154	167.61	167.01	169.51	169.51	168.97	X

Index

DISCARDED
CONCORDIA UNIV. LIBRARY

CONCORDIA UNIVERSITY LIBRARY
MONTREAL